Common Core
Interactive Student Edition

Reveal MATH™

Course 1 • Volume 2

Cover: (l to r, t to b) guvendemir/E+/Getty Images, Andrey Prokhorov/E+/Getty Images, anatols/iStock/Getty Images, iava777/iStock/Getty Images

my.mheducation.com

Send all inquiries to:
McGraw-Hill Education
STEM Learning Solutions Center
8787 Orion Place
Columbus, OH 43240

ISBN: 978-0-07-899714-3
MHID: 0-07-899714-3

Reveal Math, Course 1
Interactive Student Edition, Volume 2

Printed in the United States of America.

4 5 6 7 LMN 23 22 21 20 19

Contents in Brief

Reveal Math™ Makes Math Meaningful...

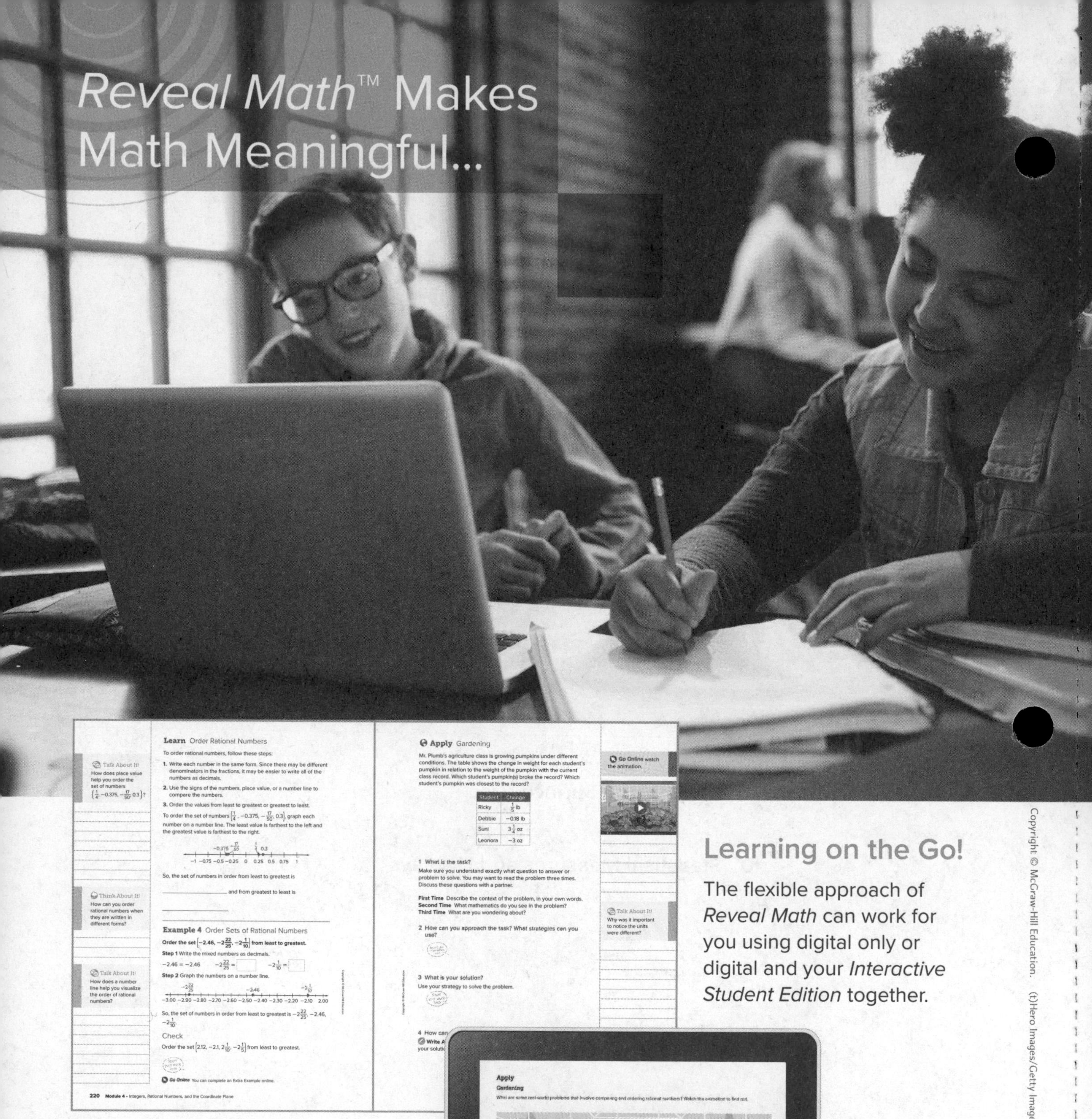

Learn Order Rational Numbers

To order rational numbers, follow these steps:

1. Write each number in the same form. Since there may be different denominators in the fractions, it may be easier to write all of the numbers as decimals.
2. Use the signs of the numbers, place value, or a number line to compare the numbers.
3. Order the values from least to greatest or greatest to least.

To order the set of numbers $\{\frac{1}{4}, -0.375, -\frac{17}{50}, 0.3\}$, graph each number on a number line. The least value is farthest to the left and the greatest value is farthest to the right.

−1 −0.75 −0.5 −0.25 0 0.25 0.5 0.75 1

So, the set of numbers in order from least to greatest is

________________ and from greatest to least is

____

Talk About It! How does place value help you order the set of numbers $\{\frac{1}{4}, -0.375, -\frac{17}{50}, 0.3\}$?

Think About It! How can you order rational numbers when they are written in different forms?

Example 4 Order Sets of Rational Numbers

Order the set $\{-2.46, -2\frac{22}{25}, -2\frac{1}{10}\}$ from least to greatest.

Step 1 Write the mixed numbers as decimals.

$-2.46 = -2.46$ $-2\frac{22}{25} = \square$ $-2\frac{1}{10} = \square$

Step 2 Graph the numbers on a number line.

−3.00 −2.90 −2.80 −2.70 −2.60 −2.50 −2.40 −2.30 −2.20 −2.10 2.00

So, the set of numbers in order from least to greatest is $-2\frac{22}{25}, -2.46, -2\frac{1}{10}$.

Talk About It! How does a number line help you visualize the order of rational numbers?

Check

Order the set $\{2.12, -2.1, 2\frac{1}{10}, -2\frac{1}{5}\}$ from least to greatest.

Go Online You can complete an Extra Example online.

220 Module 4 • Integers, Rational Numbers, and the Coordinate Plane

Apply Gardening

Mr. Plumb's agriculture class is growing pumpkins under different conditions. The table shows the change in weight for each student's pumpkin in relation to the weight of the pumpkin with the current class record. Which student's pumpkin(s) broke the record? Which student's pumpkin was closest to the record?

Student	Change
Ricky	$\frac{1}{5}$ lb
Debbie	−0.18 lb
Suni	$3\frac{1}{4}$ oz
Leonora	−3 oz

1 What is the task?

Make sure you understand exactly what question to answer or problem to solve. You may want to read the problem three times. Discuss these questions with a partner.

First Time Describe the context of the problem, in your own words.
Second Time What mathematics do you see in the problem?
Third Time What are you wondering about?

2 How can you approach the task? What strategies can you use?

3 What is your solution?

Use your strategy to solve the problem.

4 How can

Write A

your soluti

Go Online watch the animation.

Talk About It! Why was it important to notice the units were different?

Interactive Student Edition

Learning on the Go!

The flexible approach of *Reveal Math* can work for you using digital only or digital and your *Interactive Student Edition* together.

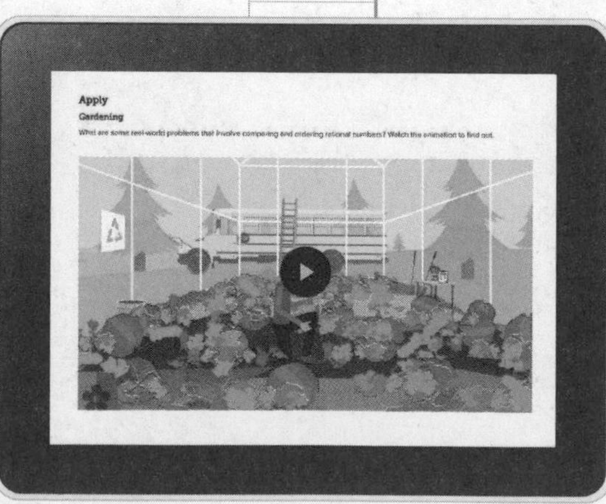

Student Digital Center

...to Reveal YOUR Full Potential!

Reveal Math™ Brings Math to Life in Every Lesson

Reveal Math is a blended print and digital program that supports access on the go. You'll find the *Interactive Student Edition* aligns to the Student Digital Center, so you can record your digital observations in class and reference your notes later, or access just the digital center, or a combination of both! The Student Digital Center provides access to the interactive lessons, interactive content, animations, videos, and technology-enhanced practice questions.

Write down your username and password here

Username: ______________________

Password: ______________________

Go Online! my.mheducation.com

Web Sketchpad® Powered by The Geometer's Sketchpad®- Dynamic, exploratory, visual activities embedded at point of use within the lesson.

Animations and Videos – Learn by seeing mathematics in action.

Interactive Tools – Get involved in the content by dragging and dropping, selecting, and completing tables.

Personal Tutors – See and hear a teacher explain how to solve problems.

eTools – Math tools are available to help you solve problems and develop concepts.

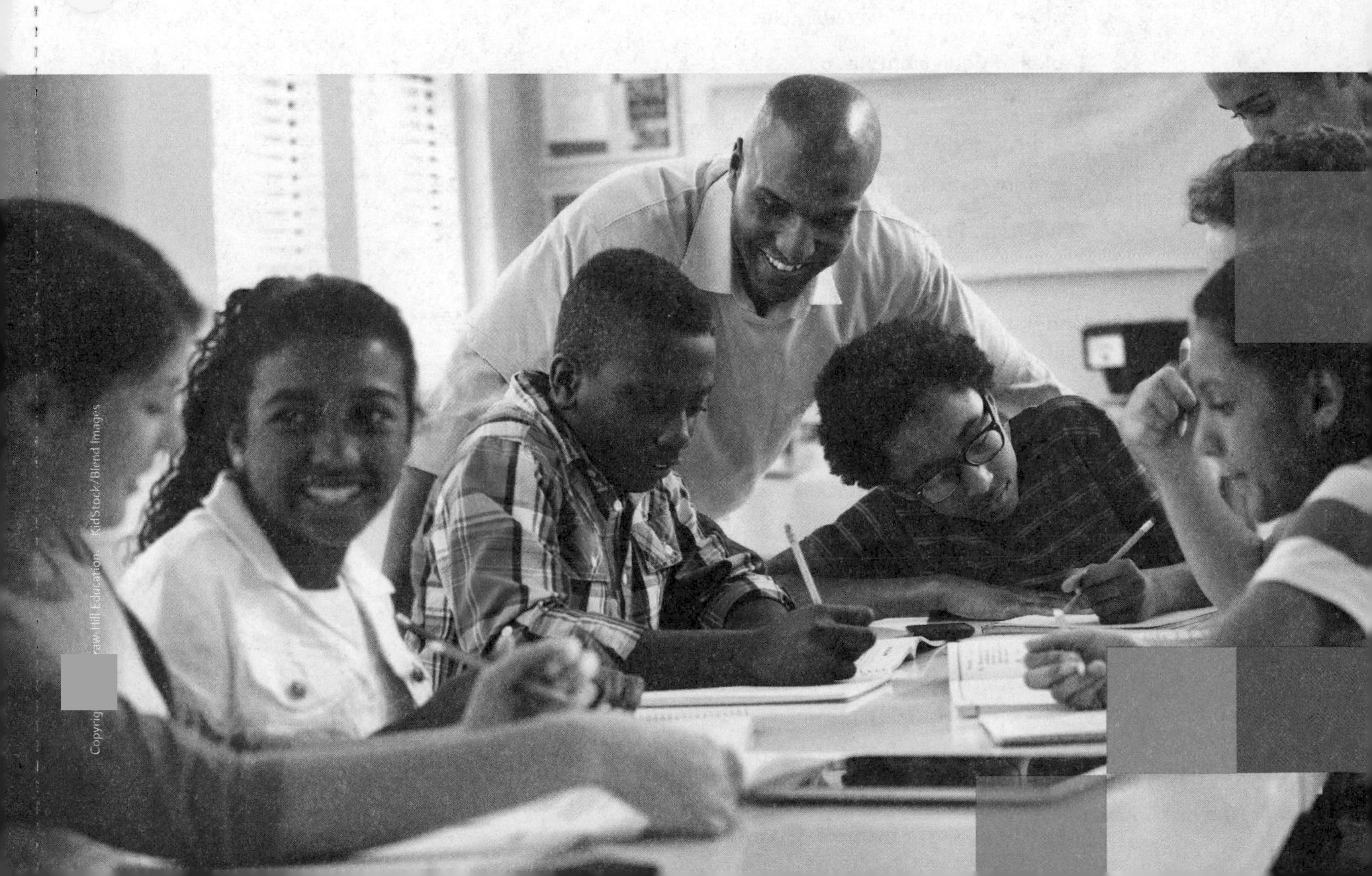

TABLE OF CONTENTS

Module 1

Ratios and Rates

Essential Question

How can you describe how two quantities are related?

Module 2

Fractions, Decimals, and Percents

Essential Question

How can you use fractions, decimals, and percents to solve everyday problems?

TABLE OF CONTENTS

Module 3

Compute with Multi-Digit Numbers and Fractions

Essential Question

How are operations with fractions and decimals related to operations with whole numbers?

Module 4

Integers, Rational Numbers, and the Coordinate Plane

e Essential Question

How are integers and rational numbers related to the coordinate plane?

TABLE OF CONTENTS

Module 5

Numerical and Algebraic Expressions

Essential Question

How can we communicate algebraic relationships with mathematical symbols?

Module 6

Equations and Inequalities

Essential Question

How are the solutions of equations and inequalities different?

TABLE OF CONTENTS

Module 7

Relationships Between Two Variables

Essential Question

What are the ways in which a relationship between two variables can be displayed?

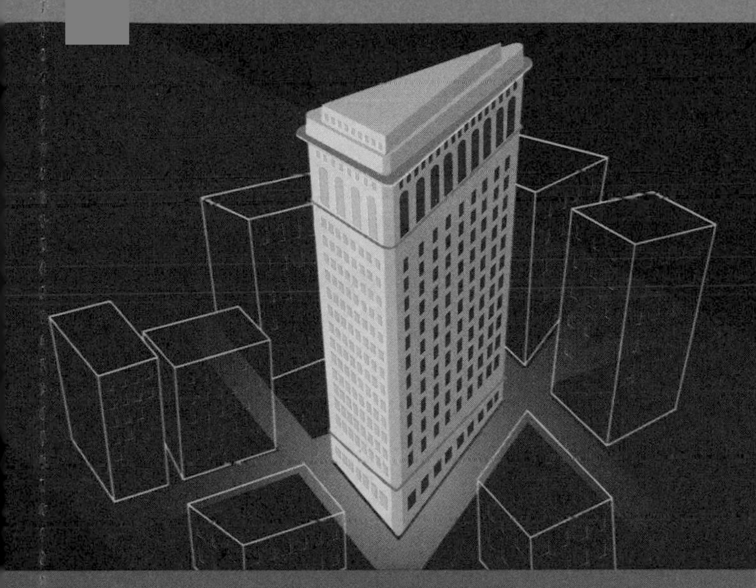

Module 8

Area

Essential Question

How are the areas of triangles and rectangles used to find the areas of other polygons?

Module 9

Volume and Surface Area

Essential Question

How can you describe the size of a three-dimensional figure?

Module 10

Statistical Measures and Displays

Essential Question

Why is data collected and analyzed and how can it be displayed?

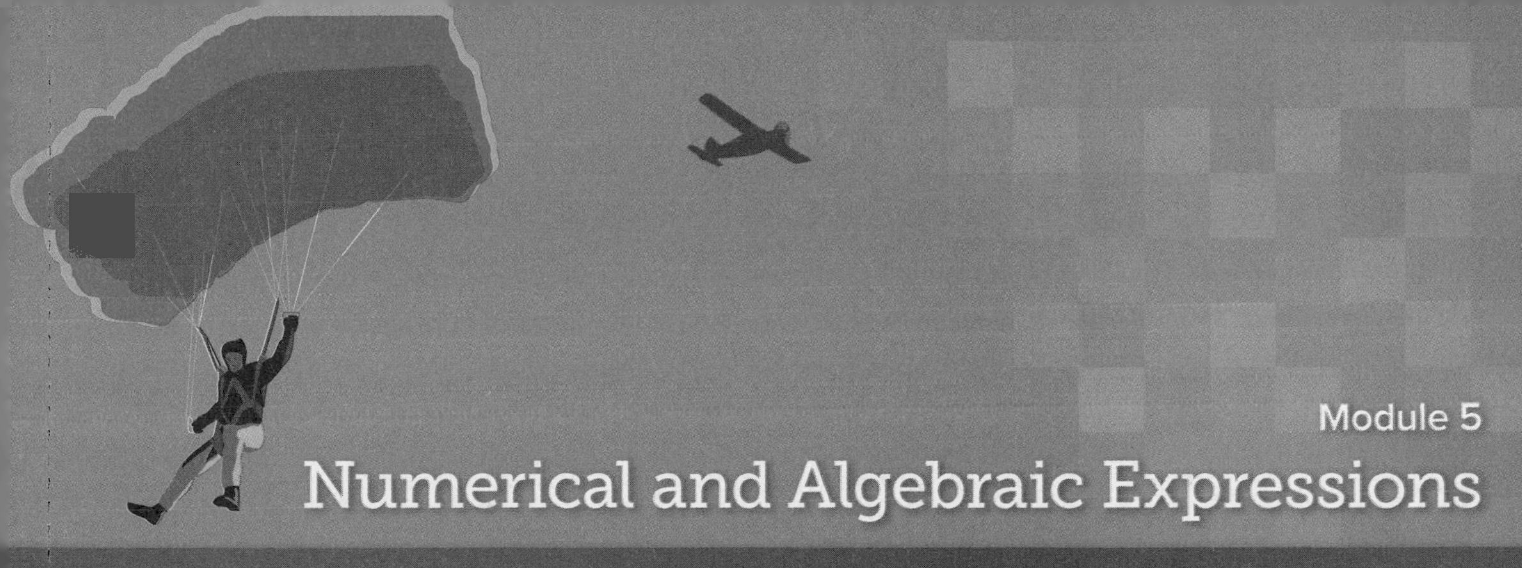

Module 5

Numerical and Algebraic Expressions

Essential Question

How can we communicate algebraic relationships with mathematical symbols?

What Will You Learn?

Place a checkmark (✓) in each row that corresponds with how much you already know about each topic **before** starting this module.

KEY

□ — I don't know. ◇ — I've heard of it. ☆ — I know it!

	Before			After		
	□	◇	☆	□	◇	☆
writing products as powers						
evaluating powers						
evaluating numerical expressions						
writing numerical expressions						
writing algebraic expressions						
evaluating algebraic expressions						
finding the greatest common factor of two whole numbers						
finding the least common multiple of two whole numbers						
using the Distributive Property						
using the greatest common factor to factor numerical expressions						
identifying equivalent expressions						
simplifying expressions by combining like terms						

Foldables Cut out the Foldable and tape it to the Module Review at the end of the module. You can use the Foldable throughout the module as you learn about numerical and algebraic expressions.

What Vocabulary Will You Learn?

Check the box next to each vocabulary term that you may already know.

- ☐ algebra
- ☐ algebraic expression
- ☐ Associative Property
- ☐ base
- ☐ coefficient
- ☐ Commutative Property
- ☐ constant
- ☐ defining the variable
- ☐ Distributive Property
- ☐ equivalent expressions
- ☐ evaluate
- ☐ exponent
- ☐ factoring the expression
- ☐ greatest common factor
- ☐ Identity Property
- ☐ least common multiple
- ☐ like terms
- ☐ numerical expression
- ☐ order of operations
- ☐ power
- ☐ simplest form
- ☐ term
- ☐ variable

Are You Ready?

Study the Quick Review to see if you are ready to start this module.
Then complete the Quick Check.

Quick Review

Example 1

Multiply repeated factors.

Multiply 5 × 5 × 5 × 5.

The number 5 is used as a factor four times.

$5 \times 5 \times 5 \times 5 = 625$

Example 2

Subtract fractions and mixed numbers.

Find $3\frac{7}{8} - 1\frac{1}{2}$.

$3\frac{7}{8} - 1\frac{1}{2}$

$= 3\frac{7}{8} - 1\frac{4}{8}$ Rewrite using the LCD, 8.

$= 2\frac{3}{8}$ Subtract.

Quick Check

1. Multiply 7 × 7 × 7.

2. Multiply 2 × 2 × 2 × 2 × 2.

3. Find $\frac{4}{5} - \frac{1}{2}$.

4. Find $3\frac{1}{10} - 2\frac{5}{6}$.

How Did You Do?

Which exercises did you answer correctly in the Quick Check?
Shade those exercise numbers at the right.

Powers and Exponents

I Can... write a product of whole numbers, fractions, or decimals as a power and write a power as a product of factors.

What Vocabulary Will You Learn?
base
exponent
power

Learn Products as Powers

A product of like factors can be written in exponential form using an exponent and a base. A number expressed using an exponent is called a **power**. The **base** is the number used as a factor. The **exponent** tells how many times a base is used as a factor.

Go Online Watch the animation to see how an expression involving repeated factors can be written as a power.

The animation explains how to write the product of the repeated factor, 3, as a power.

$$\underbrace{3 \times 3 \times 3 \times 3}_{\text{4 factors}} = 3^4$$

Use an exponent to express the number of times 3 is used as a factor.

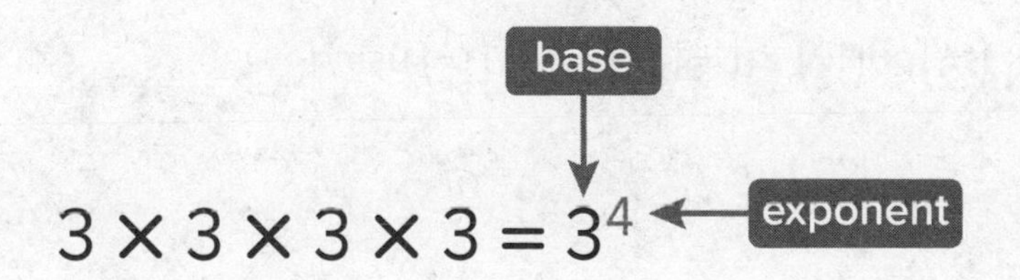

The base is the common factor that is being multiplied, and the exponent tells how many times the base is used as a factor.

Label each part of the equation using the words below.

factors   exponent   base   power

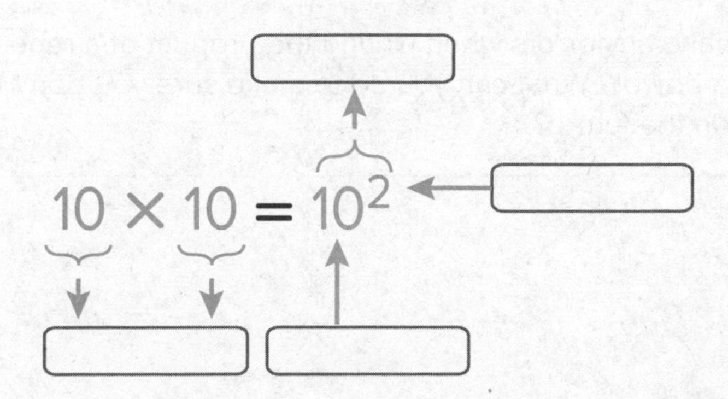

Talk About It!
Explain the difference between a base and an exponent.

Your Notes

Example 1 Write Products as Powers

Write $7 \times 7 \times 7 \times 7 \times 7$ using an exponent.

The base ______ is used as a factor ______ times.

So, $7 \times 7 \times 7 \times 7 \times 7$ can be written using an exponent as 7^5.

Check

Write $8 \times 8 \times 8 \times 8 \times 8 \times 8 \times 8 \times 8 \times 8 \times 8$ using an exponent.

Example 2 Write Products as Powers

Write $\frac{2}{5} \times \frac{2}{5} \times \frac{2}{5}$ using an exponent.

The base $\frac{2}{5}$ is used as a factor ______ times.

So, $\frac{2}{5} \times \frac{2}{5} \times \frac{2}{5}$ can be written as $\left(\frac{2}{5}\right)^3$.

Check

Write $\left(1\frac{1}{2}\right) \times \left(1\frac{1}{2}\right) \times \left(1\frac{1}{2}\right) \times \left(1\frac{1}{2}\right) \times \left(1\frac{1}{2}\right) \times \left(1\frac{1}{2}\right) \times \left(1\frac{1}{2}\right)$ using an exponent.

Go Online You can complete an Extra Example online.

Think About It!
What information do you have that will help you use the correct exponent?

Talk About It!
Why are parentheses used in $\left(\frac{2}{5}\right)^3$?

Pause and Reflect

Did you make any errors when writing the product of a repeated factor as a power? What can you do to make sure you don't repeat that error in the future?

Learn Powers as Products

To write powers as products, determine the base and the exponent. The base of 3^2 is 3 and the exponent is 2. To read powers, consider the exponent. The power 3^2 is read as *three to the second power* or *three squared*, the power 3^3 is read *three to the third power* or *three cubed*, and 3^5 is read *three to the fifth power*.

To evaluate powers, find the value of the power after multiplying.

Complete the table for the first four powers of 5.

Powers			
Power	Words	Factors	Value
5^1	5 to the first power	5	5
5^2	5 to the second power	5×5	
5^3			125
5^4	5 to the fourth power		

Example 3 Evaluate Powers

Evaluate 4^5.

$4^5 = 4 \times 4 \times 4 \times 4 \times 4$ Write 4^5 as a product.

$= 1{,}024$ Simplify.

So, 4^5 is ______.

Check

Evaluate 8^4.

Go Online You can complete an Extra Example online.

Talk About It!

What are some mistakes that could be made when evaluating 5^3?

Talk About It!

A friend evaluates the expression 4^5 and arrives at a value of 20 for the solution. Describe the mistake.

Think About It!

What does the exponent of 4 mean?

Talk About It!

A friend evaluates the expression $\left(\frac{1}{3}\right)^4$ and arrives at a value of $\frac{1}{12}$. Describe the likely mistake.

Example 4 Evaluate Powers

Evaluate $\left(\frac{1}{3}\right)^4$.

$\left(\frac{1}{3}\right)^4 = \frac{1}{3} \times \frac{1}{3} \times \frac{1}{3} \times \frac{1}{3}$ Write $\left(\frac{1}{3}\right)^4$ as a product.

$= \frac{1}{81}$ Simplify.

So, $\left(\frac{1}{3}\right)^4$ is ________.

Check

Evaluate $\left(\frac{2}{5}\right)^3$.

Example 5 Evaluate Powers

Evaluate $(2.5)^3$.

$(2.5)^3 = (2.5) \times (2.5) \times (2.5)$ Write $(2.5)^3$ as a product.

$= 15.625$ Simplify.

So, $(2.5)^3$ is ________.

Check

Evaluate $(0.2)^4$.

Go Online You can complete an Extra Example online.

Apply Biology

Delmar is studying the growth rate of a specific type of bacteria. He places 3 cells in a Petri dish and records the number of bacteria over time. He records the results over 20 hours in the table shown and notices a pattern. At this rate, how many bacteria are expected to be present in the Petri dish after 30 hours?

Number of Hours	Number of Bacteria
5	3×3
10	$3 \times 3 \times 3$
15	$3 \times 3 \times 3 \times 3$
20	$3 \times 3 \times 3 \times 3 \times 3$

Go Online watch the animation.

1 What is the task?

Make sure you understand exactly what question to answer or problem to solve. You may want to read the problem three times. Discuss these questions with a partner.

First Time Describe the context of the problem, in your own words.
Second Time What mathematics do you see in the problem?
Third Time What are you wondering about?

2 How can you approach the task? What strategies can you use?

Talk About It!

Suppose Delmar originally placed 4 cells in the Petri dish. Could you use the same method to determine the total cells after 30 hours? Explain.

3 What is your solution?

Use your strategy to solve the problem.

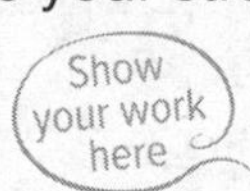

4 How can you show your solution is reasonable?

Write About It! Write an argument that can be used to defend your solution.

Check

Faith is turning 12 this year. She asks her parents to give her $2 on her birthday and to double that amount for her next birthday. If she continues with this pattern, how much money will Faith receive on her 20th birthday?

Birthday	Amount ($)
12th	2
13th	2×2
14th	$2 \times 2 \times 2$
15th	$2 \times 2 \times 2 \times 2$

Go Online You can complete an Extra Example online.

Pause and Reflect

How well do you understand the process of evaluating powers? What questions do you still have? How can you get those questions answered?

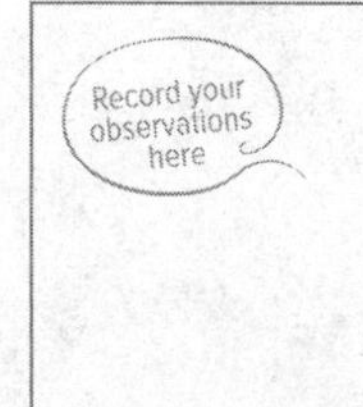

Name ______________________ Period __________ Date __________

Practice

Go Online You can complete your homework online.

Write each product using an exponent. (Examples 1 and 2)

1. $4 \times 4 \times 4$

2. $3 \times 3 \times 3 \times 3 \times 3$

3. $15 \times 15 \times 15 \times 15$

4. $\frac{3}{4} \times \frac{3}{4} \times \frac{3}{4} \times \frac{3}{4} \times \frac{3}{4} \times \frac{3}{4}$

5. $\frac{1}{3} \times \frac{1}{3} \times \frac{1}{3} \times \frac{1}{3} \times \frac{1}{3} \times \frac{1}{3} \times \frac{1}{3}$

6. 1.625×1.625

Evaluate each power. (Examples 3–5)

7. $5^5 =$ ________

8. $6^3 =$ ________

9. $10^4 =$ ________

10. $\left(\frac{1}{2}\right)^2 =$ ________

11. $\left(\frac{2}{5}\right)^3 =$ ________

12. $\left(\frac{1}{4}\right)^4 =$ ________

13. $(1.5)^3 =$ ________

14. $(0.2)^2 =$ ________

15. $(0.4)^3 =$ ________

16. The table shows the approximate area in square miles of the largest and smallest states in the United States. What is the difference between the areas in square miles?

State	Area (mi^2)
Alaska	87^3
Rhode Island	39^2

Test Practice

17. **Multiselect** Select all expressions that are equivalent to $7 \times 7 \times 7 \times 7$.

- ☐ 4^7
- ☐ 7^4
- ☐ 7^7
- ☐ 28
- ☐ 2,401
- ☐ 16,384

Apply

18. Willa is studying the growth rate of a specific type of organism called a ciliate. She places 2 cells in a dish and records the number of cells over time. The table shows her results. If the pattern continues, how many cells will be in the dish after 12 hours?

Number of Hours	Number of Cells
2	2×2
4	$2 \times 2 \times 2$
6	$2 \times 2 \times 2 \times 2$
8	$2 \times 2 \times 2 \times 2 \times 2$

19. Christiano is performing a science experiment and studying the growth rate of a certain type of onion root cell under different conditions. He places a cell in a dish and records the number of cells each day. Based on the pattern shown in the table, predict the number of cells in the dish after 5 days.

Number of Days	Number of Cells
1	4
2	4×4
3	$4 \times 4 \times 4$

20. Write a power whose value is greater than 500 but less than 1,000.

21. MP **Find the Error** A student was evaluating 2^3. Find the student's mistake and correct it.

$$2^3 = 3 \times 3$$
$$= 9$$

22. MP **Reason Inductively** Suppose the world population is about 8 billion. Is 8 billion closer to 10^{10} or 10^{11}? Explain.

23. MP **Be Precise** Explain how exponential form is similar to multiplication being the process of repeated addition.

Numerical Expressions

I Can... write and evaluate a numerical expression using the correct order of operations.

What Vocabulary Will You Learn?
evaluate
numerical expression
order of operations

Learn Numerical Expressions

A **numerical expression** is a combination of numbers and at least one operation, such as $4^2 + 7 \div (3 - 1)$.

To **evaluate** a numerical expression, you find its value. A numerical expression can be evaluated using the order of operations. The **order of operations** are the rules that tell which operation to perform first when more than one operation is used. This guarantees that the same value of a numerical expression is found each time the expression is evaluated.

Order of Operations

1. Simplify the expressions inside of grouping symbols, such as parentheses.
2. Find the value of all powers.
3. Multiply and divide in order from left to right.
4. Add and subtract in order from left to right.

Talk About It!
Use the expression $12 + 3 \times 4$ to explain why we need rules for evaluating expressions.

Go Online Watch the animation to learn how to simplify a numerical expression using the order of operations.

The animation explains how to evaluate the expression below.

$6^2 - 12 \div 3 + (14 - 9) \times 2$

$= 6^2 - 12 \div 3 + 5 \times 2$	Simplify the expression inside the parentheses.
$= 36 - 12 \div 3 + 5 \times 2$	Evaluate the exponent.
$= 36 - 4 + 5 \times 2$	Divide 12 by 3.
$= 36 - 4 + 10$	Multiply 5 by 2.
$= 32 + 10$	Subtract 4 from 36.
$= 42$	Add 32 and 10.

So, the value of the expression is 42.

Your Notes

Think About It!

What does the order of operations tell you to do first?

Talk About It!

Explain why subtraction was performed before addition when the expression showed $100 - 90 + 2$.

Example 1 Evaluate Numerical Expressions

Evaluate $100 - 3^2 \times (6 + 4) + 2$.

$100 - 3^2 \times (6 + 4) + 2$	Write the expression.
$100 - 3^2 \times (6 + 4) + 2 = 100 - 3^2 \times 10 + 2$	Simplify parentheses.
$= 100 - 9 \times 10 + 2$	Evaluate the exponent.
$= 100 - 90 + 2$	Multiply.
$= 10 + 2$	Subtract.
$= 12$	Add.

So, the value of the expression is ______.

Check

Evaluate $[23 - (8 + 2^3)] \times 2 + 10$.

Example 2 Evaluate Numerical Expressions

Evaluate $5 + \left(8^2 \div \frac{2}{5}\right) \times 2$.

$5 + \left(8^2 \div \frac{2}{5}\right) \times 2$	Write the expression.
$5 + \left(8^2 \div \frac{2}{5}\right) \times 2 = 5 + \left(64 \div \frac{2}{5}\right) \times 2$	Evaluate the exponent inside of the parentheses.
$= 5 + 160 \times 2$	Simplify parentheses.
$= 5 + 320$	Multiply.
$= 325$	Simplify.

So, the value of the expression is ______.

Check

Evaluate $8.2 \times (2^4 - 3) + 8$.

Go Online You can complete an Extra Example online.

Learn Write Numerical Expressions

In a real-world situation where one or more operations occur, you can write an expression to represent the situation.

Suppose Mariana and her friends are buying snacks at a hockey game. **Hot dogs cost $4**, **boxes of popcorn cost $2**, and **drinks cost $2.50**. The expression below represents the total cost of **4 hot dogs**, **3 boxes of popcorn**, and **2 drinks**.

The different colored text represents each part of the expression.

hot dogs + **popcorn** + **drinks**

($4 × 4) + **($2 × 3)** + **($2.50 × 2)**

Talk About It!

How else can you represent the part of the expression written as (4 × 4)?

Example 3 Write and Evaluate Numerical Expressions

Paula is shopping for the items shown in the table.

Item	lotion	candle	lip balm
Cost ($)	5.00	7.80	2.49

Write an expression to represent the total cost of 5 lotions, 2 candles, and 4 lip balms. Then find the total cost.

Part A Write an expression.

cost of lotions + cost of candles + cost of lip balms

(5^2) + (2 × 7.80) + (4 × 2.49)

Part B Find the total cost.

(5^2) + (2 × 7.80) + (4 × 2.49) = ☐ + ☐ + ☐

= 50.56

So, the total cost is $_______.

Talk About It!

In this situation, does the placement of the parentheses have an effect on the evaluation of the expression? Explain.

Math History Minute

Mary G. Ross (1908–2008) is considered the first known Native American female mathematician and engineer. After her retirement, Ross was an advocate of women studying STEM fields (Science, Technology, Engineering, and Mathematics). She earned a place in the Silicon Valley Engineering Council's Hall of Fame.

Check

Tickets to a play cost \$10.50 for adult members of the theater, \$19.95 for adult non-members, and \$8 for students.

Part A

Suppose 4 non-members, 2 members, and 8 students are buying tickets for the play. Which expression could be used to find the total cost of the tickets?

Ⓐ $4(19.95 + 10.50 + 8)$

Ⓑ $(4 \times 19.95) + (2 \times 10.50) + 8^2$

Ⓒ $(2 \times 19.95) + (4 \times 10.50) + 8^2$

Ⓓ $(4 \times 19.95) + (2 \times 10.50) + 8^8$

Part B

What is the total cost of the tickets?

Go Online You can complete an Extra Example online.

Pause and Reflect

Describe some examples of when writing an expression can help you solve problems in everyday life. How does understanding the order of operations help you to evaluate those expressions?

Apply Art Supplies

An art store sells different-sized art kits that include crayons and a sketch pad. The table shows the number of boxes of crayons and sketch pads in each kit. A school buys 30 small, 35 medium, and 10 large art kits. Then they return 11 medium art kits. How many boxes of crayons and sketch pads do they have in all?

Art Kit Size	Boxes of Crayons	Sketch Pads
Small	16	20
Medium	24	40
Large	68	100

1 What is the task?

Make sure you understand exactly what question to answer or problem to solve. You may want to read the problem three times. Discuss these questions with a partner.

First Time Describe the context of the problem, in your own words.
Second Time What mathematics do you see in the problem?
Third Time What are you wondering about?

2 How can you approach the task? What strategies can you use?

3 What is your solution?

Use your strategy to solve the problem.

4 How can you show your solution is reasonable?

Write About It! Write an argument that can be used to defend your solution.

Talk About It!
How could you solve this problem another way?

Check

At a summer camp, you can buy clothing embroidered with the camp logo. Short sleeve T-shirts are $15 each, shorts are $18 each, and long sleeve T-shirts are $25 each. During the final week of summer, the clothing is marked down to $\frac{3}{4}$ the original price. You buy two short sleeve T-shirts, one pair of shorts, and three long sleeve T-shirts. What was the total cost of the purchase?

Go Online You can complete an Extra Example online.

Pause and Reflect

Where did you encounter difficulty in this lesson, and how did you deal with it? Write down any questions you still have.

Name ______________________ Period __________ Date __________

Practice

Go Online You can complete your homework online.

Evaluate each expression. **(Examples 1 and 2)**

1. $64 \div (15 - 7) \times 2 - 9$

2. $9 + 8 \times 3 - (5 \times 2)$

3. $4 \times (5^2 - 12) - 6$

4. $78 - 2^4 \div (14 - 6) \times 2$

5. $9 + 7 \times (15 + 3) \div 3^2$

6. $13 + (4^3 \div 2) \times 5 - 17$

7. $4 + \left(6^2 \div \frac{1}{4}\right) \times 3$

8. $12 + \left(2^3 \div \frac{2}{3}\right) - 2$

9. $36 \div \left(3^2 \div \frac{3}{4}\right) - 2.4$

10. $80 \div \left(4^2 \div \frac{2}{5}\right) + 3.75$

11. Mei is shopping for the items shown in the table. Write an expression to represent the total cost of 6 bubbles, 2 beach balls, and 3 sand buckets. Then find the total cost. **(Example 3)**

Item	bubbles	beach ball	sand buckets
Cost	\$1.49	\$2.00	\$3.50

12. Roy and 2 friends are at a game center. Each person buys a hot dog for \$3, fries for \$2.49, and a drink for \$2.50. They also have a coupon for \$1 off each drink. Write an expression to represent the total cost. Then find the total cost. **(Example 3)**

Test Practice

13. Multiselect Alice takes an art class after school once a week. At each class, she buys a bottle of juice for \$1.25 and a bag of pretzels for \$0.85. Select all expressions that represent the amount of money, in dollars, Alice spends after attending 8 art classes.

- ☐ 8(1.25)(0.85)
- ☐ 8(1.25 + 0.85)
- ☐ 8(1.25) + 8(0.85)
- ☐ 8 + 1.25 + 0.85
- ☐ (8 + 1.25) + (8 + 0.85)

Apply

14. An art teacher is ordering colored pencils for the new school year. The table shows the number of colored pencils per box size. The teacher buys 24 small boxes, 12 medium boxes, and 5 large boxes. She had to return 3 small boxes due to defects. How many colored pencils does the teacher have in all?

Box Size	Number of Colored Pencils
Small	12
Medium	64
Large	100

15. A bakery sells boxes of muffins in the sizes shown in the table. On Monday, the bakery sold 15 minis, 8 dozens, and 6 jumbos. However, 6 of the minis sold were free with a coupon. How many total muffins were paid for on Monday?

Size	Number of Muffins
Mini	6
Dozen	12
Jumbo	24

16. MP **Persevere with Problems** Refer to the expression $2 + 6 \div 2 + 4 \times 3$.

a. Place parentheses in the expression so that the value of the expression is 16.

b. Place parentheses in the expression so that the value is *not* equal to 16. Then find the value of the new expression.

17. MP **Find the Error** A student is evaluating the expression $42 + 6 \div 2$. Find the student's mistake and correct it.

$$\begin{aligned} 42 + 6 \div 2 &= 48 \div 2 \\ &= 24 \end{aligned}$$

18. Write an expression that contains parentheses, 5 numbers, two different operations, and has a value of 20.

19. Create Write about a real-world situation that could be represented by a numerical expression. Then write and evaluate the expression.

Write Algebraic Expressions

I Can... identify parts of an expression from a verbal description in order to write an algebraic expression, using variables for unknown quantities, that models a real-world or mathematical problem.

What Vocabulary Will You Learn?
algebra
algebraic expression
coefficient
constant
defining the variable
like terms
term
variable

Explore Write Algebraic Expressions

Online Activity You will use algebra tiles to write algebraic expressions.

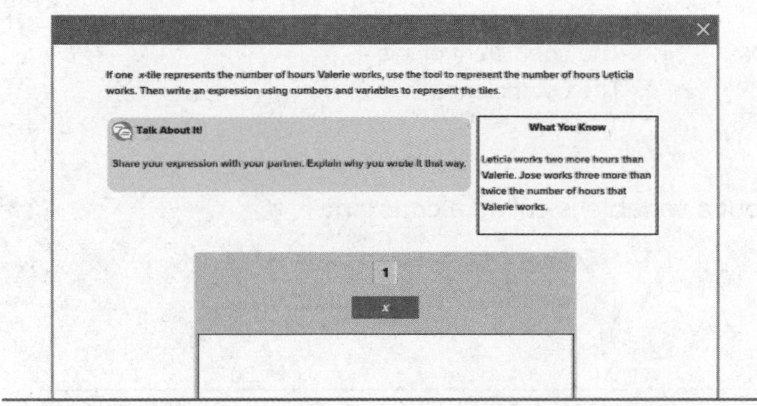

Learn Structure of Algebraic Expressions

Algebra is a branch of mathematics that uses symbols. A **variable** is a symbol, usually a letter, used to represent a number. An **algebraic expression** is a combination of variables, numbers, and at least one operation. For example, the expression $n + 2$ represents the phrase *the sum of an unknown number and two*. In this case, n is the variable.

Any letter can be used as a variable, but the letter x is commonly used. To avoid confusion with × as a multiplication symbol, multiplication is shown in other ways. Division can also be written in different ways.

Words	Variables
five times the variable x	$5 \cdot x$, $5(x)$, $5x$
five times x divided by 3	$5x \div 3$, $\frac{5x}{3}$
five times twice the value of x	$5 \cdot 2x$, $5(2x)$, $5 \cdot 2 \cdot x$ or $10x$

Talk About It!
A classmate said that $4(3x) = 12x$. Is the student correct? Justify your reasoning.

(continued on next page)

Your Notes

Talk About It!

In the expression $2x^2y + 4xy^2$, explain why $2x^2y$ and $4xy^2$ are not like terms.

Algebraic expressions can contain like terms, coefficients, variables, and constants. When addition or subtraction signs separate an algebraic expression into parts, each part is called a **term**.

$4x + 12 + 2x$ — $4x$, 12, and $2x$ are terms.

Like terms contain the same variables to the same powers.

$4x + 12 + 2x$ — $4x$ and $2x$ are like terms.

The numerical factor of each term that contains a variable is called the **coefficient** of the variable.

$4x + 12 + 2x$ — The coefficient of x is 4.
The coefficient of x is 2.

A term without a variable is called a **constant**.

$4x + 12 + 2x$ — The number 12 is a constant.

Pause and Reflect

How is an algebraic expression similar to a numerical expression? How is it different? How do you think knowing these differences will help you as you progress through this lesson?

Record your observations here

Example 1 Identify Parts of Algebraic Expressions

Identify the terms, like terms, coefficients, and constants in the expression $6n + 7n + 4 + 2n$.

Terms are parts of the expression that are separated by addition and subtraction, so the terms are:

$$6n,\ 7n,\ 4,\ 2n$$

Circle the like terms above.

Write the coefficients of the terms and the constants in the appropriate bin.

Coefficients	Constants

So, the terms are $6n$, $7n$, 4, and $2n$.

The like terms are $6n$, $7n$, and $2n$.

The coefficients are 6, 7, and 2.

The constant is 4.

Check

Identify the *terms, like terms, coefficients*, and *constants* in the expression $3x + 2 + 10 + 4x$.

terms ______________

like terms ______________

coefficients ______________

constants ______________

Go Online You can complete an Extra Example online.

Think About It!

How would you begin identifying the parts of the expression?

Talk About It!

Are there any like terms in this expression that are constants? Explain.

Talk About It!

Suppose that an additional term, n, is added to the end of the expression. What is the coefficient for n?

Learn Write One-Step Algebraic Expressions

To write verbal phrases as algebraic expressions, use the table below. When **defining the variable**, choose a variable and decide what it represents.

Words
Describe the mathematics of the problem.
Variable
Define a variable to represent the unknown quantity.
Expression
Translate the words into an algebraic expression.

In order to translate a situation into an expression, it is important to correctly identify operations that are described in words.

Write each phrase below the operation that it describes.

the product of   increased by   less than a number
the quotient of   the sum of

Addition

Subtraction

Multiplication

Division

Talk About It!
Make a list of additional phrases that could be represented by mathematical operations. Share your list and explain how those phrases represent that operation.

Example 2 Write One-Step Algebraic Expressions

Define a variable to represent the unknown in the phrase *ten dollars more than Anthony earned*. Then write the phase as an algebraic expression.

Words
ten dollars more than Anthony earned
Variable
Let d represent the number of dollars Anthony earned.
Expression
$d + 10$

So, the expression ________ can be used to model the phrase *ten dollars more than Anthony earned*.

Check

Define a variable to represent the unknown in the phrase *twelve dollars less than the original price*. Then write the phrase as an algebraic expression.

Go Online You can complete an Extra Example online.

Pause and Reflect

Why is it important to define the variable when writing an algebraic expression? What possible errors might be made if the variable is not correctly defined?

Record your observations here

Example 3 Write One-Step Algebraic Expressions

Define a variable to represent the unknown in the phrase *four and one-half times the number of gallons*. Then write the phrase as an algebraic expression.

Words
four and one-half times the number of gallons
Variable
Let g represent the number of gallons.
Expression
$4\frac{1}{2}g$ or $4.5g$

So, the expression ________ or ________ can be written to model the phrase *four and one-half times the number of gallons*.

Talk About It!

The expression $4\frac{1}{2}g$ can also be written as $4.5g$. What is another way that the expression could be written and still be equivalent to the original expression?

Check

Define a variable to represent the unknown in the phrase *six times more money than Eliot saved*. Then write the phrase as an algebraic expression.

Go Online You can complete an Extra Example online.

Learn Write Two-Step Algebraic Expressions

Two-step expressions contain two different operations. The table shows how to translate a verbal phrase into an algebraic expression.

Words
Describe the mathematics of the problem.
Variable
Define a variable to represent the unknown quantity.
Expression
Translate the words into an algebraic expression.

Talk About It!

How can you write an algebraic expression for the phrase *two more than three times a number*?

Example 4 Write Two-Step Algebraic Expressions

Define a variable to represent the unknown in the phrase *five less than three times the number of points*. Then write the phrase as an algebraic expression.

Words
five less than three times the number of points
Variable
Let p represent the number of points.
Expression
$3p - 5$

So, the expression ________ can be written to model the phrase *five less than three times the number of points.*

Think About It!

How would you begin writing the expression?

Talk About It!

Why is the expression $5 - 3p$ not correct?

Check

Define a variable to represent the unknown in the phrase *two less than one-third of the points that the Panthers scored.* Then write the phrase as an algebraic expression.

Go Online You can complete an Extra Example online.

Pause and Reflect

Did you make any errors when writing the two-step algebraic expressions? Were the errors the same or different from any errors you made while writing one-step algebraic expressions? What can you do to make sure you do not repeat that error in the future?

Example 5 Write Algebraic Expressions

A rectangle has a length that is twice its width.

Define a variable to represent the unknown quantity. Then write an expression to represent the perimeter of the rectangle.

Words
The length of the rectangle is ________ its width.
Variable
Let w represent the width of the rectangle. Because the length of the rectangle is twice the width, use the expression ________ to represent the length.
Expression
$w + w + \square + \square$

So, the perimeter of the rectangle can be represented by the expression $w + w + 2w + 2w$, where w represents the width and $2w$ represents the length.

Think About It!

What will your variable represent?

Talk About It!

Does the order in which you write the terms in the expression matter? Explain your reasoning.

Check

A rectangle has a length that is three times its width. Define a variable to represent the unknown quantity. Then write an expression to represent the perimeter of the rectangle.

Go Online You can complete an Extra Example online.

Pause and Reflect

Describe an instance in which the order you write the terms in the expression matters. Why is this important to recognize?

Name ____________________ Period ________ Date ________

Practice

Go Online You can complete your homework online.

Identify the terms, like terms, coefficients, and constants in each expression. (Example 1)

1. $4e + 7e + 5 + 2e$

2. $5a + 2 + 7 + 6a$

3. $4 + 4y + y + 3$

For each verbal phrase, define a variable to represent the unknown quantity. Then write the phrase as an algebraic expression. (Examples 2–4)

4. three more pancakes than Hector ate

5. twelve fewer questions than were on the first test

6. two and one-half times the number of minutes spent exercising

7. one-third the number of yards

8. four less than seven times Lynn's age

9. $2.50 more than one-fourth the cost of a pizza

10. A plumber charges $50 to visit a house plus $40 for every hour of work. Define a variable to represent the unknown quantity. Then write an expression to represent the total cost of hiring a plumber. (Example 4)

11. A gymnastics studio charges an annual fee of $35 plus $20 per class. Define a variable to represent the unknown quantity. Then write an expression to represent the total cost of taking classes. (Example 4)

12. A rectangle has a length that is half its width. Define a variable to represent the unknown quantity. Then write an expression to represent the perimeter of the rectangle. (Example 5)

13. In a triangle there are two sides that have the same length and the third side is 1.5 times longer than the length of the other two. Define a variable to represent the unknown quantity. Then write an algebraic expression to represent the perimeter of the triangle. (Example 5)

14. Open Response Nate scored 5 more than twice the number of points as Jake scored. Write an expression that represents the relationship of the number of points Nate scored in terms of the number of points Jake scored, *p*.

15. MP **Identify Structure** Write an expression that has four terms and at least one constant. Identify the like terms, coefficients, and constants in your expression.

16. If x represents the number of questions on a test, analyze the meaning of each expression: $x + 4$, $x - 5$, $2x$, and $x \div 3$.

17. MP **Persevere with Problems** Norman earns $8 for every dog he washes plus 25% of the cost of the dog wash. Write an expression that represents the total amount of money Norman earns for one dog wash with a cost, *c*.

18. Create Write about a real-world situation that can be represented with an algebraic expression. Then represent the situation with the expression.

Evaluate Algebraic Expressions

I Can... use the order of operations to evaluate algebraic expressions for given values.

Explore Algebraic Expressions

Online Activity You will use Web Sketchpad to explore how to evaluate algebraic expressions.

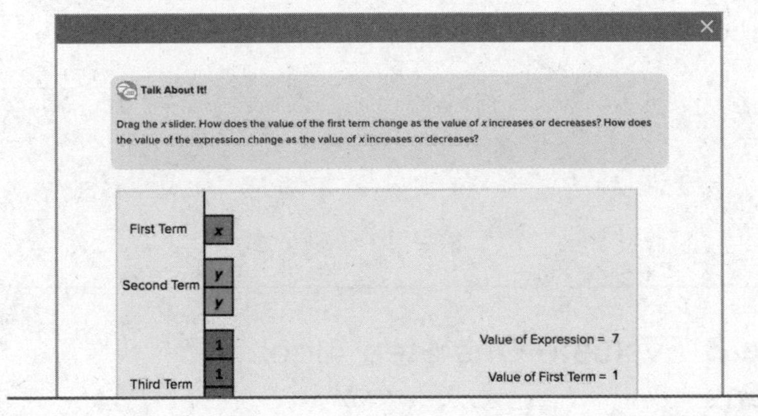

Learn Evaluate Algebraic Expressions

The variables in an algebraic expression can be replaced with a number. Once the variables have been replaced, you can evaluate, or find the value of, the algebraic expression.

Suppose that $x = 5$ in the expression $4x + 2$. The expression can be evaluated by replacing the x with 5 and simplifying according to the order of operations as shown.

$4x + 2 = 4(5) + 2$	Replace x with 5.
$= 20 + 2$	Multiply.
$= 22$	Add.

The expression $4x + 2$ is equal to ________ when $x = 5$.

Talk About It!

Can you evaluate the expression $2x + 5y - 1$ if you know that $x = 3$? Explain your reasoning.

Your Notes

Think About It!

What does it mean to evaluate an expression?

Talk About It!

Why is the value of the expression not equal to $6\frac{1}{2}$?

Example 1 Evaluate One-Step Algebraic Expressions

Evaluate the expression 6*b* when $b = \frac{1}{2}$.

$6b$	Write the expression.
$6b = 6 \cdot \frac{1}{2}$	Replace *b* with $\frac{1}{2}$.
$= 3$	Multiply.

So, the value of the expression is ________.

Check

Evaluate $\frac{x}{6}$ when $x = 33$.

Example 2 Evaluate One-Step Algebraic Expressions

Evaluate the expression $x + y$ when $x = \frac{3}{4}$ and $y = \frac{2}{3}$.

$x + y$	Write the expression.
$x + y = \frac{3}{4} + \frac{2}{3}$	Replace *x* with $\frac{3}{4}$ and *y* with $\frac{2}{3}$.
$= \frac{9}{12} + \frac{8}{12}$	Rewrite the fractions with common denominators.
$= \frac{17}{12}$ or $1\frac{5}{12}$	Add.

So, the value of the expression is ________.

Check

Evaluate $a + b$ when $a = \frac{5}{6}$ and $b = 3\frac{1}{4}$.

Go Online You can complete an Extra Example online.

Example 3 Evaluate Multi-Step Algebraic Expressions

Evaluate $(5x - 4y) \div z^2$ when $x = 4$, $y = \frac{1}{2}$, and $z = 3$.

$(5x - 4y) \div z^2$ — Write the expression.

$(5x - 4y) \div z^2 = \left(5 \cdot 4 - 4 \cdot \frac{1}{2}\right) \div 3^2$ — Replace x with 4, y with $\frac{1}{2}$, and z with 3.

$= (20 - 2) \div 9$ — Multiply.

$= 18 \div 9$ — Subtract.

$= 2$ — Divide.

So, the value of the expression is ________.

Check

Evaluate $\frac{x}{4} + 2(y^2 - 3z)$ when $x = 12$, $y = 7$, and $z = 8$.

Go Online You can complete an Extra Example online.

Pause and Reflect

Compare and contrast evaluating one-step and multi-step algebraic expressions. Do the differences affect your approach to evaluating the expressions? If yes, what do you do differently? If no, then explain why your approach remains the same.

Record your observations here

Think About It!

How would you begin solving the problem?

Talk About It!

Explain how the Commutative Property allows you to multiply $\frac{1}{2} \cdot 19 \cdot 9.8$.

Example 4 Use Algebraic Expressions

The expression $\frac{1}{2}a(b + c)$ can be used to find the area of the trapezoid.

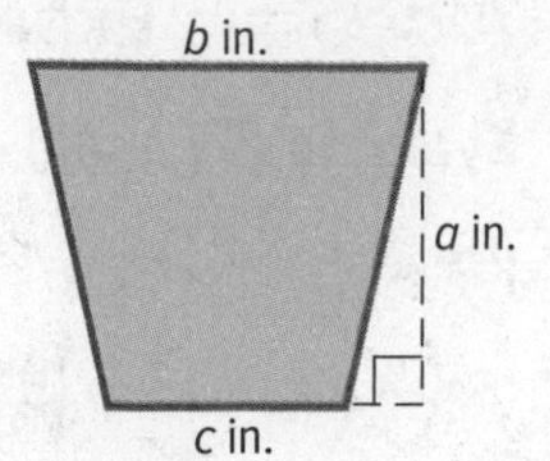

What is the area of the trapezoid when $a = 9.8$, $b = 12$, and $c = 7$?

$\frac{1}{2}a(b + c)$	Write the expression.
$\frac{1}{2}a(b + c) = \frac{1}{2} \cdot 9.8(12 + 7)$	Replace a with 9.8, b with 12, and c with 7.
$= \frac{1}{2} \cdot 9.8(19)$	Simplify inside the parentheses.
$= 4.9(19)$	Multiply.
$= 93.1$	Multiply.

So, the area of the trapezoid is _______ square inches.

Check

The expression $\frac{1}{2}a(b + c)$ can be used to find the area of the trapezoid. What is the area of the trapezoid when $a = 3.7$, $b = 6.4$, and $c = 3.6$?

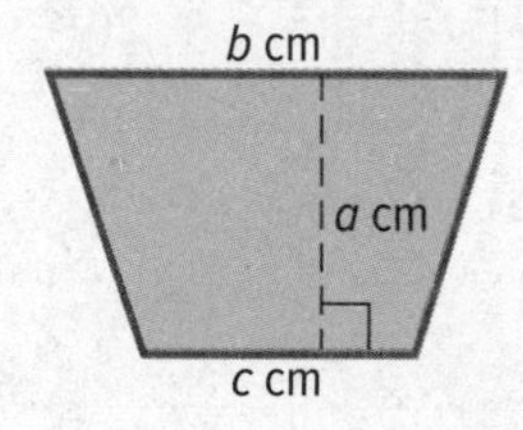

Go Online You can complete an Extra Example online.

Pause and Reflect

Did you encounter difficulty when using algebraic expressions in this Example and Check? What are some helpful tips you could give a classmate who encountered difficulty when using algebraic expressions?

Apply Woodworking

The table shows the dimensions of three rectangular picture frame sizes that Martina is making. How much wood is needed to make two small frames and three large frames? The perimeter of a rectangle is $2\ell + 2w$, where ℓ is the length and w is the width.

Picture Frame Size	Length (in.)	Width (in.)
Small	3	5
Medium	5	7
Large	8	10

1 What is the task?

Make sure you understand exactly what question to answer or problem to solve. You may want to read the problem three times. Discuss these questions with a partner.

First Time Describe the context of the problem, in your own words.
Second Time What mathematics do you see in the problem?
Third Time What are you wondering about?

2 How can you approach the task? What strategies can you use?

3 What is your solution?

Use your strategy to solve the problem.

4 How can you show your solution is reasonable?

Write About It! Write an argument that can be used to defend your solution.

Talk About It!
Suppose the dimensions of the frames were doubled. How would this affect the perimeter of the frames?

Check

At a garage sale, Georgia found some used DVDs and CDs that she wanted to buy. Each DVD costs \$3 and each CD costs \$2. She also has the option of paying \$25 for the entire box of DVDs and CDs. Evaluate the expression $3d + 2c$ when $d = 4$ and $c = 7$ to find the cost of 4 DVDs and 7 CDs. What is the difference between the cost of buying the entire box and buying the items individually?

Show your work here

Go Online You can complete an Extra Example online.

Pause and Reflect

Describe a real-world scenario when it would be advantageous to use an algebraic expression to solve a problem. How will the concepts you learned in this lesson help you to evaluate any algebraic expression you encounter?

Record your observations here

Name ____________ Period ________ Date ________

Practice

Go Online You can complete your homework online.

Evaluate each expression when $x = \frac{3}{4}$ and $y = 2.5$. (Example 1)

1. $8x$

2. y^2

3. $\frac{10}{y}$

Evaluate each expression when $a = \frac{2}{3}$, $b = \frac{4}{5}$, and $c = 6$. Write in simplest form. (Example 2)

4. $a + b$

5. $c - b$

6. $b - a$

Evaluate each expression when $a = 4$, $b = 3$, and $c = \frac{1}{3}$. (Example 3)

7. $(3a + 18c) \div b^2$

8. $(a^2 + 12c) \div (7b - 1)$

9. $(2b + 3a)(c^2)$

10. The expression $\frac{1}{2}a(b + c)$ can be used to find the area of the trapezoid. What is the area of the trapezoid when $a = 5.5$, $b = 5$, and $c = 7.2$? (Example 4)

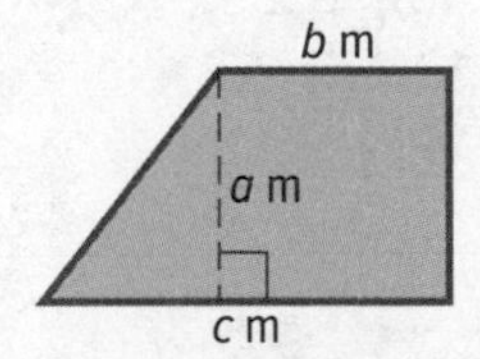

11. The expression $\frac{1}{2}a(b + c)$ can be used to find the area of the trapezoid. What is the area of the trapezoid when $a = 4.4$, $b = 8$, and $c = 3$? (Example 4)

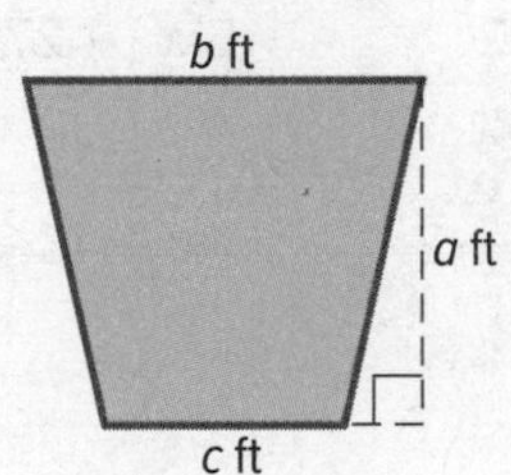

12. The perimeter of a rectangle can be found using the expression $2\ell + 2w$, where ℓ represents the length and w represents the width. Find the perimeter when $\ell = 6.2$ units and $w = 3.5$ units.

Test Practice

13. Equation Editor What is the value of the expression when $x = 7$, $y = \frac{1}{2}$, and $z = 8$? (Example 3)

$(24y + 2x) \div \left(\frac{1}{4}z\right)$

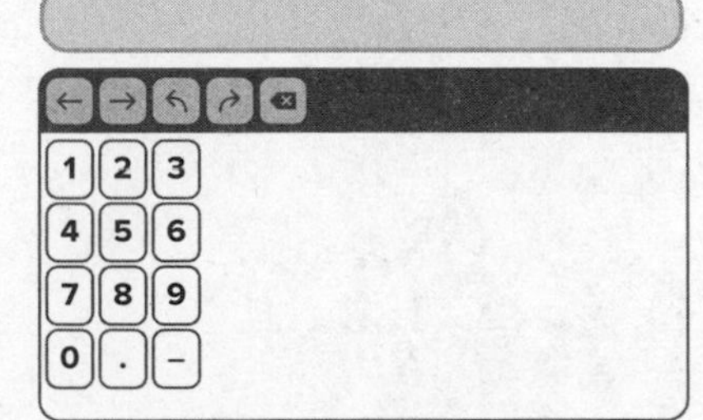

Apply

14. Mr. Young is replacing the fencing around his rabbit pens and garden. The table shows the dimensions of the different areas. How many feet of fencing will he need to replace two rabbit pens and his garden? The perimeter of a rectangle is $2\ell + 2w$, where ℓ is the length and w is the width.

Item	Length (ft)	Width (ft)
Rabbit Pen	3.5	4.5
Garden	12	10

15. Angel is comparing the price to print shirts for summer camp at two companies. Company A charges an initial fee of $50 and $12 per shirt. Company B charges an initial fee of $10 and $15 per shirt. Evaluate the expressions $50 + 12x$ and $10 + 15x$ for $x = 40$ to find the total cost to print 40 shirts at each company. What is the difference in cost between the companies?

16. Which One Doesn't Belong? Circle the expression that does not equal 13 when $x = 3$.

$5x - 2$ $\quad$ $5x^2 - 27 + 5$

$(x^3 - 1) \div 2$ $\quad$ $x + 13 - x$

17. MP **Find the Error** A student was evaluating $4b + c$ for $b = 2$ and $c = 3$. Find the student's mistake and correct it.

$$4b + c = 4(3) + 2$$
$$= 12 + 2$$
$$= 14$$

18. MP **Be Precise** Compare and contrast algebraic expressions and numerical expressions.

19. Give an example of an algebraic expression and a numerical expression that have the same value when evaluated.

Factors and Multiples

I Can... find the greatest common factor and least common multiple of two whole numbers.

What Vocabulary Will You Learn?
common factor
greatest common factor
least common multiple

Explore Greatest Common Factor

Online Activity You will find the greatest common factor of two whole numbers.

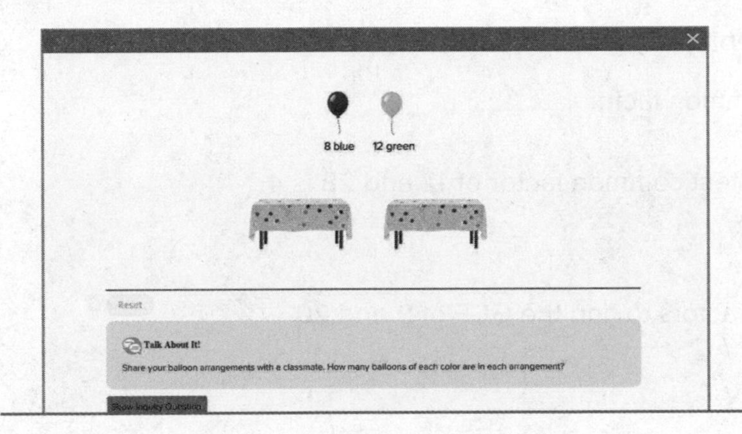

Learn Greatest Common Factor

A **common factor** is a number that is a factor of two or more numbers. The greatest of the common factors of two or more numbers is the **greatest common factor** (GCF).

You can find the GCF of two or more numbers using different methods. Some of these methods include:

- listing the factors
- making a factor tree

Go Online Watch the animation to learn how to find the GCF by listing the factors.

The animation shows the lists of factors of each number used to find the GCF of 9, 15, and 18.

factors of 9: 1, 3, 9

factors of 15: 1, 3, 5, 15

factors of 18: 1, 2, 3, 6, 9, 18

The common factors are 1 and 3.

Since 3 is greater than 1, the greatest common factor is 3.

Talk About It!
When is making a list of the factors difficult to do?

Example 1 Find the GCF by Using a List of Factors

Use a list of factors to find the greatest common factor of 12 and 28.

Step 1 List the factors of each number.

factors of 12: 1, 2, 3, 4, 6, 12

factors of 28: 1, 2, 4, 7, 14, 28

Step 2 Identify the common factors.

common factors: 1, 2, 4

Step 3 Identify the greatest common factor.

greatest common factor: ________

So, the greatest common factor of 12 and 28 is 4.

Check

Use a list of factors to find the GCF of 9 and 20.

Go Online You can complete an Extra Example online.

Example 2 Find the GCF by Using a Factor Tree

Use factor trees to find the greatest common factor of 52 and 78.

For greater numbers, listing all of the factors can be inefficient. A factor tree is another method you can use to find the GCF.

Complete each factor tree.

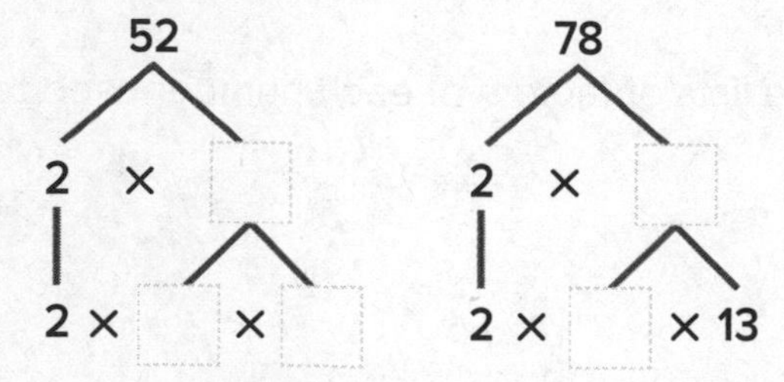

(continued on next page)

Look at the bottom row of the factor trees. The common prime factors

are ________ and ________.

Because 2 and 13 are factors of both 52 and 78, the product of 2 and 13 is also a factor of both numbers. Multiply the common prime factors to find the GCF.

So, the GCF of 52 and 78 is 2×13, or 26.

Check

Use factor trees to find the GCF of 45 and 75.

Show your work here

Go Online You can complete an Extra Example online.

Explore Least Common Multiple

Online Activity You will use Web Sketchpad to find the least common multiple of two whole numbers.

Talk About It!

Does changing the order of the factors result in a different GCF? Explain.

Learn Least Common Multiple

The least nonzero number that is a multiple of two or more whole numbers is the **least common multiple** (LCM) of the numbers.

You can find the least common multiple of a set of whole numbers by:

- listing the multiples
- using a number line

Go Online Watch the animation to learn how to find the LCM of 4 and 6 by listing the nonzero multiples.

The animation shows the lists of the first six nonzero multiples of each number.

multiples of 4: 4, 8, 12, 16, 20, 24, ...

multiples of 6: 6, 12, 18, 24, 30, 36, ...

The common multiples in the list are 12 and 24.

Since 12 is less than 24, the least common multiple is 12.

You can also find the least common multiple of a set of whole numbers by using a number line.

0 1 2 3 4 5 6 7 8 9 10 11 12 13 14 15 16 17 18 19 20 21 22 23 24 25

The least number with both an X and a dot is 12.

So, the least common multiple is 12.

Think About It!
When will Ernesto be back at the community center? Will Kamala be there? How do you know?

Example 3 Find the LCM by Using a List of Multiples

Ernesto is at the community center every 8 weeks for his painting class. Kamala is at the community center every 6 weeks for her pottery class. They were both at the center for their classes this week.

How many weeks will it be until they both have their classes in the same week again?

Step 1 List the multiples of each number.

multiples of 8: ______________________

multiples of 6: ______________________

(continued on next page)

Step 2 Identify the least common multiple.

Circle the multiples that 8 and 6 have in common.

multiples of 8: 8, 16, 24, 32, 40, 48, 56, 64...

multiples of 6: 6, 12, 18, 24, 30, 36, 42, 48...

least common multiple: ________

So, Ernesto and Kamala have their classes in the same week again in 24 weeks.

Check

Every 10th week Tamika visits the zoo. Every 12th week she visits the local pet rescue. If she visited both this week, how many weeks will it be until she visits both in the same week again?

Show your work here

Pause and Reflect

Describe the differences between finding the greatest common factor and finding the least common multiple. How will knowing these differences be helpful when checking the accuracy of your answers?

Record your observations here

Talk About It!

Can the least common multiple of two numbers ever be one of the given numbers? If so, give an example.

Example 4 Find the LCM by Using a Number Line

Use a number line to find the least common multiple of 2 and 3.

Place an X above each multiple of 2.

Place a dot above each multiple of 3.

0 1 2 3 4 5 6 7 8 9 10 11 12

What is the least number with both an X and a dot? ______

So, the least common multiple of 2 and 3 is 6.

Check

Use a number line to find the LCM of 2 and 8.

0 1 2 3 4 5 6 7 8 9 10 11 12

Show your work here

Go Online You can complete an Extra Example online.

Pause and Reflect

Compare and contrast using a number line and using a list of multiples to find the least common multiple. When might using a number line be more advantageous? When might using a list be more advantageous?

Record your observations here

Apply School Supplies

The table shows the supplies a school supply store has left at the end of the week. The store manager wants to put pencils and notepads together in bags to sell as a combo pack, and he wants to make the greatest number of bags possible. If all of the pencils and notepads are distributed evenly among all of the bags, and the store charges $4 per bag, how much money will the store bring in if they sell all of the bags?

Item	Number
Pencils	48
Pens	32
Erasers	60
Notepads	36

Go Online watch the animation.

1 What is the task?

Make sure you understand exactly what question to answer or problem to solve. You may want to read the problem three times. Discuss these questions with a partner.

First Time Describe the context of the problem, in your own words.
Second Time What mathematics do you see in the problem?
Third Time What are you wondering about?

2 How can you approach the task? What strategies can you use?

3 What is your solution?

Use your strategy to solve the problem.

4 How can you show your solution is reasonable?

Write About It! Write an argument that can be used to defend your solution.

Talk About It!

Suppose the manager wanted to distribute the erasers evenly to the combo packs in addition to the pencils and notepads. Would adding the erasers alter the number of combo packs that can be made? Explain your reasoning.

Check

A gardener has 27 pansies and 36 daisies to plant in identical rows in a community flower garden. It costs $5 to plant each row. How much will it cost if he plants the greatest number of rows possible with no flowers leftover?

Show your work here

Go Online You can complete an Extra Example online.

Pause and Reflect

Create a graphic organizer that will help you study the concepts of greatest common factor and least common multiple. You might want to consider including multiple methods of finding each.

Record your observations here

Name ____________________ Period __________ Date __________

Practice

Go Online You can complete your homework online.

Use any method to find the greatest common factor of each pair of numbers. (Examples 1 and 2)

1. 12, 30

2. 4, 16

3. 9, 36

4. 35, 63

5. 42, 56

6. 54, 81

7. On every fourth visit to the hair salon, Margot receives a discount of $5. On every tenth visit, she receives a free hair product. After how many visits will Margot receive the discount and a free product at the same time? (Example 3)

8. The table shows the city bus schedule for certain bus lines. Both buses are at the bus stop right now. In how many minutes will both buses be at the bus stop again? (Example 3)

Bus Line	Arrives at the bus stop every...
A	25 minutes
B	15 minutes

Use any method to find the least common multiple of each pair of numbers. (Example 4)

9. 4, 6

10. 3, 5

11. Monique has the flowers shown in the table. She wants to put all the flowers into decorative vases. Each vase must have the same number of flowers in it. Without mixing flowers, what is the greatest number of flowers that Monique can put in each vase?

Flower Type	Number
Daisies	20
Roses	25

Test Practice

12. Equation Editor What is the greatest common factor of 35 and 28?

← → ↶ ↷ ⌫

1 2 3
4 5 6
7 8 9
0 . −

Apply

13. The table shows the number of each type of cookie a bakery has left at the end of the day. The baker wants to make the greatest number of cookie boxes to sell, using chocolate chip and sugar cookies together. If all of the chocolate chip and sugar cookies are distributed evenly among the boxes and the baker charges $5 per box, how much money will the bakery bring in if they sell all of the boxes?

Type of Cookie	Number
Chocolate Chip	26
Oatmeal Raisin	34
Peanut Butter	18
Sugar	39

14. A teacher needs to purchase notebooks and pencils for her students. Notebooks come in packages of 6 and pencils in packages of 10. The table shows the cost of the items. What is the least amount of money the teacher can spend and have the same number of notebooks and pencils?

Item	Cost ($)
Folder Packages	3.50
Notebook Packages	5.00
Pencil Packages	2.00

15. MP **Identify Structure** Explain how the Commutative Property is applied when finding the GCF using factor trees.

16. MP **Make a Conjecture** A student is finding the GCF of 6 and 12. Without computing, will the GCF be odd or even? Explain.

17. MP **Use a Counterexample** Determine if the statement is *true* or *false*. If true, support with an example. If false, give a counterexample.

If one number is a multiple of another number, the LCM is the lesser of the two numbers.

18. MP **Make a Conjecture** Can two different pairs of numbers have the same LCM? Explain.

Use the Distributive Property

I Can... use the Distributive Property to evaluate numerical expressions, to rewrite algebraic expressions, and to factor numerical and algebraic expressions.

What Vocabulary Will You Learn?
Distributive Property
factoring the expression

Explore Use Algebra Tiles to Model the Distributive Property

Online Activity You will use algebra tiles to investigate the Distributive Property.

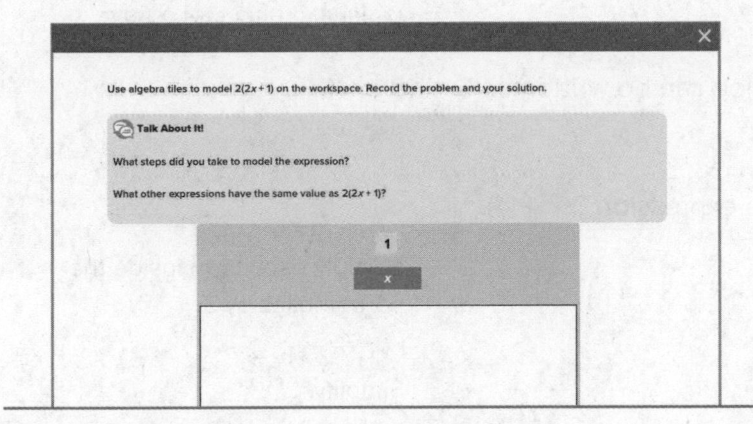

Learn The Distributive Property

The **Distributive Property** states that to multiply a sum by a number, multiply each term inside the parentheses by the number outside the parentheses.

Words
To multiply a sum by a number, multiply each addend by the number outside the parentheses.
Numbers
$2(5 + 6) = 2(5) + 2(6)$
Variables
$a(b + c) = ab + ac$

(continued on next page)

Your Notes

Talk About It!

When there are only numbers in an expression, such as 3(4 + 9), is the Distributive Property the only way to evaluate the expression?

Talk About It!

Does the Distributive Property apply to subtraction? For example, does $3(8-2) = 3(8) - 3(2)$? Does it apply to all numbers? Explain.

Go Online Watch the animation to see how to use the Distributive Property to expand an expression.

The animation shows how to expand the expression $a(b + c)$.

$a(b + c) = a \cdot b + a \cdot c$ — Multiply each term inside the parentheses by a. Then simplify.

The expression can be written as $ab + ac$.

Consider the expression $2(3 + 5)$.

$2(3 + 5) = 2 \cdot 3 + 2 \cdot 5$ — Multiply each term inside the parentheses by 2.

$6 + 10$ — Multiply 2 by 3 and 2 by 5.

The expression can be written as $6 + 10$ or 16.

Consider the expression $3(x + 4)$.

$3(x + 4) = 3 \cdot x + 3 \cdot 4$ — Multiply each term inside the parentheses by 3.

$3x + 12$ — Simplify.

The expression can be written as $3x + 12$.

Pause and Reflect

What questions do you have about the Distributive Property as a result of this Learn? Can you begin to think of an instance where the Distributive Property could be beneficial?

Example 1 Use the Distributive Property

Use the Distributive Property to expand $2(x + 3)$.

$2(x + 3) = 2(x) + 2(3)$ Distributive Property

$= 2x + 6$ Multiply.

So, $2(x + 3)$ can be written as ______.

Check

Use the Distributive Property to expand $8(x + 3)$.

Example 2 Use the Distributive Property

Use the Distributive Property to find $8 \cdot 3\frac{1}{2}$.

$8 \cdot 3\frac{1}{2}$ Write the expression.

$8 \cdot 3\frac{1}{2} = 8\left(3 + \frac{1}{2}\right)$ Write $3\frac{1}{2}$ as $\left(3 + \frac{1}{2}\right)$.

$= 8(3) + 8\left(\frac{1}{2}\right)$ Distributive Property

$= 24 + 4$ Multiply.

$= 28$ Add.

So, $8 \cdot 3\frac{1}{2}$ is ______.

Check

Use the Distributive Property to find $12 \cdot 2\frac{1}{4}$.

Go Online You can complete an Extra Example online.

Talk About It!

How can you use algebra tiles to verify you expanded the expression correctly?

Think About It!

How can you rewrite $3\frac{1}{2}$ as a sum of two terms?

Talk About It!

Can you use the Distributive Property to multiply a one-digit number by a two-digit number, such as 9×37? Explain your reasoning.

Learn Greatest Common Factor and the Distributive Property

You can use the greatest common factor (GCF) to rewrite the sum of two whole numbers with a common factor as a product. The Distributive Property allows you to write the sum as the product of the greatest common factor and the sum of the remaining factors.

When numerical or algebraic expressions are written as a product of their factors, the process is called **factoring the expression**. To factor an expression, follow these steps.

1. Find the GCF of the terms.
2. Write the terms as a product of factors.
3. Rewrite the expression as the product of two terms.

Talk About It!
How can you determine what remains in the parentheses after the GCF has been factored out of the expression?

Go Online Watch the animation to see how to use the GCF and the Distributive Property to factor an expression.

The animation explains how to factor the expression $8 + 56$.

$8 = 2 \cdot 2 \cdot 2$
$56 = 2 \cdot 2 \cdot 2 \cdot 7$ — Find the GCF of the terms.

The GCF is 8.

$8 + 56 = 8(1) + 8(7)$ — Use the GCF to write each term as a product of factors.

$= 8(1 + 7)$ — Rewrite the expression as a product of two terms.

You can also use the GCF to factor expressions containing variables, such as $45x + 6$.

$45x = 3 \cdot 3 \cdot 5 \cdot x$
$6 = 3 \cdot 2$ — Find the GCF of the terms.

The GCF is 3.

$45x + 6 = 3(15x) + 3(2)$ — Use the GCF to write each term as a product of factors.

$= 3(15x + 2)$ — Rewrite the expression as a product of two terms.

Example 3 Use GCF to Factor Numerical Expressions

Use the GCF to factor 45 + 72.

$45 + 72$	Write the expression.
$45 + 72 = 9(5) + 9(8)$	Rewrite each term as a product of the GCF, 9, and its remaining factor.
$= 9(5 + 8)$	Use the Distributive Property to write as the product of two terms.

So, 45 + 72 in factored form is ______________.

Think About It!

How can you find the GCF of 45 and 72?

Talk About It!

Are the expressions 9(5 + 8) and (5 + 8)9 equal to the same value? Explain your reasoning.

Check

Use the GCF to factor 80 + 56.

Go Online You can complete an Extra Example online.

Pause and Reflect

How did your prior knowledge of greatest common factor help you to understand the concepts in this Learn and Example?

Example 4 Use GCF to Factor Algebraic Expressions

Use the GCF to factor $6x + 15$.

$6x + 15$	Write the expression.
$6x + 15 = 3(2x) + 3(5)$	Rewrite each term as a product of the GCF, 3, and its remaining factor.
$= 3(2x + 5)$	Use the Distributive Property to write as the product of two terms.

So, $6x + 15$ in factored form is ________.

Think About It!
What is the GCF of the two terms?

Talk About It!
Is it possible to factor an expression using a factor other than the GCF? Explain.

Check

Factor $36x + 30$. Use the GCF.

Go Online You can complete an Extra Example online.

Pause and Reflect

Did you make any errors while factoring the algebraic expression in the Check exercise? If so, was it in finding the GCF or rewriting each term? If not, how could you check the accuracy of your answer?

Apply Money

Wen is buying bottles of apple juice and wants to mentally calculate how much they will cost. He buys 5 bottles of juice at $2.15 each. Use mental math and the Distributive Property to determine how much change he will receive from $20.

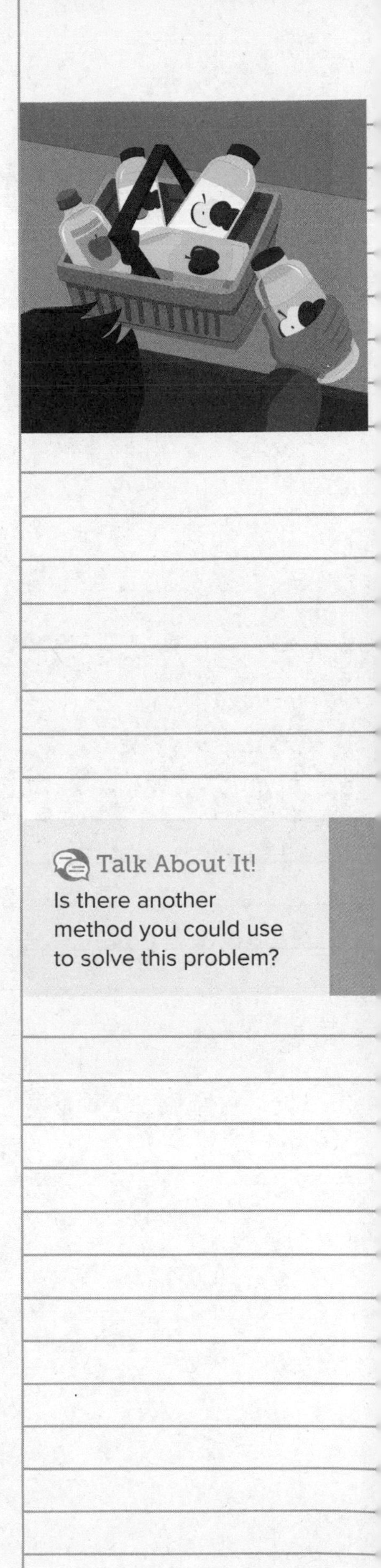

1 What is the task?

Make sure you understand exactly what question to answer or problem to solve. You may want to read the problem three times. Discuss these questions with a partner.

First Time Describe the context of the problem, in your own words.
Second Time What mathematics do you see in the problem?
Third Time What are you wondering about?

2 How can you approach the task? What strategies can you use?

3 What is your solution?

Use your strategy to solve the problem.

Talk About It!

Is there another method you could use to solve this problem?

4 How can you show your solution is reasonable?

Write About It! Write an argument that can be used to defend your solution.

Check

Martin exercised four days for 65 minutes each day. His goal is to exercise for a total of 300 minutes in 5 days. How many minutes does he need to exercise on the fifth day to meet his goal?

Go Online You can complete an Extra Example online.

Pause and Reflect

Write a real-world problem that involves writing an expression using the Distributive Property. Explain how you came up with that problem. Exchange problems with a classmate and solve each other's problem.

Name ______________________ Period __________ Date __________

Practice

Go Online You can complete your homework online.

Use the Distributive Property to expand each algebraic expression. (Example 1)

1. $3(x + 8)$

2. $5(6 + x)$

3. $9(3 + x)$

Use the Distributive Property to simplify each expression. (Example 2)

4. $12 \cdot 3\frac{3}{4}$

5. $15 \cdot 2\frac{2}{3}$

6. $8 \cdot 4\frac{1}{2}$

Use the GCF to factor each numerical expression. (Example 3)

7. $16 + 48$

8. $35 + 63$

9. $26 + 39$

Use the GCF to factor each algebraic expression. (Example 4)

10. $8x + 16$

11. $24 + 6x$

12. $42 + 7x$

13. Five friends each bought a shirt and a pair of shoes. The table shows the cost of the items. The expression $5(x + 24)$ shows the total amount of money they spent. Expand the expression using the Distributive Property.

Item	Cost ($)
Shirt	x
Shoes	24.00

Test Practice

14. Multiple Choice Which expression has the same value as $9 + 24$?

Ⓐ $3(3 + 24)$

Ⓑ $3(3 + 8)$

Ⓒ $3(9 + 8)$

Ⓓ $9(1 + 24)$

Apply

15. The table shows the cost of snacks at a basketball game. Mrs. Cooper buys 6 nachos for her daughter and 5 friends. Use mental math and the Distributive Property to determine how much change she will receive from \$30.

Item	Cost
Nachos	\$4.10
Popcorn	\$2.85

16. Jeffery is making 4 batches of chocolate chip cookies. Each batch of cookies needs $2\frac{3}{4}$ cups of chocolate chips. If he has 96 ounces of chocolate chips, how many ounces will be left over? Use mental math and the Distributive Property.

17. MP **Identify Structure** Write two equivalent numerical expressions involving fractions that illustrate the Distributive Property.

18. MP **Justify Conclusions** A student rewrote the expression $4(5 + x)$ as $20 + x$. Did the student rewrite the expression correctly? Justify your reasoning.

19. MP **Construct an Argument** Is the expression $2(6x)$ equivalent to $(2 \cdot 6)(2 \cdot x)$? Explain why or why not.

20. Are the expressions $4(x + 5) + 1$ and $(2x + 16) + 2x + 5$ equivalent? Explain.

Equivalent Algebraic Expressions

I Can... use the properties of operations to write expressions in simplest form and check to see if two expressions are equivalent.

What Vocabulary Will You Learn?
Associative Property
Commutative Property
Distributive Property
equivalent expressions
Identity Property
simplest form

Explore Properties and Equivalent Expressions

Online Activity You will use algebra tiles and mathematical properties to identify equivalent expressions.

If $4 \div 2$ means "How many groups of 2 are in 4?" what does $4 \div \frac{1}{2}$ mean? Use the model to simplify the expression.

Talk About It!

What does $4 \div \frac{1}{2}$ mean? How can you use the model to find the answer?

Follow the dashed line to draw the first half. Begin by drawing a line from the top to the bottom of the bar.

Previous Stroke Next Stroke

Learn Use Properties to Identify Equivalent Expressions

Equivalent expressions are expressions that have the same value. Algebraic expressions are equivalent when they have the same value, no matter what value is substituted for the variable(s). You can write an equivalent expression by applying the properties of operations to an expression.

Commutative Property	Associative Property
Words	**Words**
The order in which numbers are added or multiplied does not change the sum or product.	The order in which numbers are grouped when added or multiplied does not change the sum or product.
Numbers	**Numbers**
$7 + 9 = 9 + 7$ $8 \cdot 4 = 4 \cdot 8$	$3 + (4 + 7) = (3 + 4) + 7$ $7 \cdot (6 \cdot 2) = (7 \cdot 6) \cdot 2$
Variables	**Variables**
$a + b = b + a$ $a \cdot b = b \cdot a$	$a + (b + c) = (a + b) + c$ $a \cdot (b \cdot c) = (a \cdot b) \cdot c$

(continued on next page)

Your Notes

Talk About It!

Are the expressions that are written before and after a property is applied equivalent? Explain your reasoning.

Distributive Property

Words
To multiply a sum by a number, multiply each addend by the number outside the parentheses.
Numbers
$2(7 + 9) = 2(7) + 2(9)$ $4(5 - 2) = 4(5) - 4(2)$
Variables
$a(b + c) = a(b) + a(c)$ $a(b - c) = a(b) - a(c)$

Identity Property

Words
The sum of an addend and 0 is the addend. The product of a factor and 1 is the factor.
Numbers
$13 + 0 = 13$ $7 \cdot 1 = 7$
Variables
$a + 0 = a$ $a \cdot 1 = a$

Go Online Watch the animation to see how to use properties to identify equivalent expressions.

The animation explains how to use properties to determine whether or not $4(2x + 3) + 5$ and $(4x + 3) + 5$ are equivalent expressions.

Simplify the expressions and draw a conclusion.

$4(2x + 3) + 5 = 8x + 12 + 5$ Distributive Property

$= 8x + (12 + 5)$ Associative Property

$= 8x + 17$

$(4x + 3) + 5 = 4x + (3 + 5)$ Associative Property

$= 4x + 8$

Because the simplified expressions have different terms, they will *never* have the same value. So, the expressions are *not* equivalent.

The animation also explains how to use properties to determine whether or not $8 + 3n + 2$ and $3(n + 1) + 7$ are equivalent expressions.

Simplify the expressions and draw a conclusion.

$8 + 3n + 2 = 3n + 8 + 2$ Commutative Property

$= 3n + (8 + 2)$ Associative Property

$= 3n + 10$

$3(n + 1) + 7 = 3n + 3 + 7$ Distributive Property

$= 3n + (3 + 7)$ Associative Property

$= 3n + 10$

Because the simplified expressions have the same terms, they will *always* have the same value. So, the expressions are equivalent.

Example 1 Identify Equivalent Expressions

Use the properties of operations to determine whether or not $3(x + 7) + 2$ and $5 + 3(x + 6)$ are equivalent.

Step 1 Simplify the first expression.

$3(x + 7) + 2 = 3x + 21 + 2$	Distributive Property
$= 3x + (21 + 2)$	Associative Property
$= 3x + 23$	Add.

Step 2 Simplify the second expression.

$5 + 3(x + 6) = 5 + 3x + 18$	Distributive Property
$= 3x + 18 + 5$	Commutative Property
$= 3x + (18 + 5)$	Associative Property
$= 3x + 23$	Add.

So, the expressions $3(x + 7) + 2$ and $5 + 3(x + 6)$ ______ equivalent.

Check

Use the properties of operations to determine whether or not $\frac{1}{2}a + \frac{1}{2}b$ and $\frac{1}{2}(a + b)$ are equivalent.

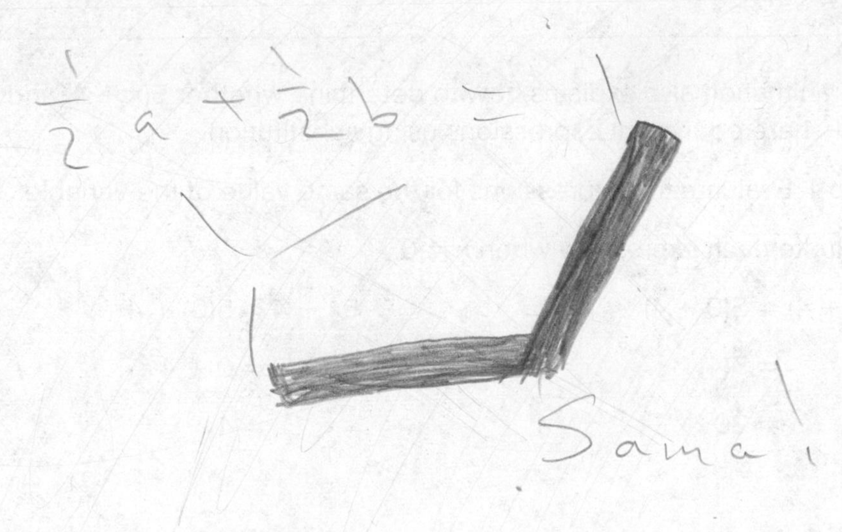

Think About It!

What determines whether two expressions are equivalent?

Talk About It!

What are some expressions that are not equivalent to $3x + 23$?

Go Online You can complete an Extra Example online.

Learn Use Substitution to Identify Equivalent Expressions

Go Online Watch the animation to learn about using substitution to identify when expressions are equivalent.

The animation explains how to determine whether $y + y + y$ and $3y$ are equivalent expressions using substitution.

Step 1 Evaluate the expressions for the same value of the variable.

Evaluate each expression when $y = 0$.

$y + y + y = 0 + 0 + 0$
$= 0$

$3y = 3(0)$
$= 0$

Repeat Step 1 using a different value for the variable. Evaluate each expression when $y = 5$.

$y + y + y = 5 + 5 + 5$
$= 15$

$3y = 3(5)$
$= 15$

Talk About It!

Why is it important to substitute more than one value into an expression to help determine equivalency?

Step 2 Draw a conclusion.

Based on substitution, the expressions appear to be equivalent. If you continue substituting additional values of x, you will see that the expressions will always be equivalent, regardless of the value of the variable being substituted.

The animation also explains how to determine whether $5(x + 4)$ and $5x + 4$ are equivalent expressions using substitution.

Step 1 Evaluate the expressions for the same value of the variable.

Evaluate each expression when $x = 0$.

$5(x + 4) = 5(0 + 4)$
$= 5(4)$
$= 20$

$5x + 4 = 5(0) + 4$
$= 0 + 4$
$= 4$

Step 2 Draw a conclusion.

Based on substitution, the expressions are not equivalent.

Example 2 Determine Equivalency Using Substitution

Use substitution to determine whether or not $2x + x + x$ and $4x$ are equivalent.

Let $x = 0$, 1, and 2. Substitute those values into both expressions. Then compare to determine whether or not they are equivalent.

$2x + x + x$	Write the expression.	$4x$
$2(0) + 0 + 0 = 0$	$x = 0$	$4(0) = 0$
$2(1) + 1 + 1 = 4$	$x = 1$	$4(1) = 4$
$2(2) + 2 + 2 = 8$	$x = 2$	$4(2) = 8$

When x is replaced with different values, the results are the same for both expressions.

So, the expressions are __________ because they ______ have the same value when values are substituted in for the variable.

Talk About It!

Try substituting 3 more values for the variable. What do you notice?

Check

Use substitution to determine whether or not $3x + x + 4$ and $x + 2(x + 1) + x$ are equivalent.

Go Online You can complete an Extra Example online.

Example 3 Determine Equivalency Using Substitution

Use substitution to determine whether or not $\frac{1}{2}x + x^2 + \frac{1}{2}$ and $\frac{1}{2}x + 3x^2 + \frac{1}{2} - x^2$ are equivalent.

$\frac{1}{2}x + x^2 + \frac{1}{2}$	Write the expressions.	$\frac{1}{2}x + 3x^2 + \frac{1}{2} - x^2$
$\frac{1}{2}(0) + (0)^2 + \frac{1}{2} = \frac{1}{2}$	$x = 0$	$\frac{1}{2}(0) + 3(0)^2 + \frac{1}{2} - (0)^2 = \frac{1}{2}$
$\frac{1}{2}(1) + (1)^2 + \frac{1}{2} = 2$	$x = 1$	$\frac{1}{2}(1) + 3(1)^2 + \frac{1}{2} - (1)^2 = 3$
$\frac{1}{2}(2) + (2)^2 + \frac{1}{2} = 5\frac{1}{2}$	$x = 2$	$\frac{1}{2}(2) + 3(2)^2 + \frac{1}{2} - (2)^2 = 9\frac{1}{2}$

So, the expressions are ________________ because they ________________ have the same value when $x = 1$ and $x = 2$.

Check

Use substitution to determine whether or not $y + 2y + 3$ and $1 + y + 2(y + 1)$ are equivalent.

Go Online You can complete an Extra Example online.

Pause and Reflect

Did you encounter any difficulty when using substitution to determine equivalency? What are some important things to remember as you progress through this lesson?

Learn Combine Like Terms

An expression is in **simplest form** if it has no like terms and no parentheses. You can use the structure of an algebraic expression to combine like terms and write it in simplest form.

When you have an expression with only constants, such as $5 + 2$, you can combine these terms for a result of 7.

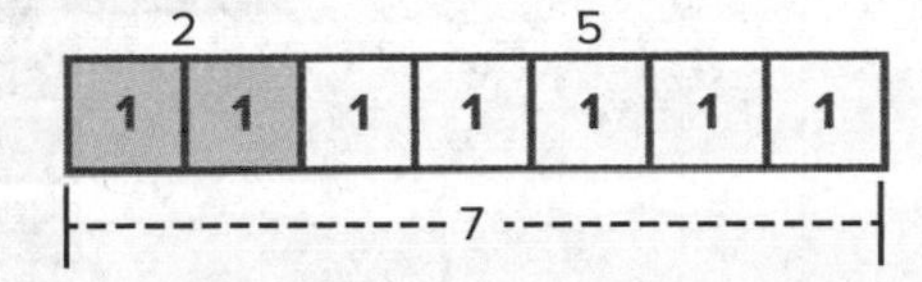

Sometimes you have an expression with like terms, such as $3x + 5x$, which means 3 groups of x plus 5 groups of x. You can combine these terms for a result of $8x$.

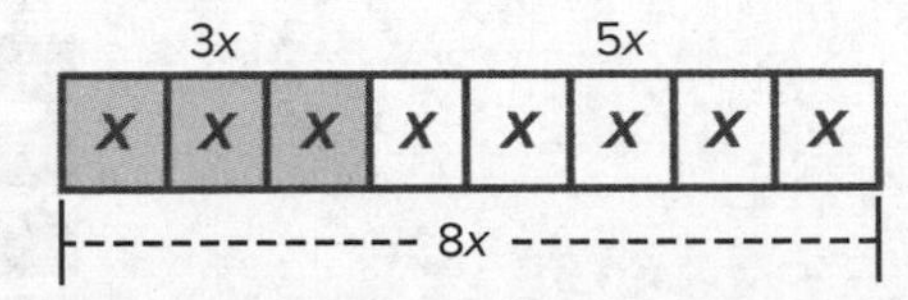

Algebra tiles can also be used to model and simplify an expression that contains like terms.

Go Online Watch the animation to learn about using algebra tiles to combine like terms in an algebraic expression.

The animation demonstrates how to simplify the expression $2x + 4 + 3x$.

To model the expression, place two x-tiles, four 1-tiles, and three more x-tiles on the integer mat.

Combine like tiles.

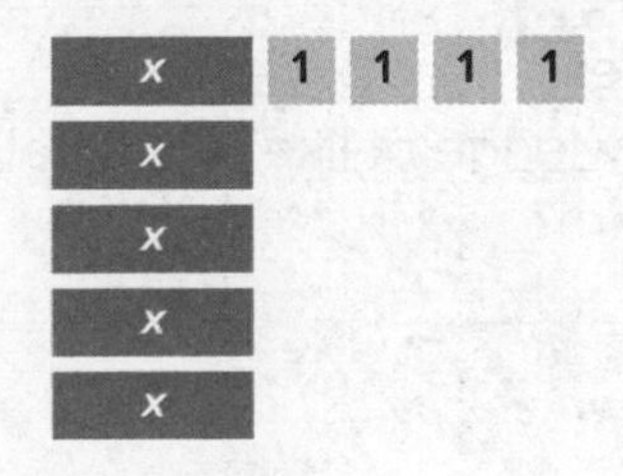

There are five x-tiles and four 1-tiles.

The simplified expression is $5x + 4$.

(continued on next page)

The animation also demonstrates how to simplify the expression $x + 5x + x$.

To model the expression, place one x-tile, then five x-tiles, and then one more x-tile on the integer mat.

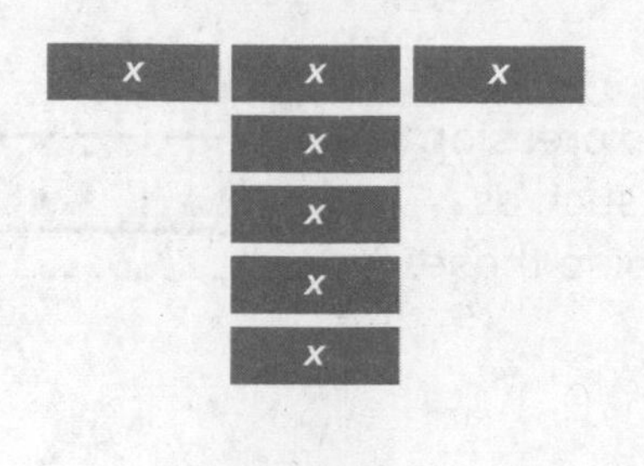

Combine like tiles.

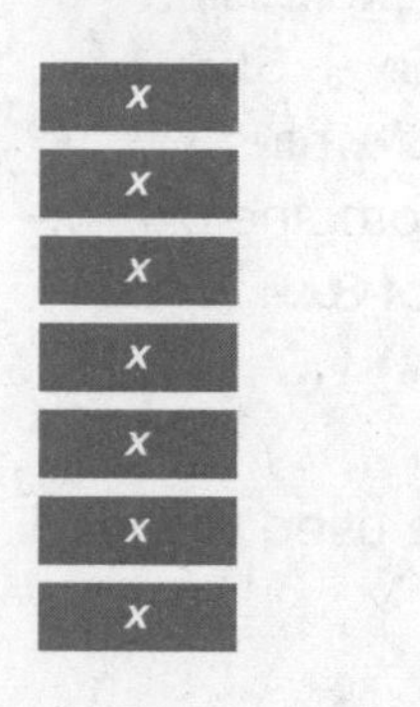

There are seven x-tiles.

The simplified expression is $7x$.

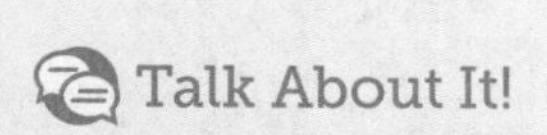
Talk About It!

How can you use the Distributive Property to combine like terms for the expression $2x + 3x + 3x^2 + 6x^2$?

You can also use the Distributive Property to combine like terms. This method allows you to simplify by adding or subtracting the coefficients of the terms.

$3x + 5x = x(3 + 5)$	Factor out the common factor in the terms, x.
$= x(8)$	Add inside parentheses.
$= 8x$	Multiply.

Pause and Reflect

How did your prior knowledge of like terms help you to understand the concepts in this Learn?

Example 4 Combine Like Terms

Simplify $2x + 5 + 4x - 2$.

$2x + 5 + 4x - 2$	Write the expression.
$2x + 5 + 4x - 2 = 2x + 4x + 5 - 2$	Commutative Property.
$= x(2 + 4) + 5 - 2$	Factor.
$= 6x + 5 - 2$	Add.
$= 6x + 3$	Subtract.

So, the simplified expression is ______.

Check

Simplify $2x + 5 + 1x - 1$.

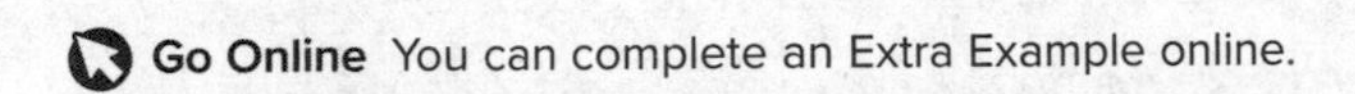

Go Online You can complete an Extra Example online.

Example 5 Combine Like Terms

Simplify $5x^2 + 2x + 2 + x^2 + 6$.

Step 1 Identify like terms.

Write the terms in the appropriate bins that describe them.

x^2 terms	x terms	constants
		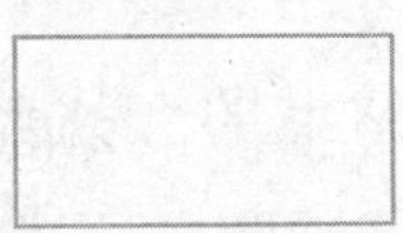

Step 2 Simplify the expression.

$5x^2 + 2x + 2 + x^2 + 6$	Write the expression.
$= 5x^2 + x^2 + 2x + 6 + 2$	Reorder using the Commutative Property.
$= 6x^2 + 2x + 8$	Combine like terms.

So, the simplified expression is ______.

Think About It!
What are the like terms in this expression and what property allows you to reorder the terms?

Talk About It!
Study the expressions $2x + 5 + 4x - 2$ and $6x + 3$. What is the relationship between the coefficients of x in each expression?

Talk About It!
Why is $2x$ not combined with any other terms when the expression $5x^2 + 2x + 2 + x^2 + 6$ is simplified?

Check

Simplify $2x^3 + x^3 + 0.5 + x^2 + 1.5$.

Learn Apply Properties to Write Equivalent Expressions

When you simplify an expression, you can apply properties and combine like terms to write equivalent expressions.

$3(4x + 1) + 2x^2 + x = 12x + 3 + 2x^2 + x$	Distributive Property
$= 2x^2 + 12x + x + 3$	Commutative Property.
$= 2x^2 + 13x + 3$	Combine like terms.

So, $3(4x + 1) + 2x^2 + x$ is equivalent to ______________.

Talk About It!
What property allows you to combine like terms?

Example 6 Write Equivalent Expressions

Simplify $\frac{1}{2}\left(2x^2 + \frac{1}{2}\right) + \frac{2}{5}x^2 + 7$.

$\frac{1}{2}\left(2x^2 + \frac{1}{2}\right) + \frac{2}{5}x^2 + 7 = x^2 + \frac{1}{4} + \frac{2}{5}x^2 + 7$	Distributive Property
$= x^2 + \frac{2}{5}x^2 + \frac{1}{4} + 7$	Commutative Property
$= 1\frac{2}{5}x^2 + 7\frac{1}{4}$	Combine like terms.

So, $\frac{1}{2}\left(2x^2 + \frac{1}{2}\right) + \frac{2}{5}x^2 + 7$ is equivalent to ______________.

Think About It!
What property should be used first to simplify the expression?

Talk About It!
What are some other expressions that are equivalent to $\frac{1}{2}\left(2x^2 + \frac{1}{2}\right) + \frac{2}{5}x^2 + 7$?

Check

Simplify $\frac{1}{4}(4x + 12) + \frac{1}{2}x + 1 + \frac{3}{2}x$.

Go Online You can complete an Extra Example online.

Apply Shipping

Dawit wants to buy some vintage comic books at a local shop and have them shipped to his cousin. The price of a comic book is based on its condition. The table shows the total cost of x number of comic books for each condition. He buys two that are in excellent condition, two that are in good condition, and two that are in fair condition. The shipping cost for the comic books is $5.00. What expression represents the total cost of buying and shipping the comic books?

Condition	Book Costs
Poor	x
Fair	$4.5x$
Good	$9.75x$
Excellent	$18x$
Like New	$25.5x$

1 What is the task?

Make sure you understand exactly what question to answer or problem to solve. You may want to read the problem three times. Discuss these questions with a partner.

First Time Describe the context of the problem, in your own words.
Second Time What mathematics do you see in the problem?
Third Time What are you wondering about?

2 How can you approach the task? What strategies can you use?

3 What is your solution?

Use your strategy to solve the problem.

4 How can you show your solution is reasonable?

Write About It! Write an argument that can be used to defend your solution.

Talk About It!
How would the expression change if you needed to include a tax rate of 8% on the comic books?

Check

Yasmin bought a case of 144 beach hats for $7.39 per hat and a case of 125 pairs of flip-flops for $2.09 per pair. She sold x number of hats for $15.75 each and y number of pairs of flip-flops for $4.95 each. Write an expression that represents Yasmin's profit.

Go Online You can complete an Extra Example online.

Foldables It's time to update your Foldable, located in the Module Review, based on what you learned in this lesson. If you haven't already assembled your Foldable, you can find the instructions on page FL1.

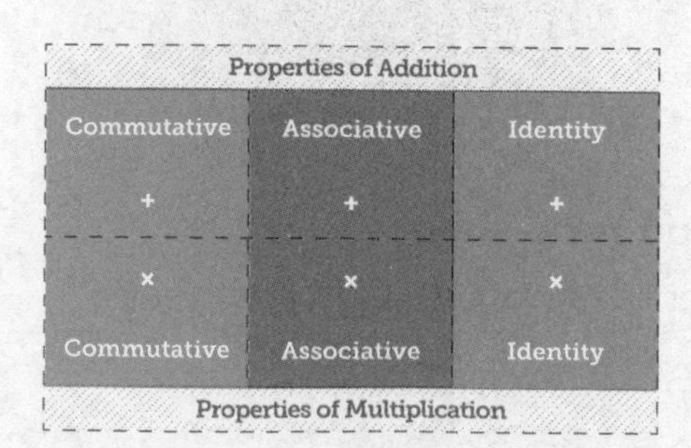

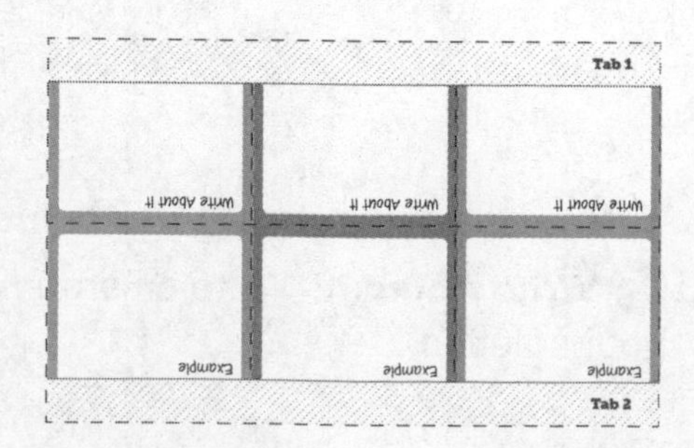

Name ______ Period ______ Date ______

Practice

Go Online You can complete your homework online.

Use properties of operations to determine whether or not the expressions are equivalent. (Example 1)

1. $(x + 10) + x + 9$ and $2(x + 7) + 5$

2. $0.5x + 1$ and $1(0.5x)$

Use substitution to determine whether or not the expressions are equivalent. (Examples 2 and 3)

3. $3x + 2x + x$ and $7x$

4. $x^2 + 1$ and $\frac{2}{3}x^2 + \frac{1}{3}x^2 + 1 + x$

Simplify each expression. (Examples 4 and 5)

5. $3x + 4 + 5x - 1$

6. $10 + 7x - 5 + 4x$

7. $4x^2 + 6x + 8 + x + 2$

8. $\frac{1}{2}x^2 + x + \frac{1}{2} + 2x + \frac{1}{2}x^2$

9. Simplify $\frac{3}{4} + \frac{2}{3}(9x + 6) + 4x + 3\frac{1}{4}$. (Example 6)

Test Practice

10. Multiselect Which of the following are equivalent to $\frac{3}{4}(8x^2 + 1) + 3x^2 + \frac{1}{4}$? Select all that apply.

- ☐ $6x^2 + \frac{3}{4} + 3x^2 + \frac{1}{4}$
- ☐ $6x^2 + 1 + 3x^2 + \frac{1}{4}$
- ☐ $9x^2 + 1\frac{1}{4}$
- ☐ $9x^2 + \frac{3}{4} + \frac{1}{4}$
- ☐ $9x^2 + 2$
- ☐ $9x^2 + 1$

Apply

11. Mrs. Watson is buying vintage records for a friend at a local record shop. The price of a record is based on its condition. The table shows the total cost of x number of records for each condition. She buys 3 that are in good condition, 2 that are in like new condition, and 1 in fair condition. The shipping cost for the records is $8.00. What expression represents the total cost of buying and shipping the records?

Condition	Total Cost
Poor	x
Fair	$5x$
Good	$10.5x$
Like New	$19.95x$

12. Jake is buying baseball cards for his brother in college. The price of a card is based on its condition. The table shows the total cost of x number of cards for each condition. He buys 6 that are in fair condition, 5 that are in good condition, and 2 that are in excellent condition. The shipping cost for the baseball cards is $4.00. What expression represents the total cost of buying and shipping the baseball cards?

Condition	Total Cost
Poor	x
Fair	$1.75x$
Good	$9.5x$
Excellent	$20.5x$
Like New	$45.65x$

13. MP **Identify Structure** Write an expression that when simplified is equivalent to $3y^2 + 2y + \frac{1}{2}$.

14. MP **Justify Conclusions** A student said the expressions $\frac{1}{2}x + 2 + 1\frac{1}{2}x$ and $2x + 2$ are equivalent. Is the student correct? Justify your reasoning.

15. Write two expressions that are equivalent because of the Identity Property of Zero.

16. MP **Reason Inductively** Are the expressions $x^2 + x^2 + x^2$ and $4x^2$ equivalent when $x = 3$? Explain your reasoning.

Review

Foldables Use your Foldable to help review the module.

Tab 1 Properties of Addition

Example	Example	Example
Write About It	Write About It	Write About It

Tab 2 Properties of Multiplication

Rate Yourself!

Complete the chart at the beginning of the module by placing a checkmark in each row that corresponds with how much you know about each topic after completing this module.

Write about one thing you learned.

Write about a question you still have.

Reflect on the Module

Use what you learned about numerical and algebraic expressions to complete the graphic organizer.

Essential Question

How can we communicate algebraic relationships with mathematical symbols?

Expression	Variable	Write a real-world example to represent the given expression. What does the variable represent?
$7x$	x	Each ticket to the school play costs \$7. The variable x represents the number of tickets purchased.
$9 + y$		
$23 - p$		
$\frac{d}{4}$		
$\frac{3}{5}c$		

Name _______________ Period _______ Date _______

Test Practice

1. Multiselect Which expression is equivalent to 5^3? Select all that apply. **(Lesson 1)**

- ☐ $3 \times 3 \times 3 \times 3 \times 3$
- ☐ $5 \times 5 \times 5$
- ☐ $5 \times 5 \times 5 \times 5 \times 5$
- ☐ 15
- ☐ 125

2. Equation Editor Market researchers are studying the effects of sending an advertisement through text messaging. On the first day of the advertisement program, the researcher sent a text message to 8 people. On the next day, each of those people will send the text message to another 8 people, and so on. The pattern of sending the advertisement through text messaging is shown in the table. **(Lesson 1)**

Number of Days	Number of People Receiving Text Message
1	8
2	8×8
3	$8 \times 8 \times 8$
4	$8 \times 8 \times 8 \times 8$

Predict the number of people who will receive the text message on the 8th day of the advertising program.

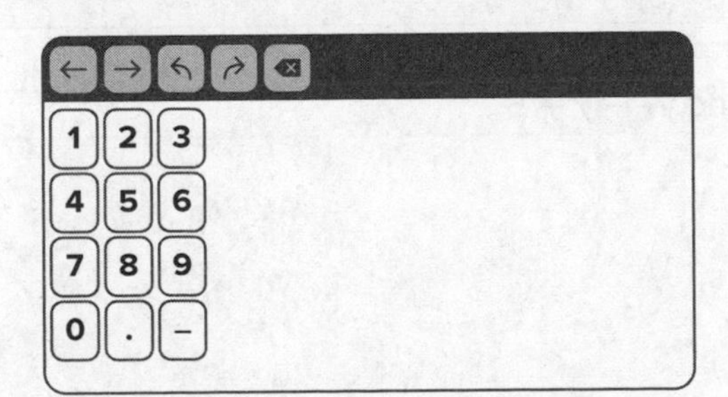

3. Open Response Roberto is buying fruit from a local farmer's market. The prices are shown in the table. **(Lesson 2)**

Item	Mango	Peach	Watermelon
Cost	$1.79	$0.75	$3.00

A. Write an expression to represent the total cost of buying 2 peaches, 5 mangoes, and 3 watermelons.

B. What is the total cost for the fruit? Round your answer to the nearest hundredth.

4. Equation Editor The local food bank is requesting donations in order to distribute meals during a holiday. Turkeys cost $18 each, a bag of potatoes cost $2.55 each, and cans of green beans cost $1.25 each. As of last week, the food bank needed 30 turkeys, 28 bags of potatoes, and 62 cans of green beans for meals. However, this week a grocery store donated 15 of the turkeys. How much money will need to be donated to distribute meals for all the families? **(Lesson 2)**

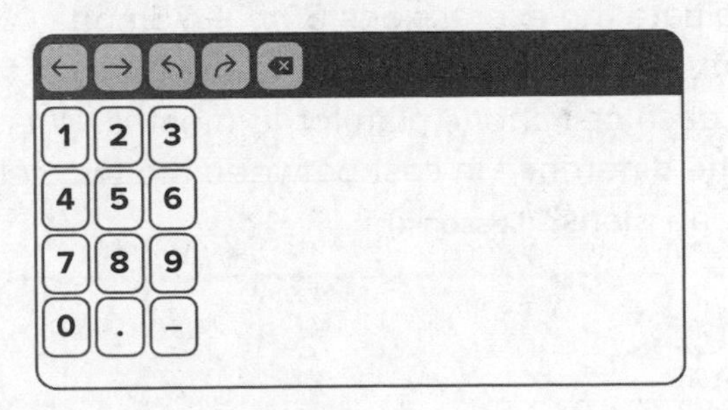

5. Open Response Identify the terms, like terms, coefficients, and constants in the expression $8p + 6q + 5 + 9q + 12p$. **(Lesson 3)**

6. Open Response Write *fifteen dollars more than the original cost* as an algebraic expression. Let c represent the original cost. Do not include dollar signs in your expression. **(Lesson 3)**

7. Multiple Choice Evaluate $(6x + 3y) - z^2 \div (2x)$ when $x = 3$, $y = 4$, and $z = 6$. **(Lesson 4)**

Ⓐ −1

Ⓑ 11

Ⓒ 24

Ⓓ 96

8. Open Response Savannah is choosing between two cell phone plans. Plan A charges $60 a month plus a one-time activation fee of $75. Plan B charges $63 a month plus a one-time activation fee of $15. Evaluate the expressions $60m + 75$ and $63m + 15$ when $m = 18$ to find the total cost for each cell phone plan for 18 months. What is the difference in cost between the two cell phone plans? **(Lesson 4)**

9. Multiple Choice Consider the expression $32x + 56$. **(Lesson 6)**

A. What is the GCF of $32x$ and 56?

Ⓐ 2

Ⓑ 4

Ⓒ 8

Ⓓ 14

B. Use the GCF to factor $32x + 56$.

10. Open Response Avery is buying cupcakes for 12 friends at the school bake sale. Each cupcake costs $1.25. **(Lesson 6)**

A. Write an expression using the Distributive Property to find the total cost of 12 cupcakes.

B. If Avery has $24, how much money will he have left?

11. Table Item Indicate whether or not the two expressions are equivalent using substitution. **(Lesson 7)**

	Equivalent	Not Equivalent
$5x + 2$ and $4x + 1 + x + 2$		
$(8y + 4x + 4y + 5)$ and $4(3y + x) + 5$		
$y^2 + 4y + 5 - 3y$ and $y^2 + y + 5$		

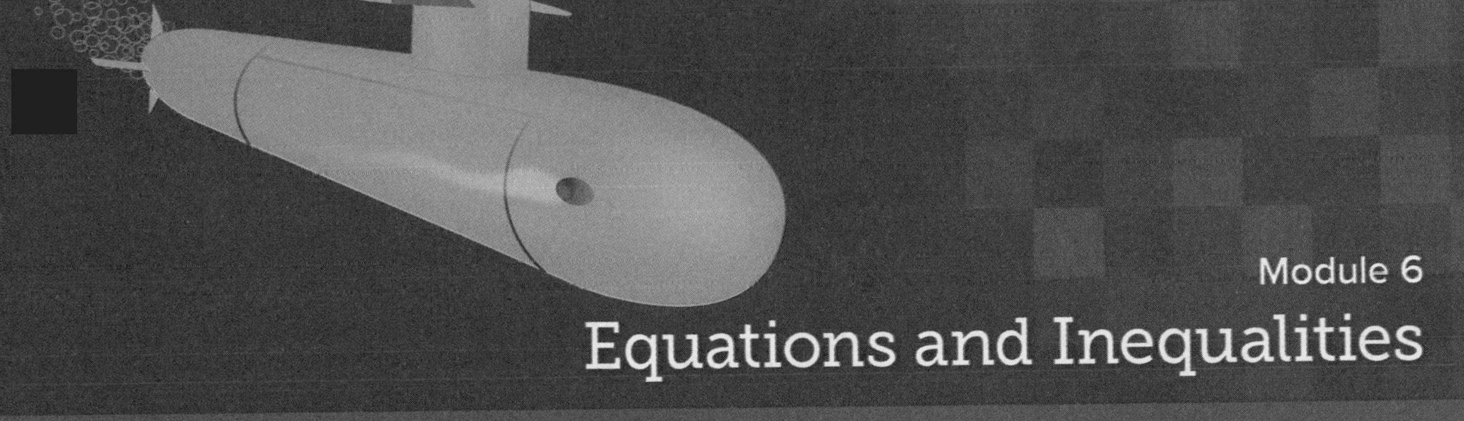

Module 6

Equations and Inequalities

Essential Question

How are the solutions of equations and inequalities different?

What Will You Learn?

Place a checkmark (✓) in each row that corresponds with how much you already know about each topic **before** starting this module.

KEY

□ — I don't know. ◇ — I've heard of it. ☆ — I know it!

	Before			After		
	□	◇	☆	□	◇	☆
solving equations using substitution						
writing and solving one-step addition equations						
writing and solving one-step subtraction equations						
writing and solving one-step multiplication equations						
writing and solving one-step division equations						
writing and graphing inequalities						
finding solutions of inequalities						

Foldables Cut out the Foldable and tape it to the Module Review at the end of the module. You can use the Foldable throughout the module as you learn about equations and inequalities.

What Vocabulary Will You Learn?

Check the box next to each vocabulary term that you may already know.

- ☐ Addition Property of Equality
- ☐ Division Property of Equality
- ☐ equals sign
- ☐ equation
- ☐ guess, check, and revise strategy
- ☐ inequality
- ☐ inverse operations
- ☐ Multiplication Property of Equality
- ☐ solution
- ☐ solve
- ☐ Subtraction Property of Equality

Are You Ready?

Study the Quick Review to see if you are ready to start this module.
Then complete the Quick Check.

Quick Review

Example 1

Subtract decimals.

Find 2.46 − 1.37.

$\begin{array}{r} 2.\overset{3}{\cancel{4}}\overset{16}{\cancel{6}} \\ -1.37 \\ \hline 1.09 \end{array}$	Line up the decimal points. Subtract.

Example 2

Add fractions.

Find $\frac{2}{3} + \frac{1}{5}$.

$\frac{2}{3} + \frac{1}{5}$	Write the problem.
$= \frac{10}{15} + \frac{3}{15}$	Rename using the LCD, 15.
$= \frac{13}{15}$	Add the numerators.

Quick Check

1. Find 14.39 – 7.45.

2. Find $\frac{3}{4} + \frac{7}{10}$.

How Did You Do?

Which exercises did you answer correctly in the Quick Check?
Shade those exercise numbers at the right.

Use Substitution to Solve One-Step Equations

I Can... use substitution to determine whether a given number is a solution of a one-step equation.

Learn Equations

An **equation** is a mathematical sentence showing that two expressions are equal. An equation contains an **equals sign,** =.

Equation	
Definition	**Example**
a mathematical sentence showing two expressions are equal	$3x = 6$

Learn Solve Equations Using Substitution

When you **solve** an equation, you find the value for the given variable that makes the equation true. This value is called the **solution** of the equation.

You may be given a specified set of values to use to find the solution of an equation. You can determine whether a value is a solution of an equation by using substitution. For example, given the equation $x + 3.5 = 7.9$, is 3.4, 4.2, or 4.4 a solution?

Value of x	$x + 3.5 = 7.9$	Is the value a solution?
3.4	$3.4 + 3.5 \stackrel{?}{=} 7.9$ $6.9 \neq 7.9$	no
4.2	$4.2 + 3.5 \stackrel{?}{=} 7.9$ $7.7 \neq 7.9$	no
4.4	$4.4 + 3.5 \stackrel{?}{=} 7.9$ $7.9 = 7.9$	yes

(continued on next page)

What Vocabulary Will You Learn?

equals sign

equation

guess, check, and revise strategy

solution

solve

Talk About It!

Describe the similarities and differences between equations and expressions.

Your Notes

Talk About It!

Is there another value that is a solution of $4.5x = 135$? Explain your reasoning.

You can also use the **guess, check, and revise strategy** to find the solution of an equation. To find the solution of the equation $4.5x = 135$, begin by choosing a reasonable value for x. For example, try $x = 20$.

Value of x	$4.5x = 135$	Is the value a solution?
20	$4.5(20) \stackrel{?}{=} 135$ $90 \neq 135$	No, because $90 < 135$, the value of x is too small. Try revising the number guessed.
25	$4.5(25) \stackrel{?}{=} 135$ $112.5 \neq 135$	No, because $112.5 < 135$, the value of x is too small. Try revising the number guessed.
30	$4.5(30) \stackrel{?}{=} 135$ $135 = 135$	Yes, because $135 = 135$, 30 is the correct solution.

Example 1 Solve Equations Using Substitution

Is 3, 4, or 5 the solution of the equation $p + 9.7 = 13.7$?

Complete the table to find the solution of the equation.

Value of p	$p + 9.7 = 13.7$	Is the value a solution?
3	$3 + 9.7 \stackrel{?}{=} 13.7$ $12.7 \neq 13.7$	NO
4	$4 + 9.7 \stackrel{?}{=} 13.7$ $13.7 = 13.7$	YES
5	$5 + 9.7 \stackrel{?}{=} 13.7$ $14.7 \neq 13.7$	NO

So, the solution is 4.

Check

Is 1, 2, or 3 the solution of the equation $m + \frac{4}{5} = 2\frac{4}{5}$?

Go Online You can complete an Extra Example online.

Example 2 Solve Equations Using Substitution

Navaeh is building a door that is 36 inches wide using wooden planks that are $4\frac{1}{2}$ inches wide.

Use the *guess, check, and revise* strategy to solve the equation $4\frac{1}{2}p = 36$ to find *p*, the number of planks Nevaeh will need to make her door.

Begin by substituting 6 into the equation.

$$4\frac{1}{2}p = 36$$

$$4\frac{1}{2}(6) \stackrel{?}{=} 36$$

$$\square \neq 36$$

Since 27 < 36, try a greater number of planks.

Substitute 7 into the equation.

$$4\frac{1}{2}p = 36$$

$$4\frac{1}{2}(7) \stackrel{?}{=} 36$$

$$\square \neq 36$$

Since $31\frac{1}{2} < 36$, try a greater number of planks.

Substitute 8 into the equation.

$$4\frac{1}{2}p = 36$$

$$4\frac{1}{2}(8) \stackrel{?}{=} 36$$

$$\square = 36$$

The sentence is true, so 8 is the solution of the equation $4\frac{1}{2}p = 36$.

So, Nevaeh needs to use ________ planks to build the door.

Think About It!

How will you make your first *guess*?

Talk About It!

How can you use mental math to solve the equation?

Check

This year, students ate 100 pounds of broccoli in the Walnut Springs Middle School cafeteria. This is $6\frac{1}{4}$ times as much as they ate in the previous year. Use the *guess, check, and revise* strategy to solve the equation $6\frac{1}{4}b = 100$ to find b, the number of pounds of broccoli the students ate the previous year.

Go Online You can complete an Extra Example online.

Pause and Reflect

Write a real-world problem that uses the guess, check, and revise strategy to solve an equation. Explain how you came up with that problem. Exchange problems with a classmate and solve each other's problem.

Name Kayla Supinski Period ______ Date ______

Practice

Go Online You can complete your homework online.

Identify the solution of each equation from the specified set. **(Example 1)**

1. $x + 5.6 = 11.6$; 5, 6, 7

2. $4.2 + z = 11.2$; 6, 7, 8

3. $b - 9.7 = 13.3$; 23, 24, 25

4. $d - 8.4 = 8.6$; 15, 16, 17

5. $4.5x = 18$; 3, 4, 5

6. $2.25c = 27$; 12, 13, 14

7. $d \div 5.5 = 4$; 22, 23, 24

8. $36.3 \div y = 12.1$; 2, 3, 4

9. Brinley is making headbands for her friends. Each headband needs $16\frac{1}{2}$ inches of elastic and she has 132 inches of elastic. Use the *guess, check, and revise* strategy to solve the equation $16\frac{1}{2}h = 132$ to find h, the number of headbands Brinley can make. **(Example 2)**

10. Maddox has \$12.25 to spend on sports drinks. Each drink costs \$1.75. Use the *guess, check, and revise* strategy to solve the equation $1.75d = \$12.25$ to find d, the number of drinks Maddox can buy. **(Example 2)**

11. Manuel has two different recipes for chocolate chip muffins. The table shows the amount of chocolate chips needed per batch for each recipe. He has $8\frac{3}{4}$ cups of chocolate chips. Use the *guess, check, and revise* strategy to solve the equation $1\frac{1}{4}b = 8\frac{3}{4}$ to find b, the number of batches of muffins he can make if he uses Recipe 2.

Recipe	Chocolate Chips (cups)
1	$\frac{3}{4}$
2	$1\frac{1}{4}$

Test Practice

12. Multiple Choice Consider the following equation.

$x + 9 = 17$

Which of the values can be substituted for x to make the equation true?

Ⓐ 7

Ⓑ 8

Ⓒ 9

Ⓓ 26

13. Create Write a real-world problem that can be solved using the equation $7.5 + x = 16$.

14. MP Justify Conclusions A student said that for $x + 7 = 11$, the value of x can be any value. Is the student correct? Write an argument that can be used to defend your response.

15. MP Be Precise Compare and contrast the expression $x + 1$ and the equation $x + 1 = 2$.

16. Give an example of an addition equation and a subtraction equation that each have a solution of 10.

One-Step Addition Equations

I Can... write and solve addition equations for real-world and mathematical problems by using the Subtraction Property of Equality.

What Vocabulary Will You Learn?
inverse operations
Subtraction Property of Equality

Explore Use Bar Diagrams to Write Addition Equations

Online Activity You will use a model to explore how to write one-step addition equations to model real-world problems.

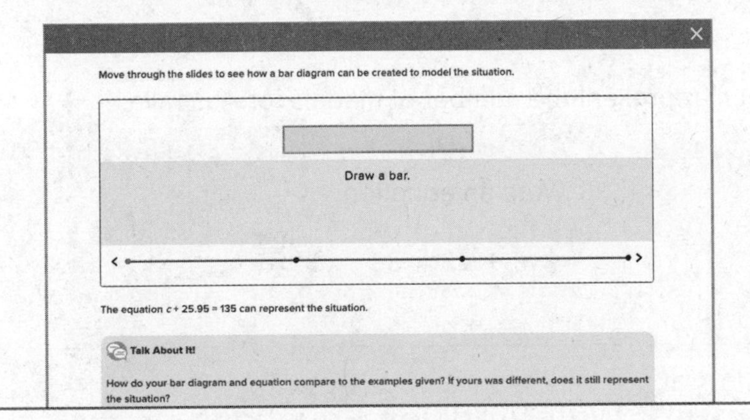

Learn Write Addition Equations

You can write equations to represent many real-world problems involving addition. The table below shows the steps for writing an equation to represent a real-world problem.

Words
Describe the mathematics of the problem. Use only the most important words in the problem.
Variable
Define a variable to represent the unknown quantity.
Equation
Translate the words into an algebraic equation.

Describing the quantity that a variable represents and selecting a letter to represent that unknown quantity is called *defining the variable*.

Talk About It!
Why is defining a variable an important step in writing the equation for a real-world problem?

(continued on next page)

Your Notes

Go Online Watch the animation to learn how to write an addition equation to represent the following real-world problem.

In a recent Summer Olympics, the United States won 23 more medals in swimming than Australia. The United States won a total of 33 swimming medals. Write an addition equation that can be used to determine the number of medals won by Australia.

Words
Describe the mathematics of the problem.
medals for Australia plus 23 equals 33 medals for the U.S.
Variable
Define the variable.
Let m represent the number of medals for Australia.
Equation
Write an equation.
$m + 23 = 33$

Example 1 Write Addition Equations

Together, Ruben and Tariq downloaded 245.5 megabytes (MB) of music. Ruben downloaded 132 MB of that total.

Write an addition equation that can be used to find how many megabytes of music Tariq downloaded.

Words
Describe the mathematics of the problem.
Ruben and Tariq downloaded a total of 245.5 MB of music.
Of this total, Ruben downloaded 132 MB
Variable
Define the variable.
Let m represent the MB that Tariq downloaded.
Equation
Write an equation.
m + 132 = 245.5

So, the equation $132 + m = 245.5$ can be used to find the MB that Tariq downloaded.

Think About It!
What is the unknown in this problem?

Talk About It!
What are some other ways to write the equation based on the real-world problem?

Check

Together, Zacharias and Paz have \$756.80. If Zacharias has \$489.50, how much does Paz have? Write an addition equation that can be used to find the amount of money that belongs to Paz.

Go Online You can complete an Extra Example online.

Explore One-Step Addition Equations

Online Activity You will use a balance to explore how to solve one-step addition equations.

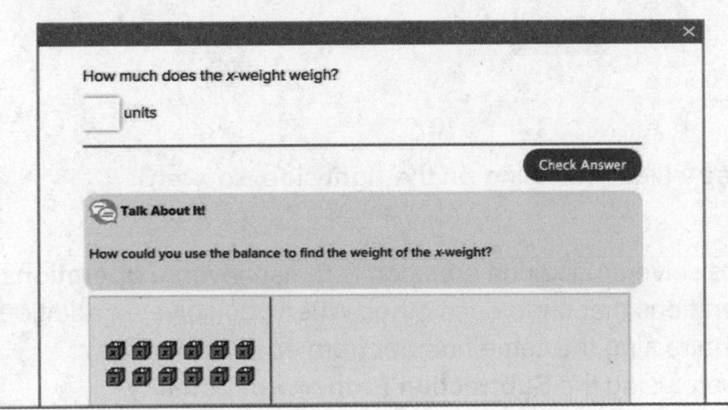

Pause and Reflect

In the Explore, you used a balance to solve equations, such as $x + 3 = 5$ and $x + 3 = 7$. Then you made a conjecture as to how to solve an addition equation without using a balance. When might a balance not be the most advantageous method to use?

Learn Solve Addition Equations

You can use substitution, models, or properties of mathematics to solve addition equations.

Go Online Watch the video to learn how to solve one-step addition equations using algebra tiles.

Talk About It!
How does using algebra tiles to solve an addition equation model the Subtraction Property of Equality?

The video demonstrates how to find the value of x for the equation $x + 4 = 7$.

To model the equation, place one x-tile and four 1-tiles on the left side of the mat. Place seven 1-tiles on the right side of the mat.

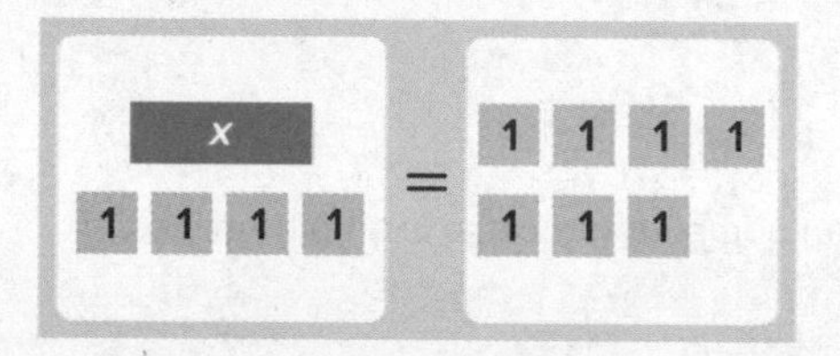

To isolate the variable, or the x-tile, remove the same number of 1-tiles from each side of the mat until the x-tile is by itself.

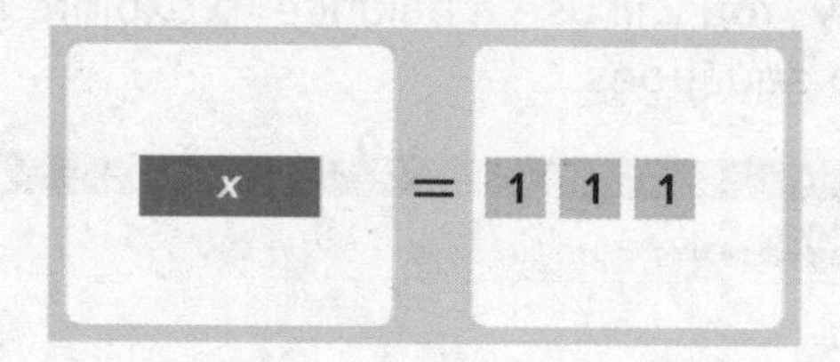

There are three 1-tiles remaining on the right side, so $x = 3$.

Another way to solve an addition equation is to use **inverse operations**, which are operations that undo each other. When you solve an addition equation by subtracting the same number from each side of the equation, you are using the **Subtraction Property of Equality**.

Words	Examples
If you subtract the same number from each side of an equation, the two sides remain equal.	If $5 = 5$, then $5 - 2 = 5 - 2$. If $x + 2 = 3$, then $x + 2 - 2 = 3 - 2$.

To solve the addition equation $x + 4 = 7$ by using inverse operations, undo the addition of 4 by subtracting 4 from each side of the equation.

$x + 4 = 7$	Write the equation.
$\underline{-4 \quad -4}$	Subtraction Property of Equality
$x = 3$	The solution is $x = 3$.

Example 2 Solve Addition Equations

Solve $8 = x + 3$. Check your solution.

Method 1 Use a model.

Step 1 Place eight 1-tiles on the left side of the mat and one x-tile and three 1-tiles on the right side of the mat.

Step 2 Remove three 1-tiles from each side of the mat.

$\square = x$

Method 2 Use the Subtraction Property of Equality.

$8 = x + 3$	Write the equation.
$-3 \quad -3$	Subtraction Property of Equality
$\square = x$	Simplify.

Method 3 Use a bar diagram.
Draw a bar diagram to represent the equation.

8
x | 3

The total length of the bar represents 8, which represents the value of the equation. What is the value of x? ________

So, the solution of the equation is 5.

Check the solution.

$8 = x + 3$	Write the equation.
$8 \stackrel{?}{=} 5 + 3$	Replace x with 5.
$8 = 8$	The sentence is true.

Think About It!
What property will you use to solve for x?

Talk About It!
Give an example of when it might be more efficient or appropriate to use Method 2 rather than Method 1. Explain your reasoning.

Check

Solve $570 = 33 + x$ for x.

Example 3 Solve Addition Equations

Think About It!
What do you notice about the equation? Does this change your approach to solving it? Why or why not?

Solve $3\frac{3}{4} + m = 7\frac{1}{2}$. Check your solution.

$3\frac{3}{4} + m = 7\frac{1}{2}$	Write the equation.
$3\frac{3}{4} + m = 7\frac{2}{4}$	Rewrite with like denominators.
$-3\frac{3}{4} \qquad -3\frac{3}{4}$	Subtraction Property of Equality
$m = 3\frac{3}{4}$	

So, the solution of the equation is ______.

Check the solution.

$3\frac{3}{4} + m = 7\frac{1}{2}$	Write the equation.
$3\frac{3}{4} + 3\frac{3}{4} \stackrel{?}{=} 7\frac{1}{2}$	Replace m with $3\frac{3}{4}$.
$7\frac{1}{2} = 7\frac{1}{2}$	The sentence is true.

Check

Solve $k + \frac{5}{8} = 2\frac{1}{4}$ for k. Check your solution.

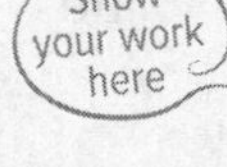

Go Online You can complete an Extra Example online.

Apply Money

An online bookstore is having a sale on mysteries. The table shows the cost of each book format. Abigail has $70 to spend. She bought two paperbacks, one hardcover, and one e-book. Write an addition equation that can be used to determine how much more money Abigail still has to spend. Then solve the equation.

Books	Hardcover	Paperback	E-book	Audio book
Cost ($)	19.49	8.25	10.99	25.19

1 What is the task?

Make sure you understand exactly what question to answer or problem to solve. You may want to read the problem three times. Discuss these questions with a partner.

First Time Describe the context of the problem, in your own words.
Second Time What mathematics do you see in the problem?
Third Time What are you wondering about?

2 How can you approach the task? What strategies can you use?

3 What is your solution?

Use your strategy to solve the problem.

4 How can you show your solution is reasonable?

Write About It! Write an argument that can be used to defend your solution.

Check

Miguel has $2\frac{1}{2}$ hours to work on his homework. The table shows how much time he spent working on his English homework and his math homework. Write an equation that can be used to find how much time, in minutes, he has left to work on his science project if he wants to take a 15-minute snack break. Then solve the equation.

Homework	Time Spent (min)
English	45
Math	28
Science Project	?

Go Online You can complete an Extra Example online.

Foldables It's time to update your Foldable, located in the Module Review, based on what you learned in this lesson. If you haven't already assembled your Foldable, you can find the instructions on page FL1.

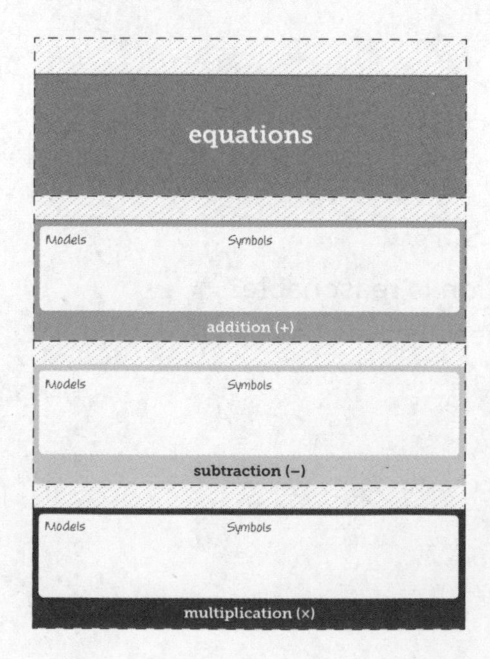

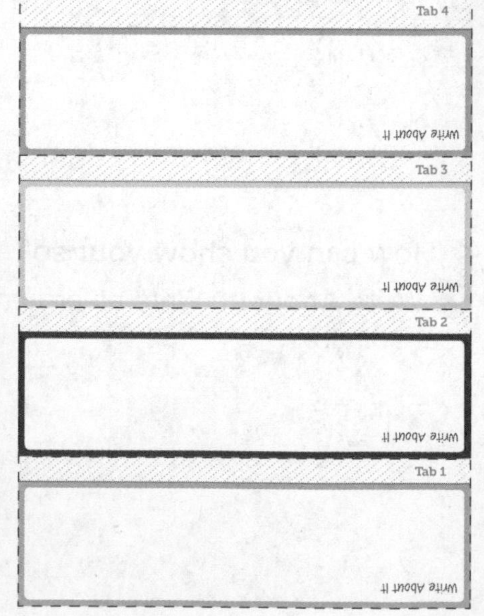

Name ________________ Period ________ Date ________

Practice

Go Online You can complete your homework online.

1. On Saturday and Sunday, Jarrod went running and burned a total of 647.5 Calories. He burned 320 of those Calories on Saturday. Write an addition equation that could be used to find the number of Calories Jarrod burned on Sunday. **(Example 1)**

2. Maggie and her sister bought a gift for their mother that cost $54.75. Maggie contributed $26 to the cost of the gift. Write an addition equation that could be used to find how much money Maggie's sister contributed to the gift. **(Example 1)**

3. A piece of material measures 38.25 inches. Courtney cuts the piece of material into two pieces. One piece measures 19.5 inches. Write an addition equation that could be used to find the length of the other piece of material. **(Example 1)**

4. On a two-day car trip, the Roberts family drove a total of 854.25 miles. On Day 1, the family drove 497.75 of those miles. Write an addition equation that could be used to find how many miles the Roberts family drove on Day 2 of their trip. **(Example 1)**

Solve each equation. Check your solution. **(Examples 2 and 3)**

5. $9 = 3 + a$

6. $5 + x = 10$

7. $3\frac{1}{4} + z = 6\frac{3}{4}$

8. $9\frac{1}{2} = b + 2\frac{1}{4}$

9. $18.35 = c + 5.1$

Test Practice

10. Equation Editor Solve $x + 5.15 = 23.85$.

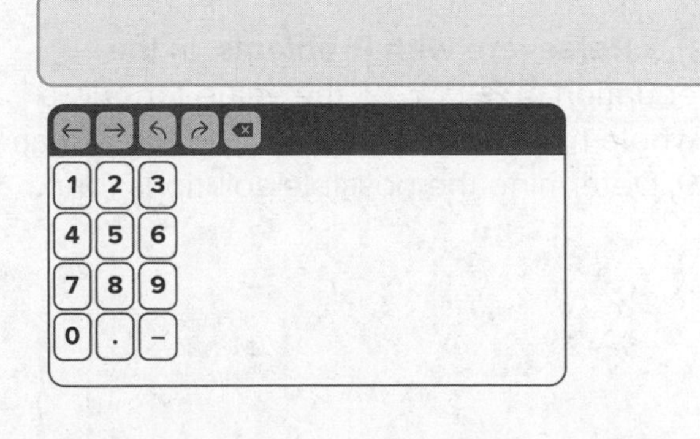

Apply

11. Jeremiah has $35 to spend on items for his dog at the pet store. The table shows the cost of the items. He bought a collar, two toys, two biscuits, and a ball. Write an addition equation that can be used to determine how much more money Jeremiah still has to spend. Then solve the equation.

Item	Cost ($)
Ball	3.45
Biscuit	1.15
Bone	2.50
Collar	8.99
Toy	5.75

12. Jasmine has $30 to spend on ice cream for a party. The table shows the cost of each size of ice cream. She bought five quarts and one gallon. Write an addition equation that can be used to determine how much more money Jasmine still has to spend. Then solve the equation.

Ice Cream Size	Cost ($)
Gallon	6.99
Pint	1.59
Quart	3.35

13. MP **Reason Abstractly** Suppose $a + b = 20$ and the value of a is increased by 1. If the sum of a and b remains the same, what must happen to the value of b?

14. MP **Find the Error** A student is solving the equation $x + 9 = 14$. Find the student's mistake and correct it.

$$\begin{aligned} x + 9 &= 14 \\ +9 &= +9 \\ \hline x &= 23 \end{aligned}$$

15. MP **Persevere with Problems** In the equation $m + n = 12$, the value for m is a whole number greater than 5 but less than 9. Determine the possible solutions for n.

16. **Create** Write and solve a real-world problem that can be solved with a one-step addition equation.

One-Step Subtraction Equations

I Can... write and solve subtraction equations for real-world and mathematical problems by using the Addition Property of Equality.

What Vocabulary Will You Learn?
Addition Property of Equality

Explore Use Bar Diagrams to Write Subtraction Equations

Online Activity You will use a model to explore how to write one-step subtraction equations to model real-world problems.

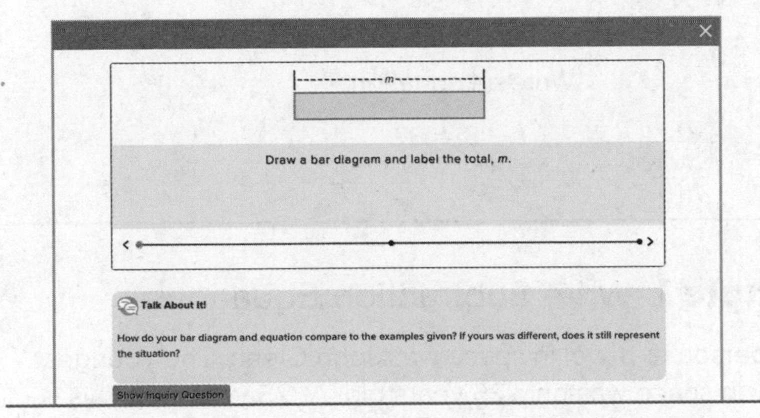

Learn Write Subtraction Equations

You can write equations to represent real-world problems involving subtraction. The table below shows the steps for writing an equation to represent a real-world problem.

Words
Describe the mathematics of the problem. Use only the most important words in the problem.
Variable
Define a variable to represent the unknown quantity.
Equation
Translate the words into an algebraic equation.

(continued on next page)

Your Notes

Talk About It!

What key words in the problem indicate subtraction?

Go Online Watch the animation to see how to write a subtraction equation to represent the following real-world problem.

Caroline gave Everly 8 beads and was left with 37 beads. Write a subtraction equation that can be used to determine the total number of beads Caroline had originally.

Words
Describe the mathematics of the problem.
The total number of beads minus the number of beads given away equals the number remaining.
Variable
Define the variable.
Let t represent the total number of beads.
Equation
Write an equation.
$t - 8 = 37$

Talk About It!

What is the unknown in this problem?

 **Example 1** Write Subtraction Equations

The oldest person to travel in space was John Glenn. The youngest person to fly in space was only 25 years old. At 25 years old, this is 52 years less than John Glenn's age.

Write a subtraction equation that can be used to find John Glenn's age when he traveled in space.

Talk About It!

The equation can be $a - 52 = 25$ or $a - 25 = 52$. Can you write any addition equations that can represent this situation? Explain.

 **Go Online** Watch the animation.

Words
Describe the mathematics of the problem.
52 years less than John Glenn's age is the youngest person's age.
Variable
Define the variable.
Let a represent the age of John Glenn.
Equation
Write an equation.
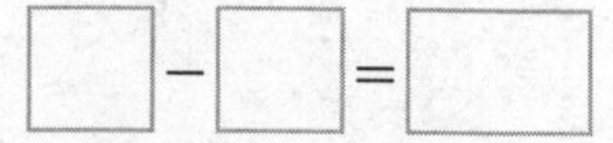

So, the equation $a - 52 = 25$ can be used to find John Glenn's age.

Check

An e-Book costs \$14.95. This is \$7.55 less than the cost of the hardback version of the same book. Write a subtraction equation that can be used to find the cost of the hardback book.

Go Online You can complete an Extra Example online.

Learn Solve Subtraction Equations

You can use substitution, models, or properties of mathematics to solve subtraction equations.

Go Online Watch the video to learn how to solve one-step subtraction equations using a bar diagram.

The video demonstrates how to find the value of x in the equation $x - 15 = 11$.

Draw a bar to represent the total. The total length of the bar represents the original amount, x. Divide the bar into two sections to show the known values, 15 and 11.

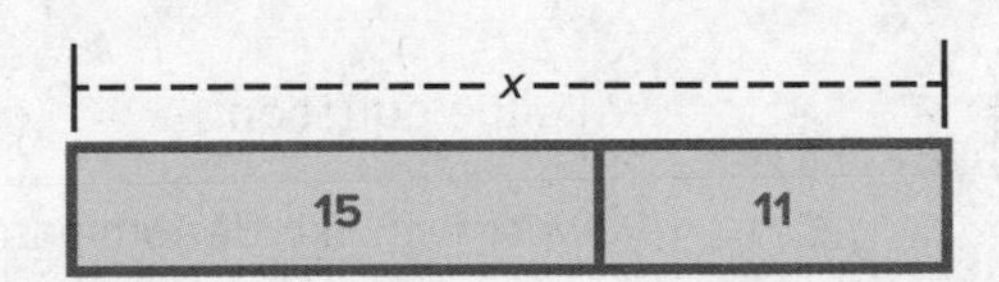

Because x represents the length of the entire bar, add 15 and 11 to find the value of x.

So, $x = 26$.

To solve a subtraction equation, use the inverse operation, which is addition. When you solve an equation by adding the same number to each side of the equation, you are using the **Addition Property of Equality**.

Words	Examples
If you add the same number to each side of an equation, the two sides remain equal.	If $10 = 10$, then $10 + 3 = 10 + 3$. If $n - 6 = 7$, then $n - 6 + 6 = 7 + 6$.

Talk About It!

Compare and contrast solving one-step addition equations and solving one-step subtraction equations.

Example 2 Solve Subtraction Equations

Solve $32 = x - 7$. Check your solution.

$32 = x - 7$	Write the equation.
$+7 = \quad +7$	Addition Property of Equality
$39 = x$	

So, the solution of the equation is ______.

Check the solution.

$32 = x - 7$	Write the equation.
$32 \stackrel{?}{=} 39 - 7$	Replace x with 39.
$32 = 32$	The sentence is true.

Check

Solve $2{,}019 = x - 731$ for x.

Example 3 Solve Subtraction Equations

Solve $m - 13\frac{2}{3} = 2\frac{1}{6}$.

$m - 13\frac{2}{3} = 2\frac{1}{6}$	Write the equation.
$m - 13\frac{4}{6} = 2\frac{1}{6}$	Rewrite with like denominators.
$+13\frac{4}{6} \quad +13\frac{4}{6}$	Addition Property of Equality
$m = 15\frac{5}{6}$	

So, the solution of the equation is ______.

Talk About It!

How can you check your solution?

Check

Solve $p - \frac{3}{4} = 4\frac{2}{5}$ for p.

Go Online You can complete an Extra Example online.

Apply Shopping

Tyson had \$302.87 in his savings account after he withdrew money to go shopping. He spent the amounts shown, and he had \$18.25 remaining. Use an equation to find how much Tyson originally had in his savings account.

Item	Total Spent (\$)
Clothes	95.21
Gifts	42.79
Soccer ball	23.75

1 What is the task?

Make sure you understand exactly what question to answer or problem to solve. You may want to read the problem three times. Discuss these questions with a partner.

First Time Describe the context of the problem, in your own words.
Second Time What mathematics do you see in the problem?
Third Time What are you wondering about?

2 How can you approach the task? What strategies can you use?

3 What is your solution?

Use your strategy to solve the problem.

4 How can you show your solution is reasonable?

Write About It! Write an argument that can be used to defend your solution.

Talk About It!
How can you solve the problem another way?

Check

Nina used $12\frac{1}{3}$ yards of ribbon to make hair bows and 5 yards of ribbon to wrap gifts. She has $17\frac{2}{3}$ yards of ribbon left. Use a subtraction equation to find how many yards of ribbon she had to start.

Go Online You can complete an Extra Example online.

Foldables It's time to update your Foldable, located in the Module Review, based on what you learned in this lesson. If you haven't already assembled your Foldable, you can find the instructions on page FL1.

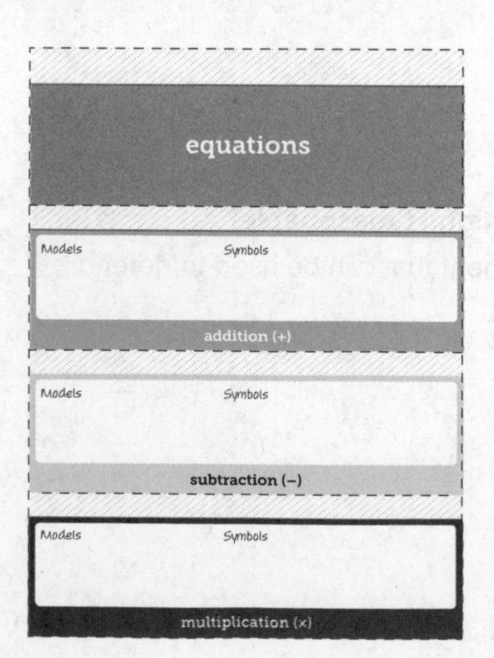

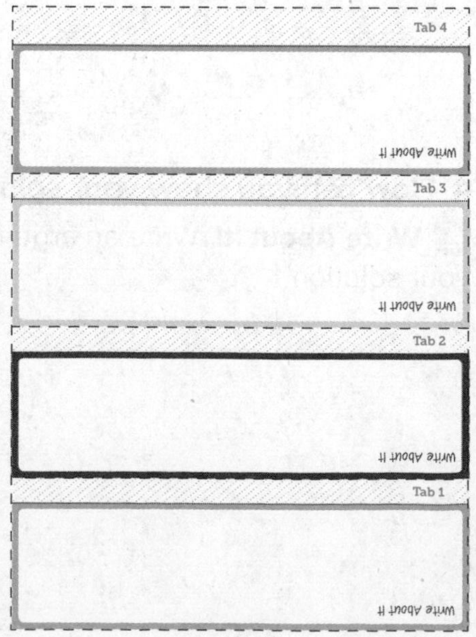

Name ____________________ Period __________ Date __________

Practice

Go Online You can complete your homework online.

1. On Monday, Homeroom 104 turned in 64 canned goods. This is 17 less than the number of canned goods turned in by Homeroom 106. Write a subtraction equation that could be used to find the number of canned goods turned in by Homeroom 106 on Monday. **(Example 1)**

2. Izan's youngest relative is 5 years old. This is 79 years less than the age of his oldest relative. Write a subtraction equation that could be used to find the age of his oldest relative. **(Example 1)**

3. To make a cake, Rose needed $1\frac{1}{2}$ cups of sugar. This is $1\frac{1}{4}$ cups less than the amount of flour she needed for the cake. Write a subtraction equation that could be used to find the amount of flour she needed for the cake. **(Example 1)**

4. On Sunday, Jax biked 10.25 miles. This is 3.5 fewer miles than the number of miles he biked on Saturday. Write a subtraction equation that could be used to find the number of miles Jax biked on Saturday. **(Example 1)**

Solve each equation. Check your solution. **(Examples 2 and 3)**

5. $24 = x - 5$

6. $z - 7 = 19$

7. $z - 9\frac{1}{3} + = 1\frac{5}{9}$

8. $5\frac{1}{2} = b - 12\frac{1}{4}$

9. $67.9 = c - 4.45$

Test Practice

10. Equation Editor Solve $x - 7.49 = 87.3$.

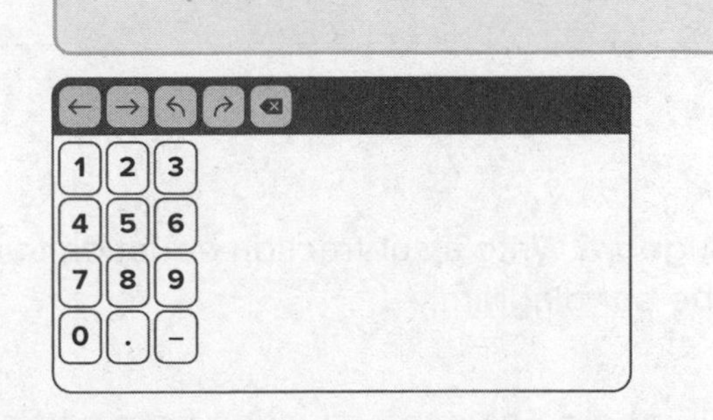

Apply

11. After spending money for a golf outing, Gus had $517.92 remaining in his checking account. The table shows how much money he spent on different items to participate in the outing. Use an equation to find how much money Gus originally had in his checking account.

Item	Cost ($)
Entry Fee	94.50
Golf Shoes	44.25
Gloves	11.25

12. Robin made two batches of every item shown in the table. At the end of the day, she had $1\frac{1}{4}$ cups of flour left. Use an equation to find how much flour Robin originally had on Saturday.

Baking Item	Amount of Flour
Bread	$1\frac{3}{4}$ cups
Muffins	2 cups
Pancakes	$1\frac{1}{2}$ cups

13. MP **Reason Abstractly** During a test flight, Jeri's rocket reached a height of 18 yards above the ground. This was 7 yards less than the height that Devon's rocket reached. Did Devon's rocket reach a height greater than 23 yards? Explain.

14. MP **Find the Error** A student is solving the equation $x - 3.2 = 5.5$. Find the student's mistake and correct it.

$$\begin{aligned} x - 3.2 &= 5.5 \\ -3.2 & -3.2 \\ \hline x &= 2.3 \end{aligned}$$

15. Multiple Representations The bar diagram represents a subtraction equation.

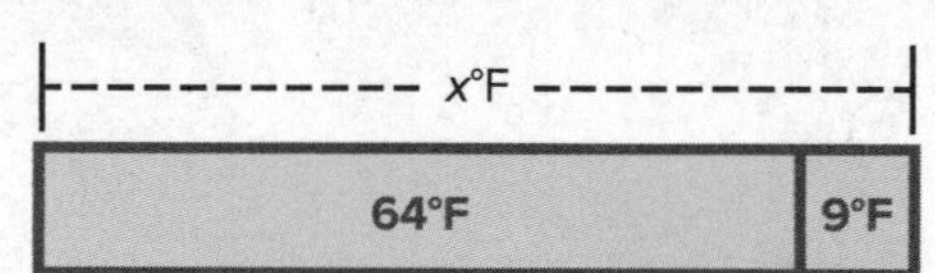

a. **Words** Write a real-world situation for the bar diagram.

b. **Algebra** Write a subtraction equation for the bar diagram.

c. **Numbers** Solve the equation from part b.

16. Create Write and solve a real-world problem involving decimals that can be solved with a one-step subtraction equation.

One-Step Multiplication Equations

I Can... write and solve multiplication equations for real-world and mathematical problems by using the Division Property of Equality.

What Vocabulary Will You Learn?
Division Property of Equality

Explore Use Bar Diagrams to Write Multiplication Equations

Online Activity You will use a model to explore how to write one-step multiplication equations to model real-world problems.

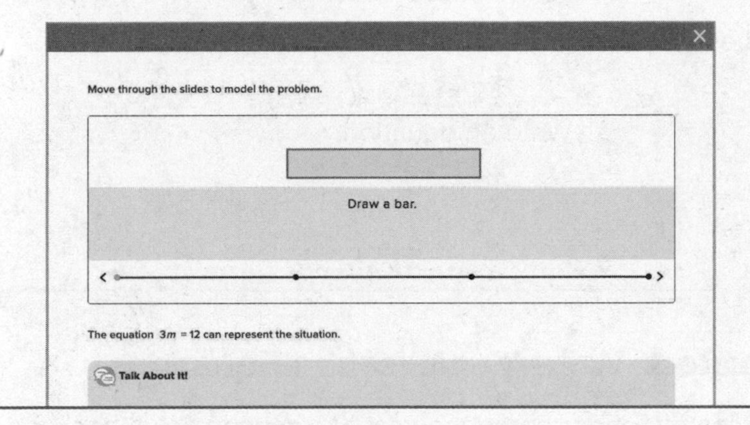

Learn Write Multiplication Equations

You can write equations to represent real-world problems involving multiplication. The table below shows the steps for writing an equation to represent a real-world problem.

Words
Describe the mathematics of the problem. Use only the most important words in the problem.
Variable
Define a variable to represent the unknown quantity.
Equation
Translate the words into an algebraic equation.

(continued on next page)

Your Notes

Talk About It!

What key words in the problem indicate multiplication?

Go Online Watch the animation to see how to write a multiplication equation to represent the following real-world problem.

Kosumi is saving an equal amount each week for 4 weeks to buy a video game for $55. Write a multiplication equation that can be used to determine the amount she is saving each week.

Words
Describe the mathematics of the problem.
The number of weeks times the amount saved each week equals the total amount saved.
Variable
Define the variable.
Let a represent the amount saved each week.
Equation
Write an equation.
$4a = 55$

Example 1 Write Multiplication Equations

Think About It!

How do you know this equation uses multiplication?

Vincent and some friends shared the cost of a season ticket package for the local football team. The package cost $745 and each person contributed $186.25.

Write a multiplication equation that can be used to find how many friends contributed to the ticket purchase.

Words
Describe the mathematics of the problem.
The number of friends times the amount each person paid equals the total cost.
Variable
Define the variable.
Let f represent the number of friends who contributed.
Equation
Write an equation.

Talk About It!

The equation can also be written as $186.25f = 745$. Does removing the multiplication symbol make it easier or more difficult to understand? Explain your reasoning.

So, the equation $f \cdot 186.25 = 745$ can be used to find the number of friends that contributed. This equation can also be written as $186.25f = 745$.

Check

A jewelry store is selling a set of 4 pairs of earrings for \$58.85 including tax. Neva and three of her friends want to buy the set so each could have one pair of earrings. Write a multiplication equation that could be used to find how much each person should pay.

Go Online You can complete an Extra Example online.

Learn Solve Multiplication Equations

You can use substitution, models, or properties of mathematics to solve multiplication equations.

Go Online Watch the video to learn how to solve one-step multiplication equations using algebra tiles.

The video demonstrates how to find the value of x for $4x = 12$.

To model the equation, place four x-tiles on the left side of the mat to represent $4x$. Place twelve 1-tiles on the right side of the mat to represent 12.

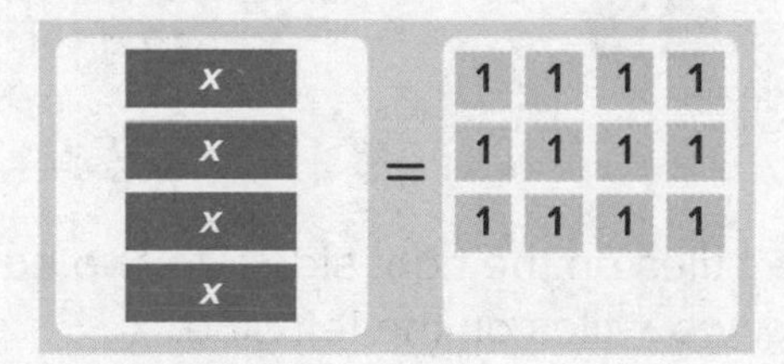

Arrange the tiles into equal groups on each side of the mat. This will allow you to group the tiles into 4 equal groups to find the value of x.

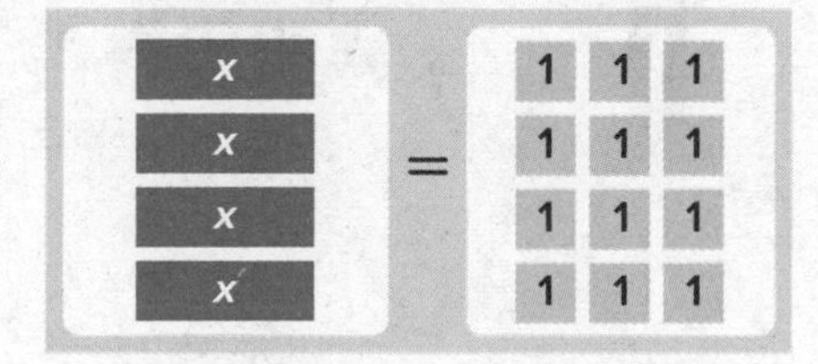

For each x-tile, there are three 1-tiles, so $x = 3$.

(continued on next page)

Talk About It!

Why is the Division Property of Equality used when solving a multiplication equation?

To solve a multiplication equation, use the inverse operation, which is division. When you solve a multiplication equation by dividing each side of the equation by the same nonzero number, you are using the **Division Property of Equality**.

Words	Examples
If you divide each side of an equation by the same nonzero number, the two sides remain equal.	If $9 = 9$, then $9 \div 3 = 9 \div 3$. If $4x = 8$, then $4x \div 4 = 8 \div 4$.

Example 2 Solve Multiplication Equations

Solve $2x = 10$. Check your solution.

Think About It!

What property will you use to solve for x?

Method 1 Use a model.

Step 1 Place two x-tiles on the left side of the mat to represent $2x$ and ten 1-tiles on the right side of the mat to represent 10.

Talk About It!

Describe a multiplication equation for which it would not be possible to use algebra tiles to solve.

Step 2 Group the 1-tiles on the right side into two equal groups because there are two x-tiles on the left side.

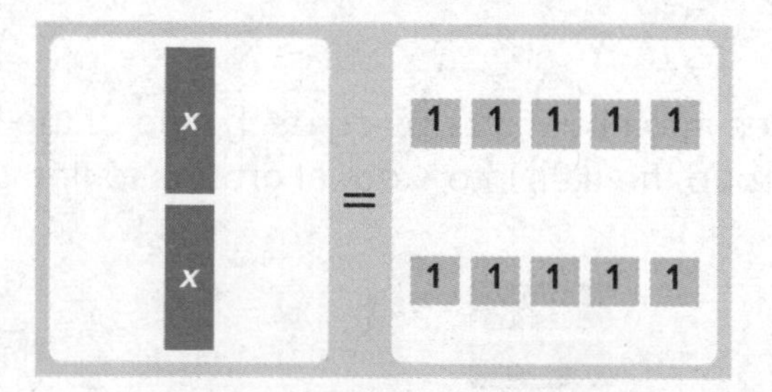

Because there are five 1-tiles for every x-tile, the value of x is 5.

$x = \square$

(continued on next page)

Method 2 Use the Division Property of Equality.

$2x = 10$ — Write the equation.

$\frac{2x}{2} = \frac{10}{2}$ — Division Property of Equality

$x = \square$ — Simplify.

So, the solution of the equation is 5.

Check the solution.

$2x = 10$ — Write the equation.

$2(5) \stackrel{?}{=} 10$ — Replace x with 5.

$10 = 10$ — The sentence is true.

Check

Solve $84 = 7x$.

Go Online You can complete an Extra Example online.

Pause and Reflect

You can also use the bar diagram shown to represent and solve the equation $2x = 10$.

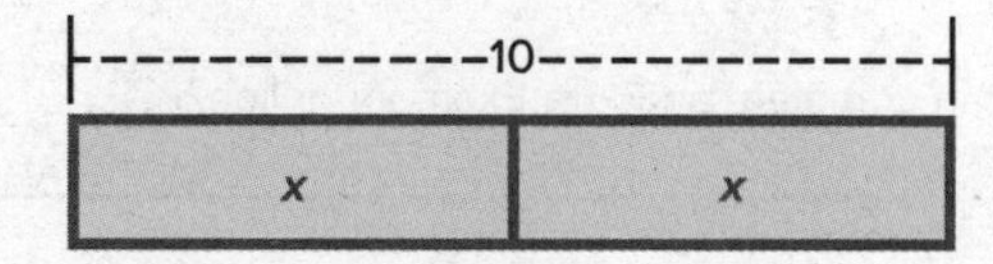

Compare and contrast each of these methods: algebra tiles, properties of equality, and bar diagrams.

Think About It!

By what number will you need to divide each side of the equation to undo multiplying by $\frac{2}{3}$?

Talk About It!

Why did you need to multiply each side of the equation by $\frac{3}{2}$, even though you used the Division Property of Equality?

Example 3 Solve Multiplication Equations

Solve $\frac{2}{3}m = \frac{5}{8}$. Check your solution.

$\frac{2}{3}m = \frac{5}{8}$ Write the equation.

$\dfrac{\frac{2}{3}m}{\frac{2}{3}} = \dfrac{\frac{5}{8}}{\frac{2}{3}}$ Division Property of Equality

$\frac{2}{3}\left(\frac{3}{2}\right)m = \frac{5}{8}\left(\frac{3}{2}\right)$ Multiply by the reciprocal.

$m = \frac{15}{16}$

So, the solution of the equation is _____.

Check the solution.

$\frac{2}{3}m = \frac{5}{8}$ Write the equation.

$\frac{2}{3}\left(\frac{15}{16}\right) \stackrel{?}{=} \frac{5}{8}$ Replace m with $\frac{15}{16}$.

$\frac{30}{48} \stackrel{?}{=} \frac{5}{8}$ Multiply.

$\frac{5}{8} = \frac{5}{8}$ Simplify. The sentence is true.

Check

Solve $\frac{4}{5}k = \frac{1}{3}$ for k.

Go Online You can complete an Extra Example online.

Pause and Reflect

Did you struggle with any Examples about solving multiplication equations? How do you feel when you struggle with math concepts? What steps can you take to understand those concepts?

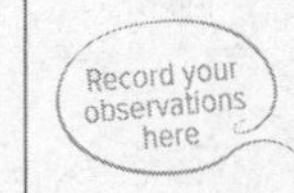

Apply Nutrition

The nutrition information for two different bottles of iced tea is shown. Alicia wants to compare the grams of sugar in a single serving for each brand. Which brand has more sugar per serving? How much more?

Aunt Maggie's Iced Tea (3 servings)	
Calories	120
Sodium (mg)	75
Sugar (g)	63

Southern Goodness Sweet Tea (4 servings)	
Calories	125
Sodium (mg)	82
Sugar (g)	74

1 What is the task?

Make sure you understand exactly what question to answer or problem to solve. You may want to read the problem three times. Discuss these questions with a partner.

First Time Describe the context of the problem, in your own words.
Second Time What mathematics do you see in the problem?
Third Time What are you wondering about?

Talk About It!
Suppose a third brand of tea has 42 grams of sugar in 2 servings. How does this compare to the brand that has more sugar per serving?

2 How can you approach the task? What strategies can you use?

we divided

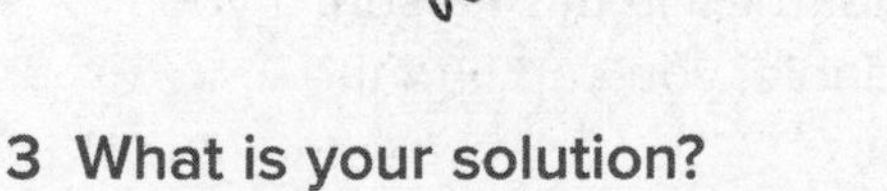

3 What is your solution?

Use your strategy to solve the problem.

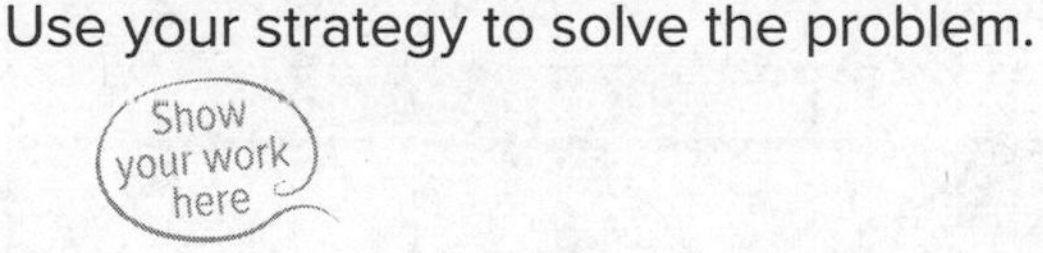

$3x = 63$ $\quad$ $4x = 74$

21 $\quad$ 18.5

4 How can you show your solution is reasonable?

Write About It! Write an argument that can be used to defend your solution.

Check

The nutrition information for two different bags of chips is shown. Frederick wants to compare the milligrams of sodium per serving in each bag of chips. Which brand has more milligrams of sodium per serving? How much more?

	Northern Grown (9 servings)	Heartland (7 servings)
Calories	1,440	1,250
Sodium	1,530 mg	1,211 mg
Sugar	9 g	5 g

Go Online You can complete an Extra Example online.

Foldables It's time to update your Foldable, located in the Module Review, based on what you learned in this lesson. If you haven't already assembled your Foldable, you can find the instructions on page FL1.

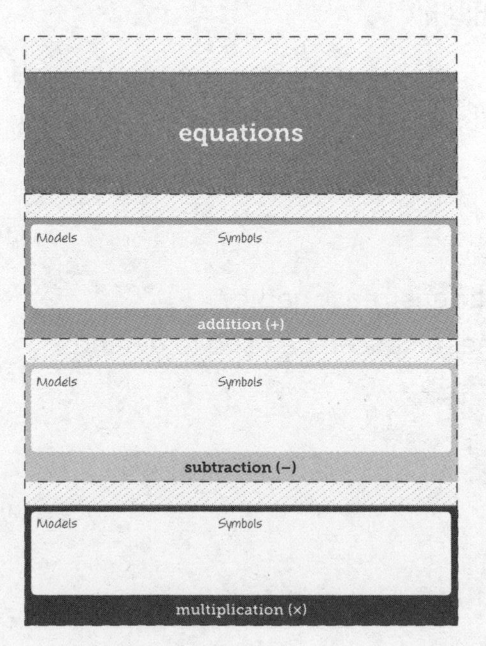

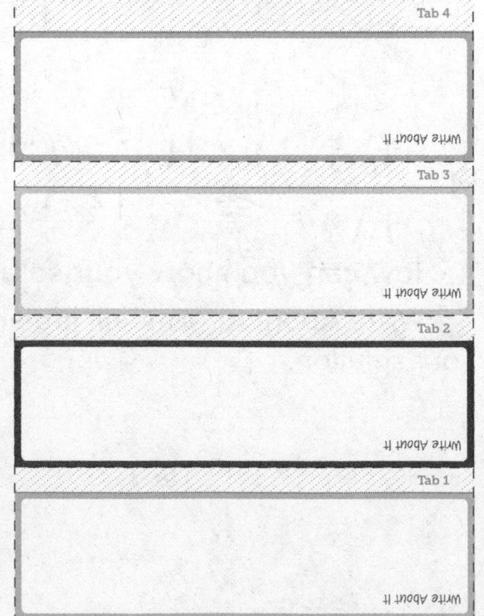

Name ______________________ Period __________ Date __________

Practice

Go Online You can complete your homework online.

1. Maribel and some friends went to an adventure park. The total cost of their tickets was $374 and each person paid $46.75. Write a multiplication equation that can be used to find how many people bought tickets to the adventure park. **(Example 1)**

2. It takes Samuel $\frac{1}{5}$ hour to walk a mile. Yesterday, Samuel walked for $1\frac{1}{2}$ hours. Write a multiplication equation that can be used to find the number of miles Samuel walked. **(Example 1)**

3. The distance around a lake is 2.6 miles. On Saturday, Doug biked a total of 18.2 miles around the lake. Write a multiplication equation that can be used to find how many times Doug biked around the lake. **(Example 1)**

4. An express delivery company charges $3.25 per pound to mail a package. Georgia paid $9.75 to mail a package. Write a multiplication equation that can be used to find the weight of the package in pounds. **(Example 1)**

Solve each equation. Check your solution. **(Examples 2 and 3)**

5. $12 = 6x$

6. $3z = 15$

7. $\frac{3}{4}z = \frac{2}{3}$

8. $\frac{1}{2} = \frac{5}{8}w$

9. $60.536 = 9.2j$

Test Practice

10. Equation Editor Solve $3.9x = 16.068$.

Apply

11. Mira is comparing two different types of popcorn. The table shows the nutritional information. She wants to compare the number of Calories per cup for each type of popcorn. Which type has more Calories per cup? How many more?

Light Popcorn $3\frac{1}{2}$ cups	Caramel Popcorn $2\frac{1}{2}$ cups
Calories: 105	Calories: 170
Carbohydrates: 21 g	Carbohydrates: 15 g
Fat: 0 g	Fat: 11 g

12. The table shows the nutritional information for two different brands of apple juice. Marcus wants to compare the number of carbohydrates in a single serving of each brand. Which brand has more carbohydrates per serving? How many more?

Brand A (4 servings)	Brand B (3 servings)
Calories: 480	Calories: 360
Carbohydrates: 120 g	Carbohydrates: 87 g
Sugars: 120 g	Sugar: 78 g

13. MP **Reason Abstractly** Earline needs to save \$367.50 for her summer vacation. She plans on saving \$52.50 per week. In 6 weeks, will she have enough money? Explain.

14. MP **Find the Error** A student is solving the equation $3x = 9$. Find the student's mistake and correct it.

$$3x = 9$$

$$3 \cdot 3x = 9 \cdot 3$$

$$x = 27$$

15. MP **Persevere with Problems** Do the equations $\frac{1}{3} = 3x$ and $\frac{1}{3} \div x = 3$ have the same solution? Explain why or why not.

16. **Create** Write and solve a real-world problem involving decimals that can be solved with a one-step multiplication equation.

One-Step Division Equations

I Can... write and solve division equations for real-world and mathematical problems by using the Multiplication Property of Equality.

What Vocabulary Will You Learn?
Multiplication Property of Equality

Explore Use Bar Diagrams to Write Division Equations

Online Activity You will use a model to explore how to write one-step division equations to model real-world problems.

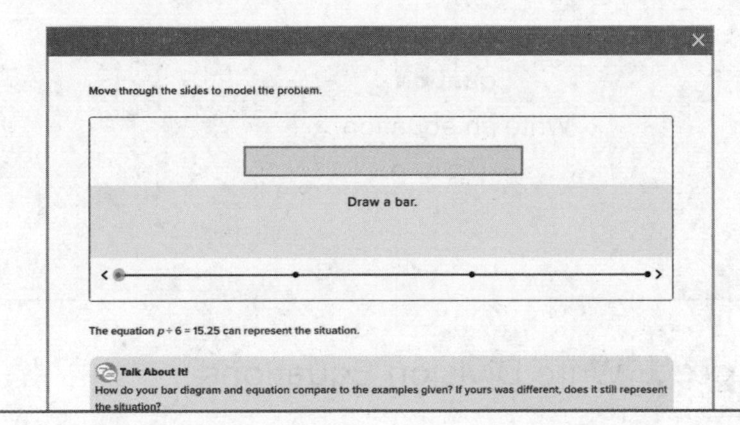

Learn Write Division Equations

You can write equations to represent real-world problems involving division. The table below shows the steps for writing an equation to represent a real-world problem.

Words
Describe the mathematics of the problem. Use only the most important words in the problem.
Variables
Define a variable to represent the unknown quantity.
Equation
Translate the words into an algebraic equation.

(continued on next page)

Your Notes

Talk About It!

Why is it important to define a variable before writing an equation?

Think About It!

How do you know that you will use division when you write the equation?

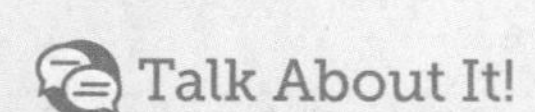

Write a multiplication equation that is equivalent to $b \div 3 = 48.5$. Construct a mathematical argument to justify your response.

Go Online Watch the animation to see how to write a division equation to represent the following real-world problem.

Cyrus, Breyon, and Michael are sharing a pack of stickers. Each student gets 9 stickers. Write a division equation that can be used to determine the total number of stickers in the pack.

Words
Describe the mathematics of the problem. The total number of stickers divided by the number of students equals the number of stickers each student receives.
Variable
Define the variable. Let s represent the total number of stickers.
Equation
Write an equation. $s \div 3 = 9$

Example 1 Write Division Equations

Benji rode his bike from Pittsburgh to Cleveland over the course of a three-day weekend. His average distance was 48.5 miles each day.

What was the total distance he rode?

Words
Describe the mathematics of the problem. The total distance divided by 3 equals 48.5 miles.
Variable
Define the variable. Let b represent the total distance he rode.
Equation
Write an equation. $\square \div \square = \square$

So, the equation $b \div 3 = 48.5$ can be used to find the total distance Benji rode.

Check

Sophia has $16.50 to spend on party favors. She wants to spend $2.75 per person. Write a multiplication equation that can be used to find the number of people Sophia can have at the party.

Go Online You can complete an Extra Example online.

Learn Solve Division Equations

You can use substitution, models, or properties of mathematics to solve division equations.

Go Online Watch the video to learn how to solve one-step division equations using bar diagrams.

The video demonstrates how to find the value of x in the equation $\frac{x}{4} = 6$.

Draw a bar to represent the total. The total length of the bar represents the original amount, x. Divide the bar into four sections to show division by 4. Then work backward to solve the equation.

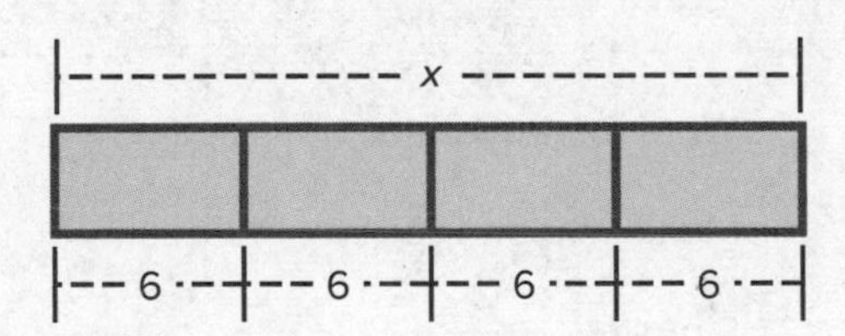

Because x represents the entire length of the bar, and there are four equal sections of 6, multiply 6 by 4 to find the value of x. So, $x = 24$.

To solve a division equation, use the inverse operation, multiplication. When you solve an equation by multiplying each side of the equation by the same number, you are using the **Multiplication Property of Equality**.

Words	Examples
If you multiply each side of an equation by the same number, the two sides remain equal.	If $6 = 6$, then $6 \times 5 = 6 \times 5$. If $x \div 3 = 4$, then $x \div 3 \times 3 = 4 \times 3$.

Talk About It!

Why might it be difficult to use algebra tiles to model a division equation, such as $\frac{x}{3} = 4$?

Think About It!

What operation is represented in the expression $\frac{x}{9}$?

Talk About It!

The equation $\frac{x}{9} = 13$ can also be written as $\frac{1}{9}x = 13$. What property of equality can you use to solve this equation? Explain your reasoning.

Talk About It!

How can you check your solution?

Example 2 Solve Division Equations

Solve $\frac{x}{9} = 13$. Check your solution.

$\frac{x}{9} = 13$	Write the equation.
$\frac{x}{9}(9) = 13(9)$	Multiplication Property of Equality
$x = 117$	Simplify.

So, the solution of the equation is ________.

Check the solution.

$\frac{x}{9} = 13$	Write the equation.
$\frac{117}{9} \stackrel{?}{=} 13$	Replace x with 117.
$13 = 13$	The sentence is true.

Check

Solve $9 = \frac{x}{17}$ for x.

Example 3 Solve Division Equations

Solve $\frac{c}{3} = \frac{2}{5}$.

$\frac{c}{3} = \frac{2}{5}$	Write the equation.
$\frac{c}{3}(3) = \frac{2}{5}(3)$	Multiplication Property of Equality
$c = \frac{6}{5}$ or $1\frac{1}{5}$	Simplify.

So, the solution of the equation is ________.

Check

Solve $\frac{k}{4} = 4\frac{2}{3}$ for k.

 **Go Online** You can complete an Extra Example online.

Apply Catering

Dario is catering a party and serves 5.5-ounce servings of chicken to twelve guests, and 5.25-ounce servings of fish to nine guests. Did Dario serve more total ounces of chicken or fish? How much more?

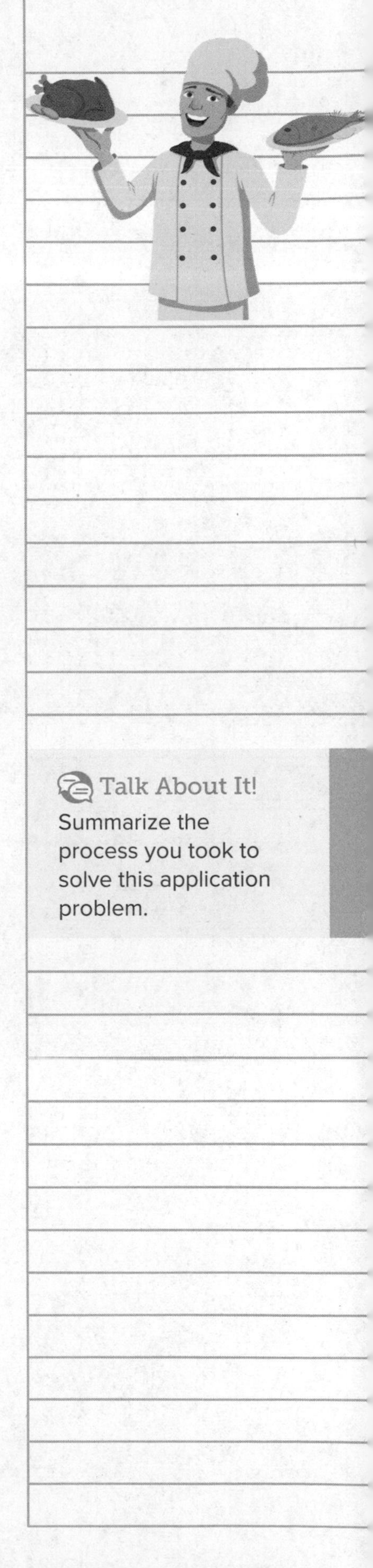

1 What is the task?

Make sure you understand exactly what question to answer or problem to solve. You may want to read the problem three times. Discuss these questions with a partner.

First Time Describe the context of the problem, in your own words.
Second Time What mathematics do you see in the problem?
Third Time What are you wondering about?

2 How can you approach the task? What strategies can you use?

3 What is your solution?

Use your strategy to solve the problem.

Talk About It! Summarize the process you took to solve this application problem.

4 How can you show your solution is reasonable?

Write About It! Write an argument that can be used to defend your solution.

Check

Marcel is purchasing boards to build a bookcase. He will use three 4.5-foot boards of pine and four 3.25-foot boards of white oak. Did Marcel use more pine or white oak to build the bookcase? How much more?

Show your work here

Go Online You can complete an Extra Example online.

Pause and Reflect

How do you feel when you are asked during class to answer a question or to explain a solution?

Record your observations here

Name ______________________ Period __________ Date __________

Practice

Go Online You can complete your homework online.

1. Jenny exercised 6 days this week. She averaged burning 284.5 Calories each day. Write a division equation that could be used to find the total number of Calories she burned this week. **(Example 1)**

2. A box of Mason's cereal contains 479.4 grams of cereal. Mason eats 28.2 grams of cereal per serving. Write a division equation that could be used to find the number of servings of cereal Mason can eat from one box of cereal. **(Example 1)**

3. On a 3-hour bike ride, Rod averaged 5.25 miles per hour. Write a division equation that could be used to find the total distance Rod biked. **(Example 1)**

4. Rowan bought a bag of jelly beans that contained 54 ounces of jelly beans. She divided the jelly beans into bags that contained 6.75 ounces each. Write a division equation that could be used to find the number of bags she made. **(Example 1)**

Solve each equation. Check your solution. **(Examples 2 and 3)**

5. $6 = \frac{j}{8}$

6. $\frac{k}{7} = 7$

7. $\frac{z}{4} = \frac{2}{3}$

8. $\frac{1}{2} = \frac{w}{8}$

9. $5.31 = \frac{p}{9.2}$

Test Practice

10. Equation Editor Solve $\frac{x}{1.3} = 1.94$.

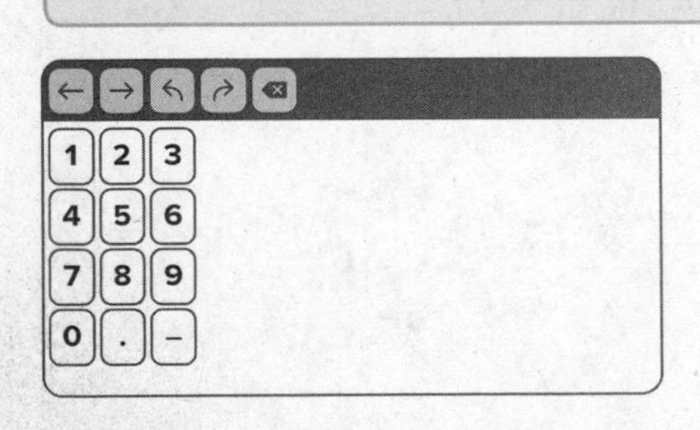

Apply

11. Each month the student council sells snack bags. The table shows the number of ounces in each bag. The first month, the student council sold 50 bags of cheese crackers and 65 bags of pretzels. How many total ounces of each snack did they sell? What is the difference in the total number of ounces?

Snack Type	Amount in Each Bag
Cheese Crackers	2.25 ounces
Pretzels	3.5 ounces

12. Jason bought two different types of boards to make picture frames. He bought a red cedar board and will cut it into eight 10.25-inch pieces. He also bought a tiger maple board that he will cut into sixteen 10.5-inch pieces. Determine the difference between the boards' total lengths.

13. MP **Reason Abstractly** Shawna noticed that the distance from her house to the ocean, which is 40 miles, was one fifth the distance from her house to the mountains. What is the distance from her house to the mountains? Explain how you solved.

14. MP **Find the Error** A student is solving the equation $\frac{x}{3} = 6$. Find the student's mistake and correct it.

$$\frac{x}{3} = 6$$

$$\frac{x}{3} \div 3 = 6 \div 3$$

$$x = 2$$

15. MP **Justify Conclusions** A model car is $\frac{1}{24}$ the size of the actual car. If a model car is 7.75 inches long, how long is the actual car? Justify your answer.

16. **Create** Write and solve a real-world problem that can be solved with a one-step division equation.

Inequalities

I Can... understand how inequalities are similar to and different from equations, and graph the solution of an inequality on a number line.

What Vocabulary Will You Learn?
inequality

Explore Inequalities

Online Activity You will use Web Sketchpad to explore inequalities using a balance and shapes that represent unknown values.

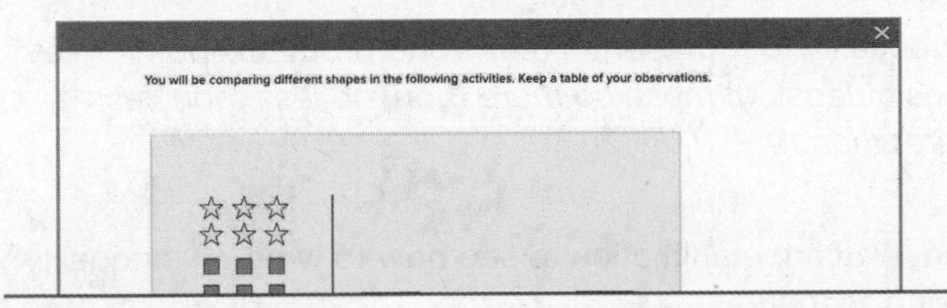

Learn Inequalities

An **inequality** is a mathematical sentence that compares quantities that may or may not be equal. The table shows the four inequality symbols, > (greater than), < (less than), ≥ (greater than or equal to), and ≤ (less than or equal to). A *solution of an inequality* is a value of the variable that makes the inequality a true statement.

Definition	Example
inequality	**inequality**
a mathematical sentence that compares quantities	$1 + x \geq 6$
symbols	**solutions of Inequality**
>, <, ≥, ≤	5, 6.5, 7, 8, 9.1...

The table compares words that are represented by the different inequality symbols.

<	≤
is less than is fewer than	is less than or equal to is at most
>	**≥**
is greater than is more than	is greater than or equal to is at least

Talk About It!
Compare and contrast an equation and an inequality.

Your Notes

Learn Write Inequalities

You can use these steps to write an inequality to represent a real-world problem.

Words
Describe the mathematics of the problem. Use only the most important words. Identify key words.
Variables
Define a variable to represent the unknown quantity.
Inequality
Translate the words into an algebraic inequality.

To write an inequality to represent a real-world problem, look for key words, such as *at least*, *at most*, *no more than*, *no less than*, *less than*, or *greater than*.

Go Online Watch the animation to see how to write an inequality for the following scenario.

A person must be at least 18 years old to vote. Write an inequality to represent the possible ages of a voter.

Words
Describe the mathematics of the problem.
The age of a voter is greater than or equal to 18 years.
Variable
Define the variable.
Let a represent the age of a voter.
Inequality
Write an inequality.
$a \geq 18$

Talk About It!

How do you know that the key words *at least* indicate using the $\geq$ symbol?

Pause and Reflect

Compare and contrast the equation $a = 18$ and the inequality $a \geq 18$. Why does the inequality represent the voter's age scenario and not the equation? Describe a scenario for which the equation might be the better representation.

Record your observations here

Example 1 Write Inequalities

In some states, you must be at least 16 years old to have a driver's license.

Write an inequality to represent the age at which you can have a driver's license.

Words
Describe the mathematics of the problem. In order to have a valid driver's license, your age must be at least 16.
Variable
Define the variable. Let a represent the age to have a license.
Inequality
Write an inequality. $a \geq 16$

So, the inequality $a \geq 16$ represents the situation.

Think About It!

What are the key words that will help you determine which inequality symbol to use?

Talk About It!

Explain why the inequality is $a \geq 16$ and not $a > 16$.

Check

A certain hotel only permits dogs that weigh less than 50 pounds to stay with hotel guests. Write an inequality that can be used to represent the weight w of dogs that are permitted to stay at the hotel.

Go Online You can complete an Extra Example online.

Pause and Reflect

Refer to the Example and Check. Why do both of these situations represent inequalities and not equations? Explain your reasoning.

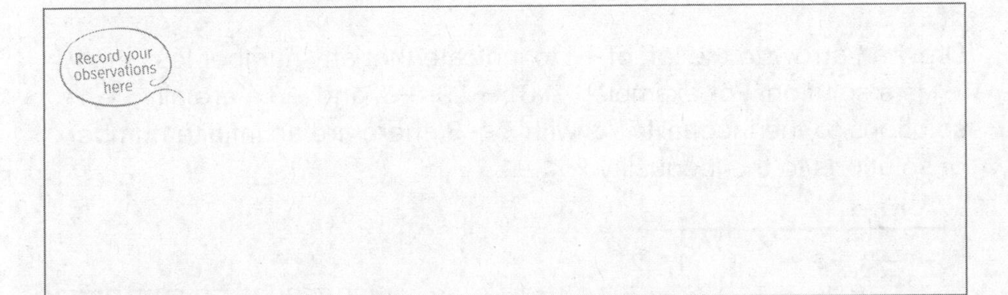

Learn Graph Inequalities

Because an inequality like $x > 5$ or $y \leq 100$ has infinitely many solutions, it is impossible to list all of them. So, inequalities can be graphed on a number line. A number line graph helps you to visualize all of the values that make the inequality true.

When you graph an inequality on a number line, place a dot at the value shown in the inequality. An open dot means the number is not included ($<$ or $>$), and a closed dot means the number is included ($\leq$ or $\geq$). Then draw an arrow in the correct direction to include all of the solutions.

Go Online Watch the video to learn more about graphing inequalities on a number line.

The video demonstrates how to graph the inequalities $x > 3$, $x < -1$, $x \leq 2$, and $x \geq -7$.

Graph the inequality $x > 3$.

Place an open dot at 3 to indicate that 3 is not a solution.

0 1 2 3 4 5 6

Draw an arrow to the right of 3 to indicate that any number greater than 3 is a solution. For example, 3.1, 3.5, 4, 4.8, and 6 are all solutions to the inequality, There are, in fact, an infinite number of solutions.

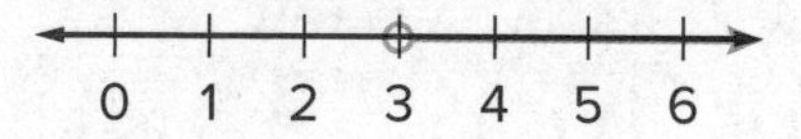

Graph the inequality $x < -1$.

Place an open dot at −1 to indicate that −1 is not a solution.

−4 −3 −2 −1 0 1 2

Draw an arrow to the left of −1 to indicate that any number less than −1 is a solution. For example, −1.01, −1.9, −3, and −3.4 are all solutions to the inequality. As with $x > 3$, there are an infinite number of solutions to the inequality $x < -1$.

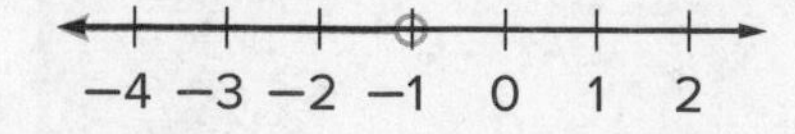

(continued on next page)

Talk About It!

Why do you think an open dot indicates the number is not a solution?

Graph the inequality $x \leq 2$.

Place a closed dot at 2 to indicate that 2 is a solution.

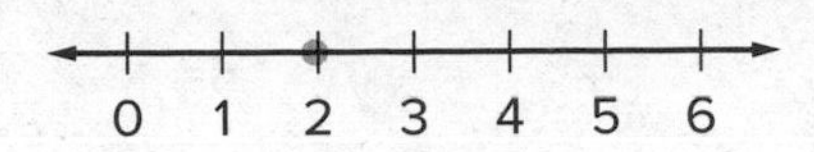

Draw an arrow to the left of 2 to indicate that any number less than 2 is also a solution.

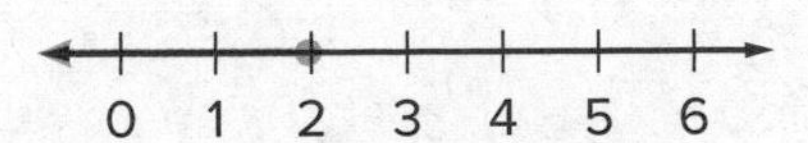

Graph the inequality $x \geq -7$.

Place a closed dot at −7 to indicate that −7 is a solution.

−10 −9 −8 −7 −6 −5 −4

Draw an arrow to the right of −7 to indicate that any number greater than −7 is also a solution.

−10 −9 −8 −7 −6 −5 −4

Example 2 Graph Inequalities

Graph the inequality $x < -5.75$.

Place an open dot at −5.75. Draw an arrow to the left of −5.75. The values that lie on the line make the inequality true.

Check

Graph the inequality $x > 1\frac{2}{5}$.

$\frac{3}{5}$ $\frac{4}{5}$ 1 $1\frac{1}{5}$ $1\frac{2}{5}$ $1\frac{3}{5}$ $1\frac{4}{5}$

Go Online You can complete an Extra Example online.

Talk About It!

Why do you think a closed dot indicates the number is a solution?

Think About It!

What does the symbol < tell you about the graph?

Talk About It!

How can you check that your graph is correct?

Think About It!

What does the symbol $\geq$ tell you about the graph?

Talk About It!

How can you check that your graph is correct?

Example 3 Graph Inequalities

Graph the inequality $x \geq \frac{2}{5}$.

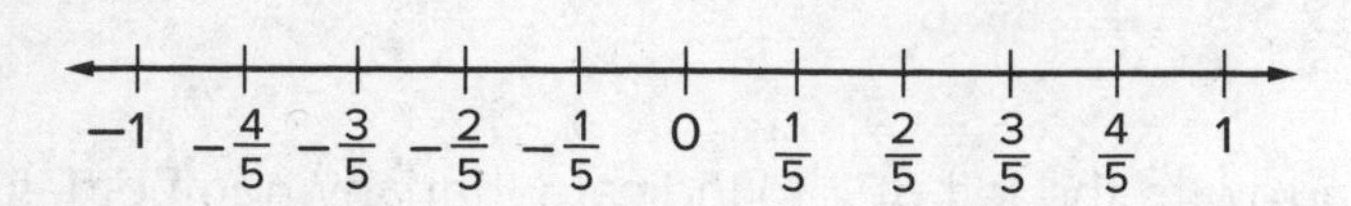

Check

Graph the inequality $x \leq -0.75$.

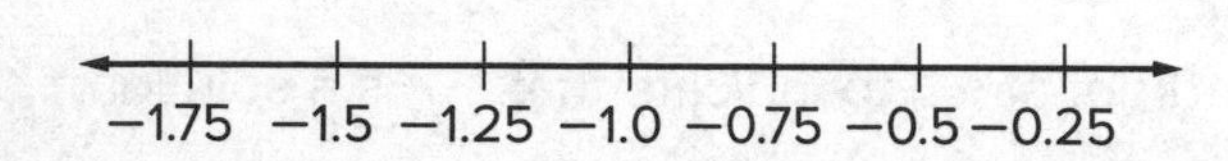

Go Online You can complete an Extra Example online.

Pause and Reflect

Did you make any errors when completing the Check exercise? What can you do to make sure you don't repeat that error in the future?

Record your observations here

Learn Find Solutions of an Inequality

When you replace a variable with a value that results in a true sentence, you solve the inequality. That value for the variable is a solution of the inequality. Some inequalities have infinitely many solutions. For example, any rational number greater than 4 will make the inequality $x > 4$ true.

Use substitution to determine if the whole numbers 5, 6, 7, 8, and 9 are solutions of the inequality $2 + x > 9$.

Value of x	$2 + x > 9$	Is the inequality true?
5	$2 + 5 > 9$ $7 > 9$	
6	$2 + 6 > 9$ $8 > 9$	
7	$2 + 7 > 9$ $9 > 9$	
8	$2 + 8 > 9$ $10 > 9$	
9	$2 + 9 > 9$ $11 > 9$	

The whole numbers ______ and ______ are solutions of the inequality.

Talk About It!

You found that 8 and 9 are solutions of the inequality $2 + x > 9$. Are there other solutions? Can you list them all? Explain your reasoning.

Example 4 Find Solutions of an Inequality

Which of the following are solutions of the inequality $a + 4 \leq 11$: 6, 7, 8?

Complete the table to determine whether or not each number is a solution of the inequality.

Value of a	$a + 4 \leq 11$	Is the inequality true?
6	$6 + 4 \overset{?}{\leq} 11$ $10 \overset{?}{\leq} 11$	
7	$7 + 4 \overset{?}{\leq} 11$ $11 \overset{?}{\leq} 11$	
8	$8 + 4 \overset{?}{\leq} 11$ $12 \overset{?}{\leq} 11$	

So, of the given values, the solutions are ______ and ______.

Talk About It!

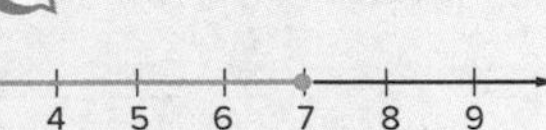

The graph shows the solution of $a + 4 \leq 11$. How can you use the graph to see if 6, 7, or 8 are solutions of the inequality?

Check

Which of the following are solutions of the inequality $c + 28 > 72$: 44, 45, 46?

Go Online You can complete an Extra Example online.

Math History Minute

The symbols < and > were first introduced in 1631 in a mathematics text. The author of the text was English mathematician **Thomas Harriot (1560 – 1621)**. While Harriot initially used triangular symbols to represent inequality, his editor changed them to < and >.

Example 5 Find Solutions of an Inequality

Which of the following are solutions of the inequality $16b > 5.6$: $\frac{1}{2}, \frac{1}{3}, \frac{1}{4}$?

Complete the table to determine if the inequality is true for each value of b.

Value of b	$16b > 5.6$	Is the inequality true?
$\frac{1}{2}$	$16 \cdot \frac{1}{2} \overset{?}{>} 5.6$ $8 \overset{?}{>} 5.6$	
$\frac{1}{3}$	$16 \cdot \frac{1}{3} \overset{?}{>} 5.6$ $5\frac{1}{3} \overset{?}{>} 5.6$	
$\frac{1}{4}$	$16 \cdot \frac{1}{4} \overset{?}{>} 5.6$ $4 \overset{?}{>} 5.6$	

So, of the given values, the solution is ________.

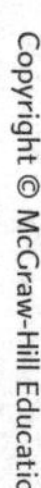

Check

Which of the following are solutions of the inequality $11.75b \leq 24.675$: 2.1, 2.3, 2.5?

Show your work here

Go Online You can complete an Extra Example online.

Example 6 Find Solutions of an Inequality

Raven has $60 to spend on matching T-shirts that cost $8.40 each for her running team. The inequality $60 \geq 8.40t$ represents the number of T-shirts t she could buy.

If there are 9 teammates on the team, how many could receive a T-shirt?

To find a solution of the inequality, substitute varying values for t. Try 9 first because that represents the number of teammates. If 9 is a solution of the inequality, then every single one of the 9 members could receive a T-shirt.

Substitute 9	Substitute 8	Substitute 7
$60 \overset{?}{\geq} 8.40t$	$60 \overset{?}{\geq} 8.40t$	$60 \overset{?}{\geq} 8.40t$
$60 \overset{?}{\geq} 8.40(9)$	$60 \overset{?}{\geq} 8.40(8)$	$60 \overset{?}{\geq} 8.40(7)$
$60 \ngeq 75.60$	$60 \ngeq 67.20$	$60 \geq 58.80$

For which values is the inequality true? ________

So, Raven can purchase no more than ________ T-shirts. This means that 7 or fewer teammates could receive a T-shirt.

Think About It!

How will the number of teammates help you choose a number to substitute?

Talk About It!

How do you know Raven has enough money to buy 1, 2, 3, 4, 5, or 6 T-shirts?

Check

At the end of his vacation, Mr. Otey has \$55 left to spend at a souvenir shop. He woud like to buy some picture frames that cost \$12.75 each to display some of his vacation photos. The inequality $12.75f < 55$ represents the number of frames f he can choose to buy. What is the greatest number of frames that he can buy?

Go Online You can complete an Extra Example online.

Pause and Reflect

Create a graphic organizer that shows the different inequality symbols and some key words that are used to indicate which symbol should be used when writing inequalities.

Apply Earnings

Several friends hope to attend a festival that costs $62.49 each to attend. To earn money, they mowed lawns for $7.50 per hour. The table shows the number of hours each person worked each day. Who earned enough money to attend the festival? What inequality can you write to represent this situation?

	Friday (hours)	Saturday (hours)
Emir	$8\frac{1}{2}$	1
Katherine	5	$2\frac{1}{2}$
Dylan	$6\frac{1}{2}$	$2\frac{1}{2}$
Anna	$3\frac{3}{4}$	$3\frac{1}{4}$

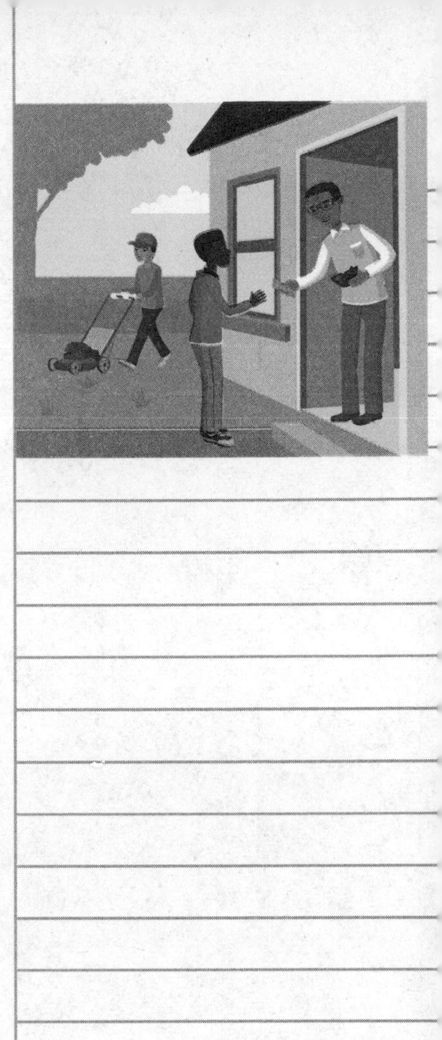

1 What is the task?

Make sure you understand exactly what question to answer or problem to solve. You may want to read the problem three times. Discuss these questions with a partner.

First Time Describe the context of the problem, in your own words.
Second Time What mathematics do you see in the problem?
Third Time What are you wondering about?

2 How can you approach the task? What strategies can you use?

3 What is your solution?

Use your strategy to solve the problem.

4 How can you show your solution is reasonable?

Write About It! Write an argument that can be used to defend your solution.

Talk About It!

How many more hours would Anna have to work to make enough money to attend?

Check

Several friends each want to buy a ticket to a football game that costs $75.99. To earn money, they worked extra hours at their job where they each earn $9.10 per hour. The table shows the number of hours each person worked each day. Who earned enough money to buy a ticket? What inequality can you write to represent this situation?

	Friday (hours)	Saturday (hours)
Aaron	$5\frac{1}{2}$	2
Cliff	$3\frac{1}{2}$	6
Missy	$7\frac{1}{2}$	$2\frac{1}{2}$
Torrance	$2\frac{1}{2}$	1

 **Go Online** You can complete an Extra Example online.

Pause and Reflect

Compare what you learned today about writing, graphing, and solving inequalities with something similar you learned about writing and solving equations. How are they similar? How are they different?

Name ______________________ Period ____________ Date ____________

Practice

Go Online You can complete your homework online.

1. The minimum deposit for a new checking account is \$75. Write an inequality to represent the amounts in dollars a that could be deposited in a new checking account. (Example 1)

2. To win a medal in a 5K race, a runner's time must be less than 22 minutes. Write an inequality to represent the times in minutes m that would win a medal. (Example 1)

Graph each inequality on the number line. (Examples 2 and 3)

3. $b < -1.5$

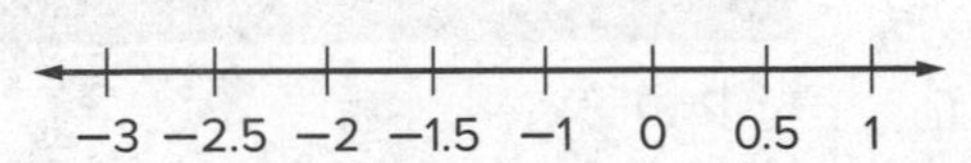

4. $d \geq 4.75$

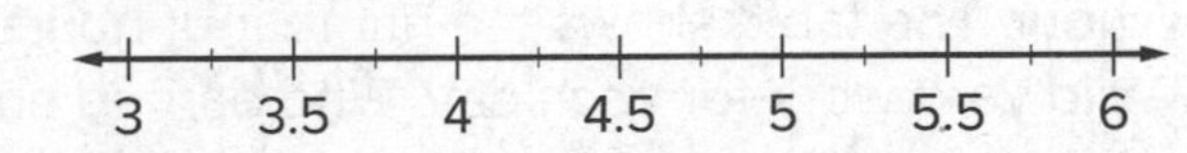

5. $a > \frac{4}{5}$

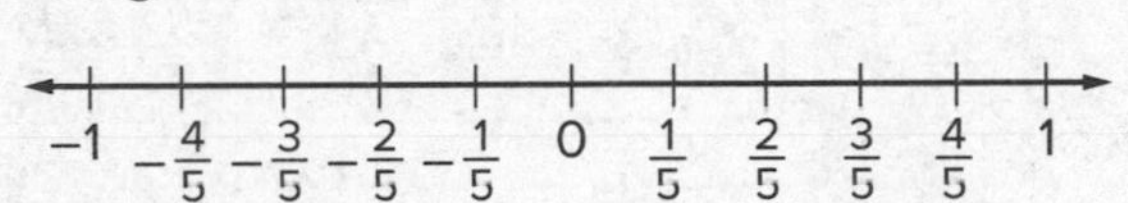

6. $d \leq -2\frac{1}{4}$

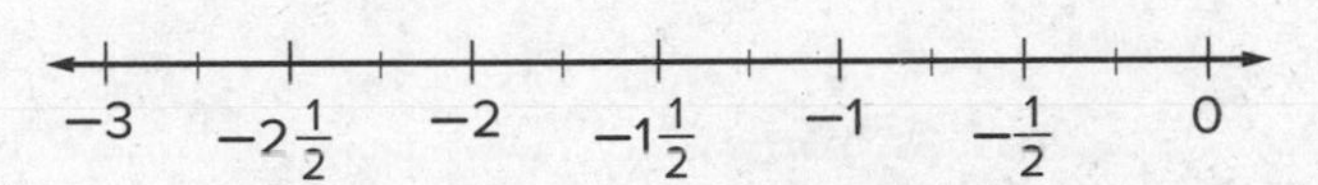

7. Which of the following are solutions of the inequality $t + 7 \leq 12$: 4, 5, 6? (Example 4)

8. Which of the following are solutions of the inequality $h - 4 > 9$: 12, 13, 14? (Example 4)

9. Which of the following are solutions of the inequality $8r \geq 1.8$: $\frac{1}{5}, \frac{1}{4}, \frac{1}{3}$? (Example 5)

10. Which of the following are solutions of the inequality $\frac{2.4}{n} < 6$: 0.25, 0.4, 0.5? (Example 5)

11. Jessica has \$32 to buy movie tickets that cost \$5.25 each for her and her friends. The inequality $32 \geq 5.25t$ represents the number of tickets t she could buy. What is the greatest number of tickets Jessica can buy? (Example 6)

Test Practice

12. **Multiselect** Stanley has \$18 to spend on packs of trading cards that cost \$1.50 each. The inequality $18 \geq 1.5p$ represents the number of packs p he can buy. Identify all the numbers of packs Stanley can buy.

☐ 10 packs ☐ 13 packs

☐ 11 packs ☐ 14 packs

☐ 12 packs ☐ 15 packs

Apply

13. Some members of a tennis team want to attend a tennis day camp that costs \$74.50 each to attend. To earn money, they washed cars for \$8.25 per hour. The table shows the number of hours each tennis player worked each day. Who earned enough money to attend the tennis day camp? What inequality can you write to represent this situation?

Tennis Player	Saturday (hours)	Sunday (hours)
Betsy	$7\frac{1}{4}$	$1\frac{1}{4}$
China	6	$3\frac{3}{4}$
Danielle	$5\frac{1}{2}$	3
Maria	$4\frac{1}{2}$	$4\frac{3}{4}$

14. Several friends each want to buy new basketball shoes that cost \$59.17. To earn money, they do yard work for \$9 an hour. The table shows the number of hours each person did yard work for each day. Who earned enough money to buy the basketball shoes? What inequality can you write to represent this situation?

Friends	Saturday (hours)	Sunday (hours)
Chad	$3\frac{1}{2}$	$3\frac{1}{2}$
Jason	4	$2\frac{3}{4}$
Martin	$3\frac{1}{2}$	3
Zek	$2\frac{1}{2}$	$3\frac{3}{4}$

15. Create Write a real-world sentence that can be represented with an inequality. Then write the inequality that represents the situation.

16. MP **Find the Error** A student is writing an inequality for the expression *a minimum donation of \$25*. Find the student's mistake and correct it.

$d \leq 25$

17. For each inequality, name a whole number that is a possible solution.

a. $18 + a > 21$

b. $7 + r \geq 18$

c. $24 - x \leq 19$

18. MP **Reason Abstractly** A roller coaster at a theme park requires children to be over 48 inches tall to ride it. Jay is 48 inches tall. Can he ride the roller coaster? Explain why or why not.

Review

Foldables Use your Foldable to help review the module.

Tab 4

Tab 3

Tab 2

Tab 1

Models Symbols

Rate Yourself!   

Complete the chart at the beginning of the module by placing a checkmark in each row that corresponds with how much you know about each topic after completing this module.

Write about one thing you learned.

Write about a question you still have.

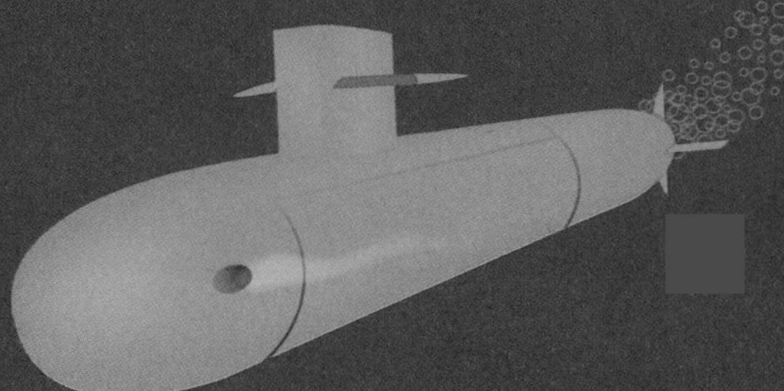

Reflect on the Module

Use what you learned about equations and inequalities to complete the graphic organizer.

Essential Question

How are the solutions of equations and inequalities different?

$x - 5 = 13$

$n + 4 = 9$

Explain how to solve each equation. Then solve the equation.

$\frac{g}{3} = 4$

$6m = 42$

What are the similarities between solving an equation and solving an inequality?

What are the differences between the solution of an equation and the solution of an inequality?

Name _______________ Period _______ Date _______

Test Practice

1. Multiple Choice Which of the following is a solution of the equation $y + \frac{1}{3} = 3\frac{2}{3}$? **(Lesson 1)**

Ⓐ $2\frac{1}{3}$

Ⓑ $2\frac{2}{3}$

Ⓒ 3

Ⓓ $3\frac{1}{3}$

2. Open Response Tonya is using $3\frac{1}{2}$ inch tiles along the 28 inch ledge of her bathroom counter. Use the *guess, check, and revise* strategy to solve the equation $3\frac{1}{2}t = 28$ to find t, the number of tiles Tonya will need. Show your work. **(Lesson 1)**

3. Multiselect Together, Rhonda and Margo saved \$478.50. If Rhonda saved \$225 of that total, how much did Margo save? Select the addition equation that could be used to find how much money m Margo saved. Select all that apply. **(Lesson 2)**

☐ $m + 478.50 = 225$

☐ $m + 225 = 478.50$

☐ $225 + 478.50 = m$

☐ $225 + m = 478.50$

☐ $478.50 + m = 225$

4. Equation Editor (Lesson 2)

A. Solve $625 = 219 + x$ for x.

$x =$ ☐

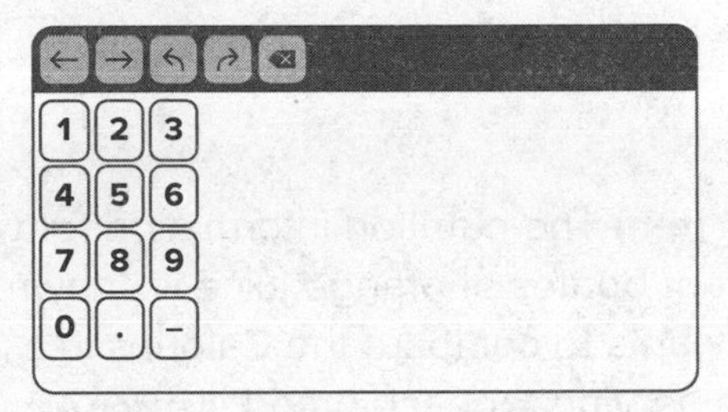

B. Check the solution.

5. Open Response A one-topping pizza costs \$12.99. This is \$6.50 less than the cost of a specialty pizza. Write a subtraction equation that could be used to find the cost c of a specialty pizza. **(Lesson 3)**

6. Equation Editor Solve $1{,}785 = x - 414$ for x. **(Lesson 3)**

$x =$ ☐

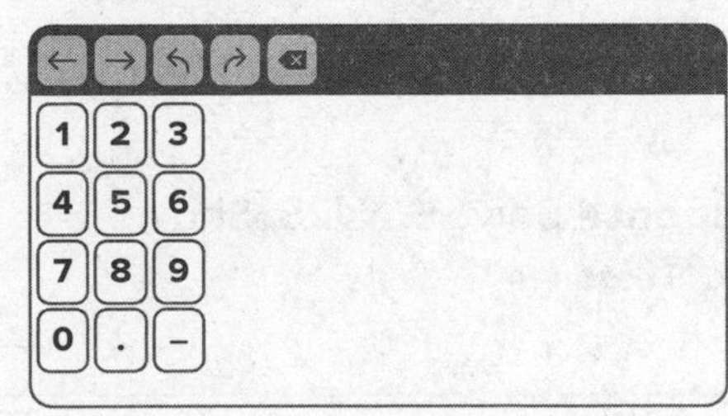

7. Multiple Choice Solve $\frac{3}{8}d = \frac{5}{24}$ for d. **(Lesson 4)**

Ⓐ $d = \frac{5}{64}$

Ⓑ $d = \frac{5}{9}$

Ⓒ $d = \frac{1}{6}$

Ⓓ $d = \frac{3}{4}$

8. Table Item The nutrition information for two different bottles of orange juice is shown. Kylie wants to compare the Calories in a single serving for each brand. **(Lesson 4)**

	Brand A (3 servings)	Brand B (2 servings)
Calories	150	220
Protein (g)	3	4
Sugar (g)	30	44

A. Find the number of Calories per serving of each brand, then indicate the correct number for each brand in the table.

Calories per Serving	Brand A	Brand B	Neither A nor B
50			
85			
110			

B. Which brand has more Calories per serving? How many more?

9. Open Response Solve $\frac{y}{11} = 28$. Show your work. **(Lesson 5)**

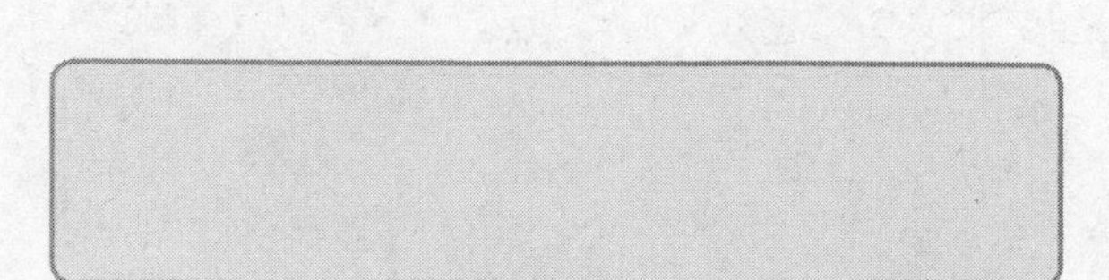

10. Multiple Choice Mr. Wolfe has a box of 144 pencils. If he wants to give 6 pencils to each of his students, which equation can be used to find the number of students s to whom Mr. Wolfe can give pencils? **(Lesson 5)**

Ⓐ $\frac{144}{s} = 6$

Ⓑ $\frac{s}{144} = 6$

Ⓒ $144s = 6$

Ⓓ $144(6) = s$

11. Grid Graph $x < \frac{1}{3}$. **(Lesson 6)**

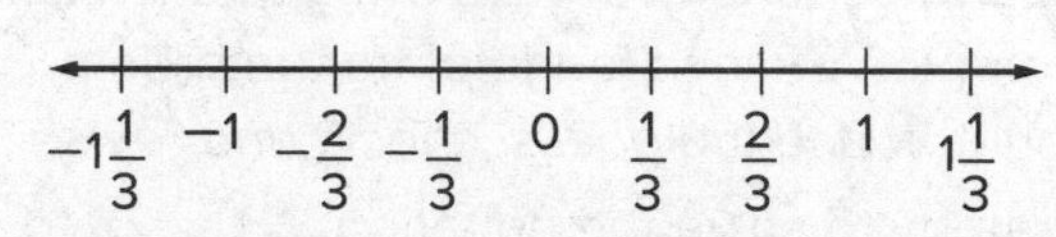

12. Table Item Use the table to indicate whether 11, 12, or 13 is a solution of the inequality $b + 8 \geq 20$. **(Lesson 6)**

Value of b	Yes	No
11		
12		
13		

13. Multiselect Brandi has $50 to spend on matching bracelets that cost $3.75 each for her volleyball team. The inequality $50 \geq 3.75b$, where b is the number of bracelets, represents the situation. If there are 16 teammates, how many will possibly receive a bracelet? **(Lesson 6)**

☐ 16 teammates

☐ 15 teammates

☐ 14 teammates

☐ 13 teammates

☐ 12 teammates

Module 7

Relationships Between Two Variables

Essential Question

What are the ways in which a relationship between two variables can be displayed?

What Will You Learn?

Place a checkmark (✓) in each row that corresponds with how much you already know about each topic **before** starting this module.

KEY

☐ — I don't know. ◇ — I've heard of it. ☆ — I know it!

	Before			After		
	☐	◇	☆	☐	◇	☆
finding dependent variable values in a table						
finding independent variable values in a table						
writing one-step and two-step equations to represent relationships between variables						
graphing relationships from equations						
writing equations from graphs						
representing relationships multiple ways						

Foldables Cut out the Foldable and tape it to the Module Review at the end of the module. You can use the Foldable throughout the module as you learn about relationships between two variables.

What Vocabulary Will You Learn?

Check the box next to each vocabulary term that you may already know.

☐ dependent variable ☐ independent variable

Are You Ready?

Study the Quick Review to see if you are ready to start this module. Then complete the Quick Check.

Quick Review

Example 1

Write algebraic expressions.

Write an algebraic expression that represents the phrase *8 less than n*. Then evaluate the expression when $n = 15$.

$n - 8$ **Write the expression.**

$n - 8 = 15 - 8$ **Replace *n* with 15.**

$= 7$ **Subtract.**

Example 2

Graph points on a coordinate plane.

Graph the point $(-3, 5)$ on a coordinate plane.

Start at (0, 0). Move 3 units left. Then 5 units up.

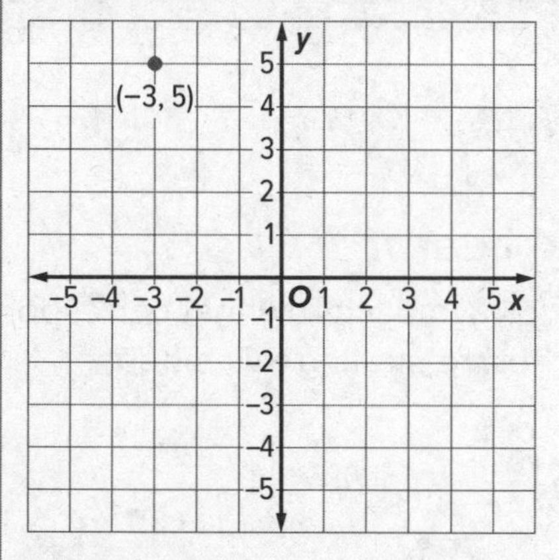

Quick Check

1. Jane spent $17 less than three times the amount Jake spent. Write an expression to find how much Jane spent if Jake spent *x* dollars. If Jake spent $15, how much did Jane spend?

2. Graph the point $(2, -4)$ on the coordinate plane.

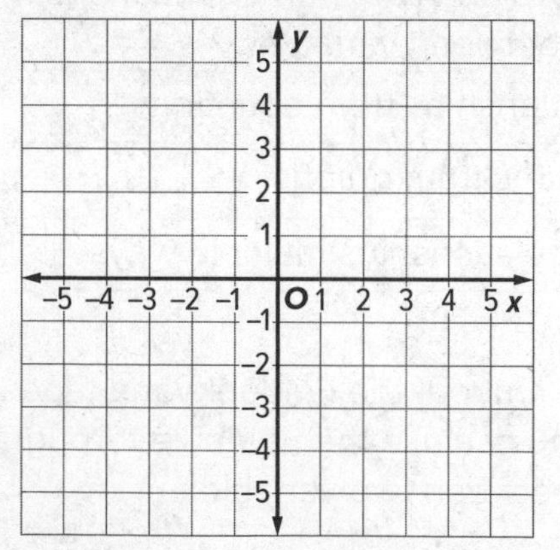

How Did You Do?

Which exercises did you answer correctly in the Quick Check? Shade those exercise numbers at the right.

Relationships Between Two Variables

I Can... use equations and rules to find missing values of independent and dependent variables in tables.

What Vocabulary Will You Learn?
dependent variable
independent variable

Explore Relationships Between Two Variables

Online Activity You will use Web Sketchpad to explore the relationship between two variables.

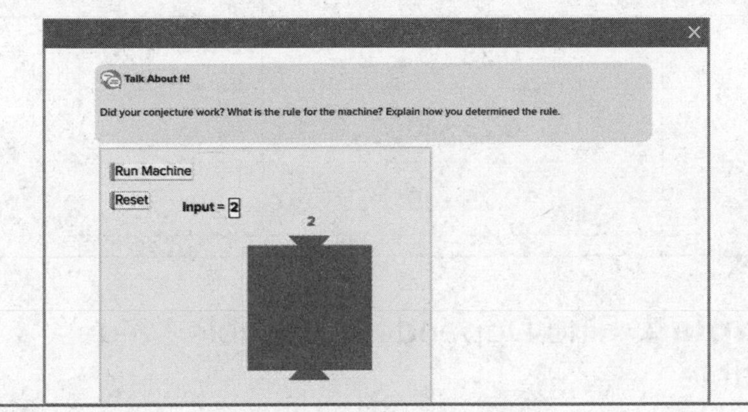

Learn Identify Independent and Dependent Variables

In a relationship between two quantities, one quantity is the independent variable and the other quantity is the dependent variable. The **independent variable**, often called the *input*, does not depend upon the other quantity. The **dependent variable**, often called the *output*, changes in response to the input for the independent variable.

Talk About It!
Use what you know about the terms *independent* and *dependent* to explain why the variables are named this way.

Consider the following situation. At top speed, a cheetah can travel about 103 feet every second. The total distance traveled at top speed t is equal to 103 times the number of seconds s.

independent variable = number of seconds

dependent variable = total distance

The total distance is the dependent variable because the cheetah's distance *depends* on the number of seconds it travels.

Suppose Jaeda earns $5 per hour for babysitting. The total amount she earns a is equal to 5 times the hours h that she babysits.

What is the independent variable? H

What is the dependent variable? A

Your Notes

Learn Find Dependent Variable Values in a Table

Suppose it costs \$0.25 to play one game at an arcade. You can use a table to show the relationship between the independent variable (input) and the dependent variable (output). In the table, the input value is the number of games played g, and the rule is $0.25g$. The output is the total cost c. To find the output, replace g with the input, and evaluate the expression.

Input (independent variable)	Rule (relationship between the input and output)	Output (dependent variable)
Number of Games Played, g	**$0.25g$**	**Total Cost (\$), c**
5	$0.25 \cdot 5$	1.25
10	$0.25 \cdot 10$	2.50
15	$0.25 \cdot 15$	3.75

Talk About It!
The unit cost is \$0.25 per game. How is this rate shown in the table? Explain your reasoning.

Example 1 Find Dependent Variable Values in a Table

Joe bought an iced coffee for \$2.95. The total cost of his breakfast c is equal to the cost of his food f plus \$2.95. The rule is $f + 2.95$.

Make a table using the rule to find the total cost of Joe's breakfast if his food costs \$5.50, \$7.75, or \$10.00.

Step 1 Identify the independent and dependent variables.

The cost of the food f is the independent variable. The total cost of his breakfast c is the dependent variable, because the total cost depends on the cost of Joe's food.

Step 2 Find each output.

Use the rule to complete the table.

Input Cost of Food (\$), f	Rule $f + 2.95$	Output Total Cost (\$), c
5.50	$5.50 + 2.95$	
7.75		
10.00		12.95

So, if his food costs \$5.50, his total cost is \$________.
If his food costs \$7.75, his total cost is \$________.
If his food costs \$10.00, his total cost is \$________.

Think About It!
How many columns will your table have? What will they be named?

Talk About It!
If Joe chooses a different drink with his breakfast, at a different price, how will it change the rule?

Check

A grocery store charges \$2.50 per gallon of fruit punch. The total cost c of g gallons of fruit punch is equal to 2.5 times g. The rule is $2.5g$. Make a table using the rule to find the total cost of buying 1, 2, or 3 gallons of fruit punch.

Input Number of Gallons, g	Rule $2.5g$	Output Total Cost (\$), c
1		
2		
3		

Go Online You can complete an Extra Example online.

Learn Find Independent Variable Values in a Table

Suppose it costs \$0.25 to play one game at an arcade. The total cost of playing any number of games can be represented by the rule $0.25g$, where g is the number of games played. You can use a table to find the independent variable (input) if you know the dependent variable (output) and the rule.

Note that in the table below, the output values are \$1.75, \$3.00, and \$4.25. Because the rule is $0.25g$, write and solve an equation to find the input g when the output is \$1.75.

$0.25g = 1.75$ The input g multiplied by 0.25 equals the output, \$1.75.

$\frac{0.25g}{0.25} = \frac{1.75}{0.25}$ Divide each side by 0.25.

$g = 7$ Simplify. The input value is 7.

Repeat this process to complete the table for the other two output values, \$3.00 and \$4.25.

Input Number of Games Played, g	Rule $0.25g$	Output Total Cost (\$), c
7	$0.25 \cdot 7$	1.75
	$0.25 \cdot$ ☐	3.00
	$0.25 \cdot$ ☐	4.25

Talk About It!

How can you use the *work backward* strategy to find each input value, instead of writing and solving an equation? How are these strategies similar and different?

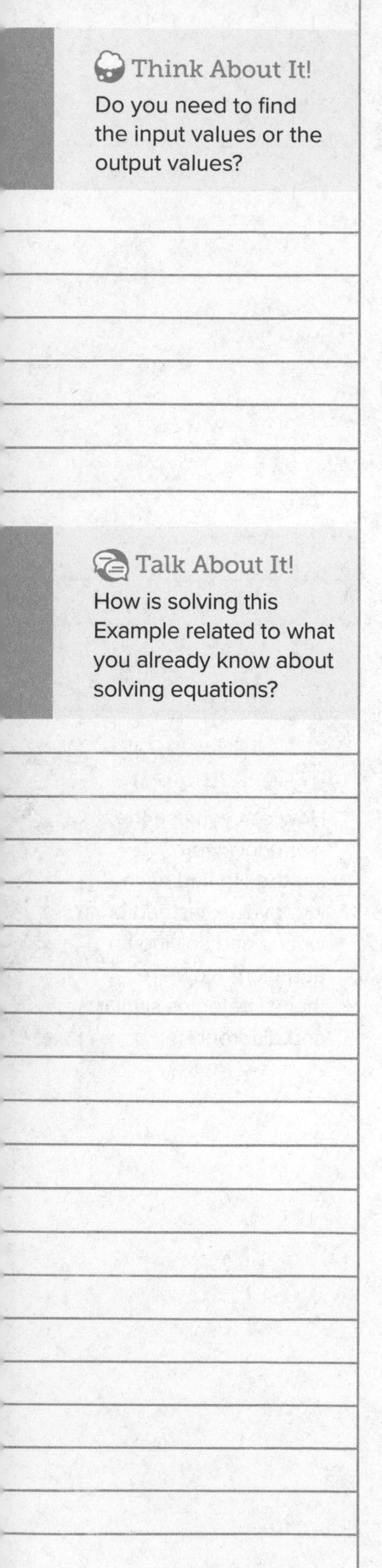

Example 2 Find Independent Variable Values in a Table

Each small pizza at the local pizza shop costs \$6.75. The total cost c of p small pizzas is equal to 6.75 times p.

Make a table to find the number of small pizzas purchased if the total cost is \$13.50, \$27, or \$33.75.

Step 1 Identify the independent and dependent variables.

The number of pizzas p is the input, or independent variable.

The total cost of the pizza c is the output, or dependent variable.

The total cost is 6.75 times p, so the rule is ________.

Step 2 Find each input.

To find the number of pizzas for each of the total costs given in the table, use the *work backward* strategy. To undo multiplication by 6.75, use the inverse operation to divide each output value by 6.75.

Complete the table.

Input Number of Pizzas, p	Rule $6.75p$	Output Total Cost (\$), c
2	6.75·2	13.50
4	6.75·4	27.00
5	6.75·5	33.75

Check

Leslie has 48 stickers to give to her friends. The number of stickers s each friend will receive is equal to 48 divided by f, the number of friends. Complete the table to find the number of friends Leslie gave stickers to if each friend receives 12, 8, or 6 stickers.

Input Number of Friends, f	Rule $48 \div f$	Output Number of Stickers, s
		12
		8
		6

Go Online You can complete an Extra Example online.

Apply Measurement

Sondra is placing sculptures on a 3-foot-tall base to display in a cabinet in the school entryway. The height including the base b is equal to the height h of the sculpture plus 3. If the cabinet is 84 inches tall, which sculpture(s) will fit in the cabinet?

Height of Sculpture (ft), h	Rule $h + 3$	Height with Base (ft), b
$2\frac{1}{4}$		
$3\frac{3}{4}$		
$5\frac{1}{8}$		

1 What is the task?

Make sure you understand exactly what question to answer or problem to solve. You may want to read the problem three times. Discuss these questions with a partner.

First Time Describe the context of the problem, in your own words.
Second Time What mathematics do you see in the problem?
Third Time What are you wondering about?

2 How can you approach the task? What strategies can you use?

3 What is your solution?

Use your strategy to solve the problem.

4 How can you show your solution is reasonable?

Write About It! Write an argument that can be used to defend your solution.

Talk About It!
Why is it helpful to convert the measurements to the same unit?

Check

Katarina wants to take four friends to an amusement park for her birthday. The total cost c is equal to the admission rate r times 5. If she can spend no more than $150 on admission tickets, which amusement park(s) can they visit?

Amusement Park	Admission Rate ($), r	Rule, $5r$	Total Cost ($), c
A	30	5 · 30	150
B	35	5 · 35	
C	40	5 · 40	

Show your work here

You can go to Amusemet part A.

Go Online You can complete an Extra Example online.

Foldables It's time to update your Foldable, located in the Module Review, based on what you learned in this lesson. If you haven't already assembled your Foldable, you can find the instructions on page FL1.

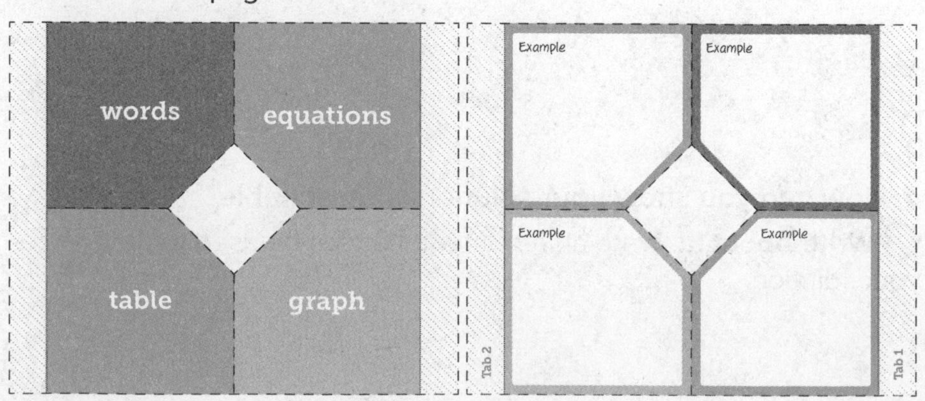

Name ______________________ Period ____________ Date ____________

Practice

Go Online You can complete your homework online.

1. Sadie ordered a pizza and had it delivered. The delivery fee is \$3.50. The total cost c is equal to the cost of her pizza p plus \$3.50. The rule is $p + 3.50$. Complete the table using the rule to find the total cost if her pizza costs \$9.75, \$12.00, or \$14.50. **(Example 1)**

Input, Cost of Pizza (\$), p	Rule $p + 3.50$	Output, Total Cost (\$), c
9.75		
12.00		
14.50		

2. Joshua has a coupon for \$1.50 off his purchase at the souvenir shop. The total cost c is equal to the cost of his purchase p minus \$1.50. The rule is $p - 1.50$. Complete the table using the rule to find the total cost if his purchase is \$5.67, \$8.34, or \$11.97. **(Example 1)**

Input, Cost of Purchase (\$), p	Rule $p - 1.50$	Output, Total Cost (\$), c
5.67		
8.34		
11.97		

3. Miranda has a coupon for \$0.75 off any salad at a restaurant. The total cost c is equal to the cost of her salad s minus \$0.75. The rule is $s - 0.75$. Complete the table using the rule to find the total cost if her salad costs \$2.79, \$3.55, or \$4.25. **(Example 1)**

Input, Cost of Salad (\$), s	Rule $s - 0.75$	Output, Total Cost (\$), c
2.79		
3.55		
4.25		

4. Avery is buying material by the yard to make bags. The material costs \$4.98 per yard. The total cost c of y yards is equal to 4.98 times y. Complete the table to find the number of yards Avery purchased if the total cost is \$14.94, \$29.88, or \$44.82. **(Example 2)**

Input, Number of Yards, y	Rule $4.98y$	Output, Total Cost (\$), c
		14.94
		29.88
		44.82

5. Each pie at a bakery costs \$9.50. The total cost c of p pies is equal to 9.50 times p. Complete the table to find the number of pies purchased if the total cost is \$19.00, \$28.50, or \$47.50. **(Example 2)**

Input, Number of Pies, p	Rule $9.50p$	Output, Total Cost (\$), c
		19.00
		28.50
		47.50

Test Practice

6. Table Item Anthony is buying plants for his garden. Each plant costs \$0.95. The total cost c of p plants is equal to 0.95 times p. Complete the table to find the number of plants Anthony purchased if the total cost is \$7.60, \$11.40, or \$15.20.

Input, Number of Plants, p	Rule $0.95p$	Output, Total Cost (\$), c
		7.60
		11.40
		15.20

Apply

7. Mara lives in a state that has no sales tax on apparel. She has a coupon for $15 off the price of one pair of shoes. The total cost c of a pair of shoes is equal to the original price of the shoes p minus 15. If she only has $60 to spend on a pair of shoes, which pair(s) could she buy?

Original Price ($), p	Rule $p - 15$	Total Cost ($), c
65		
73		
79		

8. An empty suitcase weighs 224 ounces. The total weight t of the suitcase is equal to the weight of its contents w plus 224. To not be charged an additional fee for a flight, the total weight must be no more than 50 pounds. Which suitcase(s) would be charged a fee?

Weight of Contents (oz), c	Rule $x + 224$	Weight with Suitcase (oz), t
$575\frac{1}{2}$		
576		
$576\frac{1}{2}$		

9. MP **Identify Structure** Complete the table by finding the input values.

Input, x	Rule, $2x - 2.5$	Output, y
		7.5
		10.5
		13.5

10. MP **Reason Inductively** A student said that the independent variable for the following situation is the number of days, d. Is the student correct? Explain.

Jess walks 1.5 miles every day for d days. What is the total number of miles she walks?

11. MP **Persevere with Problems** A concession stand sells soft pretzels for $2.75 each and drinks for $1.50 each. The equation $c = 2.75p + 1.50d$ can be used to represent the total cost c of p pretzels and d drinks. What is the total cost of 3 pretzels and 4 drinks? Explain how you solved.

12. Describe a real-world situation that has an independent variable and a dependent variable. Identify each variable.

Write Equations to Represent Relationships Represented in Tables

I Can... use variables, which represent independent and dependent values, to write one-step and two-step equations from real-world situations.

Learn Write One-Step Equations

Luciana earns $8 per hour walking dogs in her neighborhood. The table shows the relationship between the number of hours h she walks and the total amount d, in dollars, she earns. To write an equation that relates the variables h and d, first determine the rule that describes the relationship.

Input Number of Hours, h	Rule ?	Output Dollars Earned ($), d
1		8
2		16
3		24
4		32

The output values increase by the same number, 8, as the input values increase by 1. Because repeated addition can be written as multiplication, check each pair of input-output values to determine if the rule $8h$ accurately describes the relationship.

Input Number of Hours, h	Rule $8h$	Output Dollars Earned ($), d
1	8(1)	8
2	8(2)	16
3	8(3)	24
4	8(4)	32

So, the rule $8h$ accurately describes the relationship. Notice that the ratio of each output value to each input value is constant. This further confirms that the multiplication expression $8h$ is the rule and no other operation is involved.

$\frac{\$8}{1} = \$8 \qquad \frac{\$16}{2} = \$8 \qquad \frac{\$24}{3} = \$8 \qquad \frac{\$32}{4} = \$8$

(continued on next page)

Talk About It!

What connections do you see in this relationship that relate to what you already know about rates? Where can you see the unit rate in the table?

Your Notes

Use the rule $8h$ to write an equation relating the two variables h and d.

$d = 8h$ ← Each output value is the product of the constant ratio, 8, and the corresponding input value h.

Example 1 Write One-Step Equations

The table shows the total cost c, in dollars, of buying t souvenir T-shirts.

Number of T-shirts, t	Total Cost ($), c
1	9
2	18
3	27
4	36
5	45

Write an equation to represent the relationship between c and t.

Think About It!
How would you begin solving the problem?

Step 1 Identify the variables.

The independent variable is c.

The dependent variable is t.

Step 2 Determine the rule.

The output values increase by the same number, 9, as the input values increase by 1. Because repeated addition can be written as multiplication, check each pair of input-output values to determine if the rule $9t$ accurately describes the relationship.

Input Number of T-shirts, t	Rule $9t$	Output Total Cost ($), c
1	9(1)	9
2	9(2)	18
3	9(3)	27
4	9(4)	36
5	9(5)	45

So, the rule $9t$ accurately describes the relationship.

Talk About It!
What connections do you see in this relationship that relate to what you already know about unit price? Where can you see the unit price in the table?

Step 3 Write the equation.

Use the rule $9t$ to write an equation relating the two variables t and c.

$c = 9t$ ← Each output value is the product of the constant ratio, 9, and the corresponding input value t.

So, the equation that represents the total cost c of buying t souvenir T-shirts is ________.

Check

The table shows the total cost c of belonging to the gym for m months. Write an equation to represent the relationship between c and m.

Number of Months, m	Total Cost ($), c
1	24.95
2	49.90
3	74.85
4	99.80
5	124.75

Go Online You can complete an Extra Example online.

Explore Relationships with Rules that Require Two Steps

Online Activity You will use Web Sketchpad to explore the relationship between two variables with two-step rules.

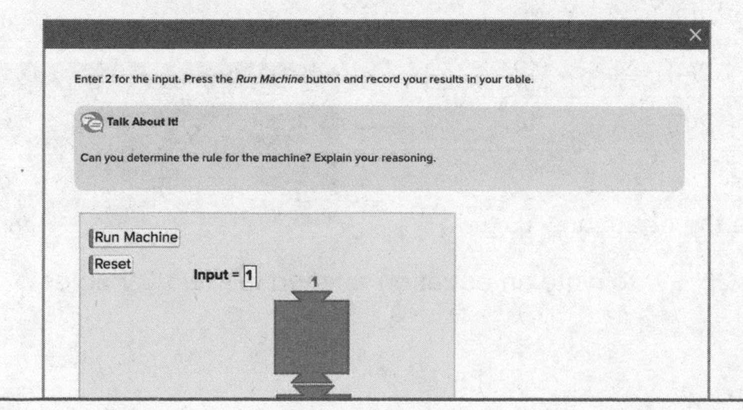

Math History Minute

In 1993, **Ellen Ochoa (1958-)** became the world's first Hispanic female astronaut. She has flown in space four times and has logged nearly 1,000 hours in orbit. Ochoa's high school calculus teacher inspired her to pursue studies in math and science.

Learn Write Two-Step Equations

Sometimes the relationship between two variables cannot be accurately described by a one-step equation. In those cases, check to see if more than one operation is involved. Consider the following scenario.

An online store sells baseball bats. You will pay for each bat that you order, plus a one-time shipping fee. The table shows the relationship between the number of bats ordered and the total cost. Write a two-step equation to represent the total cost c to ship an order of b baseball bats.

Number of Bats, b	Total Cost (\$), c
1	6
2	8
3	10

Step 1 Look for a pattern.

The output values increase by the same number, 2, as the input values increase by 1. The rule includes $2b$.

Step 2 Determine the rule.

Input, Number of Bats, b	Rule $2b$	Output, Total Cost (\$), c
1	2(1)	6
2	2(2)	8
3	2(3)	10
4	2(4)	12
5	2(5)	14

Check each pair of input-output values to determine if the rule $2b$ accurately describes the relationship. The rule $2b$ does not accurately describe the relationship.

Input, Number of Bats, b	Rule $2b + 4$	Output, Total Cost (\$), c
1	2(1) + 4	6
2	2(2) + 4	8
3	2(3) + 4	10
4	2(4) + 4	12
5	2(5) + 4	14

To obtain an output value of 6, multiply the input value 1 by 2. Then add 4.

Check each pair of input-output values. The rule $2b + 4$ accurately describes the relationship.

Step 3 Write the equation.

Use the rule $2b + 4$ to write an equation relating the two variables b and c.

$c = 2b + 4$

Talk About It!

If the store did not charge a shipping fee for the bats, how would the equation be different? What is the shipping fee?

Example 2 Write Two-Step Equations

The table shows the total number of necklaces Ari has made after a certain number of hours.

Write a two-step equation to represent the total number of necklaces *n* she will have made after *h* hours.

Time (hours), h	Number of Necklaces, n
1	5
2	7
3	9
4	11

Think About It!

Is there repeated addition in the output?

Step 1 Look for a pattern.

The output values increase by the same number, 2, as the input values increase by 1. The rule includes $2h$.

Step 2 Determine the rule.

Check each pair of input-output values to determine if the rule $2h$ accurately describes the relationship.

Time (hours), h	Rule $2h$	Number of Necklaces, n
1	2(1)	5
2	2(2)	7
3	2(3)	9
4	2(4)	11

By itself, the rule $2h$ does not describe the relationship. Check to see if this relationship involves two operations.

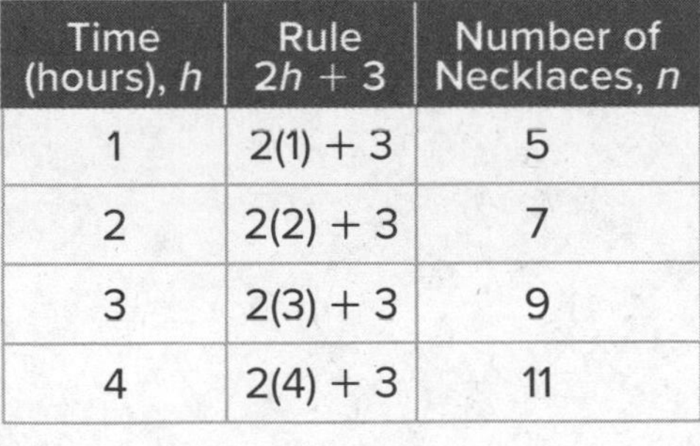

Time (hours), h	Rule $2h + 3$	Number of Necklaces, n
1	$2(1) + 3$	5
2	$2(2) + 3$	7
3	$2(3) + 3$	9
4	$2(4) + 3$	11

To obtain an output value of 5, multiply the input value 1 by 2. Then add 3.

Check each pair of input-output values. The rule $2h + 3$ describes the relationship.

Talk About It!

When a relationship can be represented by a two-step equation, is there a constant ratio between the variables? Explain.

Step 3 Write the equation.

Use the rule $2h + 3$ to write an equation relating the two variables h and n.

$n = 2h + 3$

So, the equation used to represent the total number of necklaces n Ari will have made after h hours is ____________.

Check

The table shows the total fees f for d days a library book is overdue. Write a two-step equation to represent the total fee for the number of days the book is overdue.

Time (days), d	Total Fee ($), f
1	0.30
2	0.50
3	0.70
4	0.90

Go Online You can complete an Extra Example online.

Pause and Reflect

Did you struggle more with writing two-step equations as compared to one-step equations? If so, what questions can you ask to better understand the concept? If not, how could you explain the concept to someone who is struggling?

Apply Art

Each evening, Autumn and Bennett painted signs for a school campaign. The table shows the total number of signs painted after a certain number of hours. If the pattern continues, how many more signs will Autumn have painted than Bennett after painting for 9 hours?

Hours	Total Signs: Autumn	Total Signs: Bennett
1	6	4
2	9	6
3	12	8
4	15	10

Talk About It!
What could the constant represent in the equations for Autumn and Bennett?

1 What is the task?

Make sure you understand exactly what question to answer or problem to solve. You may want to read the problem three times. Discuss these questions with a partner.

First Time Describe the context of the problem, in your own words.
Second Time What mathematics do you see in the problem?
Third Time What are you wondering about?

2 How can you approach the task? What strategies can you use?

3 What is your solution?

Use your strategy to solve the problem.

4 How can you show your solution is reasonable?

Write About It! Write an argument that can be used to defend your solution.

Check

The table shows the total number of miles Lan and Bailey have run after a certain number of days. If the pattern continues, how many more miles will Bailey have run after 6 days than Lan?

Days	Lan	Bailey
1	6	3
2	8	6
3	10	9
4	12	12

Go Online You can complete an Extra Example online.

Foldables It's time to update your Foldable, located in the Module Review, based on what you learned in this lesson. If you haven't already assembled your Foldable, you can find the instructions on page FL1.

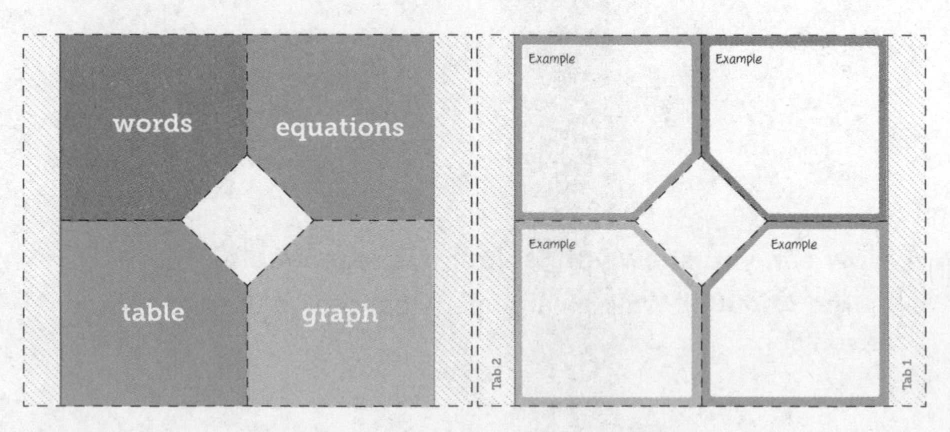

Name ______________________ Period __________ Date __________

Practice

Go Online You can complete your homework online.

1. The table shows the total cost c of buying t movie tickets. Write an equation to represent the relationship between c and t. **(Example 1)**

Number of Tickets, t	Total Cost ($), c
1	7
2	14
3	21
4	28

2. The table shows the total number of pencils p in b boxes. Write an equation to represent the relationship between p and b. **(Example 1)**

Number of Boxes, b	Total Number of Pencils, p
1	12
2	24
3	36
4	48

3. The table shows the total cost of bowling any number of games and renting bowling shoes. Write a two-step equation to represent the total cost c for bowling g games. **(Example 2)**

Number of Games, g	Total Cost ($), c
1	6
2	10
3	14
4	18

4. The table shows the total cost of renting a canoe based on the number of hours and a one-time rental fee. Write a two-step equation to represent the total cost c of renting a canoe for h hours. **(Example 2)**

Number of Hours, h	Total Cost ($), c
1	16
2	27
3	38
4	49

Test Practice

5. **Open Response** The table shows the total cost of belonging to a fitness center based on the number of months and a one-time registration fee. Write a two-step equation to represent the total cost c for belonging to the fitness center for m months.

Number of Months, m	Total Cost ($), c
1	25
2	40
3	55
4	70

Apply

6. On weekends, Peter and Aiden washed cars to raise money for a school trip. The table shows the total number of cars washed, after a certain number of hours. If the pattern continues, how many more cars will Aiden have washed than Peter after 8 hours?

Hours	Cars Washed: Peter	Cars Washed: Aiden
1	5	4
2	7	7
3	9	10
4	11	13

7. MP **Persevere with Problems** Write an equation to represent the relationship shown in the table.

Input, x	Output, y
3	4
6	5
9	6
12	7
15	8

8. MP **Reason Abstractly** A dance studio charges $45 per month, plus a $30 registration fee. Willa has $210 for dance lessons. How many months can she take lessons? Explain how you solved.

9. MP **Find the Error** A student wrote the equation $c = 20h + 12$ to represent the relationship shown in the table. Find the student's error and correct it.

Hours, h	1	2	3	4
Cost, c	$32	$44	$56	$68

10. Write about a real-world situation that can be represented with a two-step equation. Write the equation and explain the meaning of the variables.

Graphs of Relationships

I Can... graph a relationship represented by an equation and write an equation represented by a graph by identifying and using the independent and dependent variables.

Learn Graph a Relationship from an Equation

You can use an equation that represents the relationship between an independent variable (input) and a dependent variable (output) to graph the relationship on the coordinate plane. The independent variable is represented by the *x*-coordinate and the dependent variable is represented by the *y*-coordinate.

Similar to graphing ratio tables, you can make a table of values to represent the equation, use the values to generate a set of ordered pairs, and graph the relationship. Consider the following equation.

$$y = 2x + 500$$

Make a table of values.

Independent Value, x	Dependent Value, y
0	500
1	502
2	504
3	506

Write the ordered pairs.

x, y
(0, 500)
(1, 502)
(2, 504)
(3, 506)

Graph the ordered pairs. Draw a line to connect the points.

A *break* shows that the there are no values between 0 and 499.

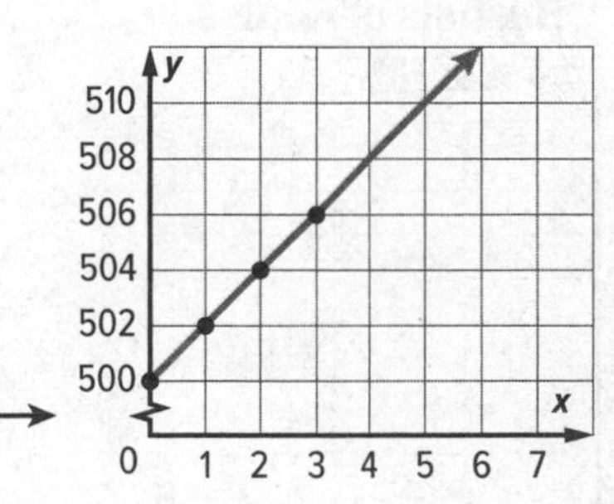

Talk About It!

How is graphing ordered pairs from an equation similar to graphing ordered pairs from a ratio table? How is it different? Explain your reasoning.

Your Notes

Think About It!
What is the dependent variable? the independent variable?

Talk About It!
How can you determine the unit rate comparing the number of apples to the number of bushels?

Example 1 Graph a Relationship from an Equation

The equation $a = 126b$ represents the approximate number of apples a in b bushels of apples. Graph the relationship on the coordinate plane.

Step 1 Determine the independent and dependent variables.

independent variable: number of ___________

dependent variable: number of ___________

Step 2 Make a table.

Use the equation $a = 126b$ to make a table of values. Place the independent variable in the first column, and the dependent variable in the second column.

Number of Bushels, b	Number of Apples, a
0	
1	
2	

Step 3 Use the values to make a list of ordered pairs.

(b, a)

Step 4 Graph the ordered pairs. Then draw the line.

Because you cannot have partial apples, the line representing the relationship should be dashed.

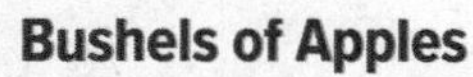

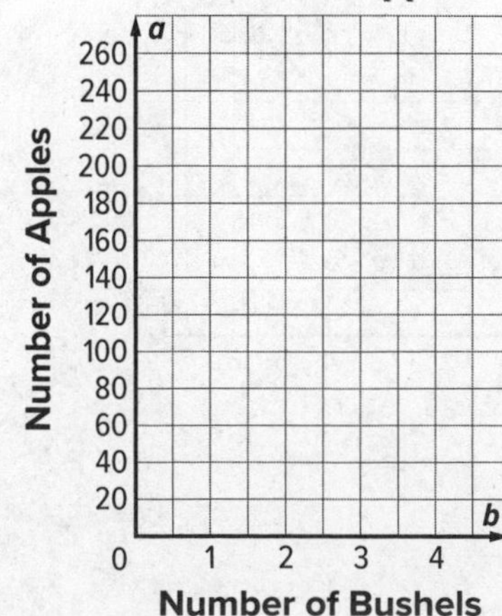

Check

The equation $c = 2p + 3$ represents the total cost c, in dollars, of ordering p pounds of personalized jelly beans from an online store with a $3 shipping fee. Graph the relationship on the coordinate plane.

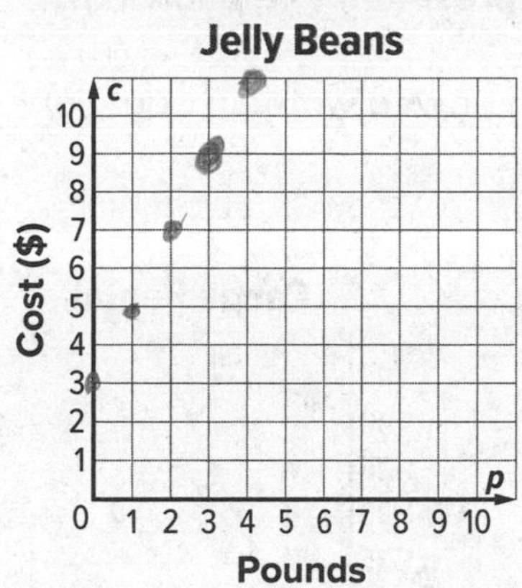

Go Online You can complete an Extra Example online.

Pause and Reflect

In Example 1, the graph of the line is dashed because you *cannot* have part of an apple in a bushel. In the Check for Example 1, the line is solid because you *can* have part of a pound. Work with a partner to give an example of two real-world relationships that could be represented using a dashed line and two real-world relationships that could be represented using a solid line.

Record your observations here

Learn Write an Equation from a Graph

An equation can be used to symbolically describe the graph of a relationship. Use the *work backward* strategy to make a table of ordered pairs, then write the equation to represent the relationship.

Talk About It!
In this situation, why does it not make sense for the graph of the line to cross the y-axis?

Go Online Watch the animation to learn how to write an equation from the following graph.

The graph shows the relationship between the cost to rent a canoe and the number of hours the canoe is rented.

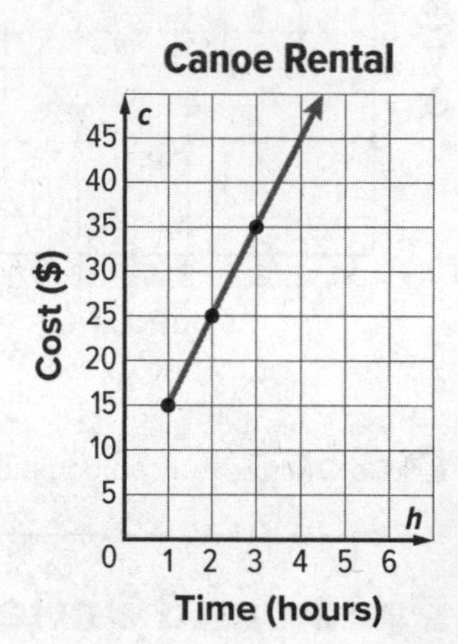

Step 1 Determine the independent and dependent variables.

The time h in hours is the independent variable, and the cost c in dollars is the dependent variable.

Step 2 Identify the ordered pairs on the graph.

The graph includes the ordered pairs (1, 15), (2, 25), and (3, 35).

Step 3 Make a table.

Time (h), h	Cost ($), c
1	15
2	25
3	35

Step 4 Write the equation.

The values of the dependent variable c increase by ________ every hour.

After multiplying by 10, you add ________ to obtain the correct value for c.

So, the equation to find the total cost c after h hours is $c = 10h + 5$.

Example 2 Write an Equation from a Graph

Martino constructed the graph that shows the height of his cactus after several years of growth.

Assuming the cactus grows at a constant rate, write an equation from the graph that could be used to find the height h of the cactus after g years.

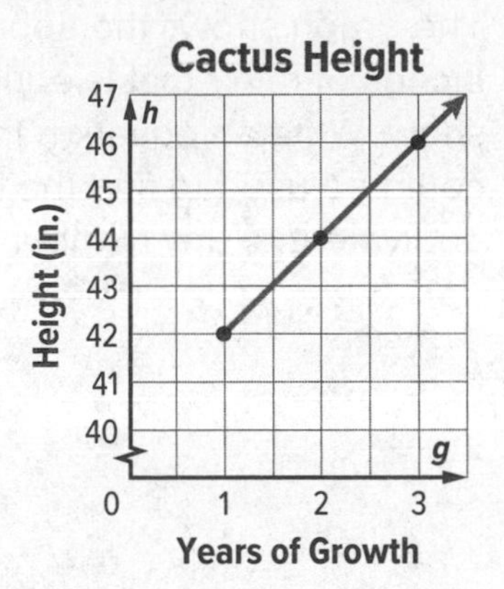

Step 1 Determine the independent and dependent variables.

The graph shows the relationship between time and the height of the cactus, where time is the __________ variable and the height is the __________ variable.

Step 2 Identify the ordered pairs on the graph.

The ordered pairs are (1, 42), (2, 44), and (3, 46).

Step 3 Make a table.

Years of Growth, g	Height (in.), h

Step 4 Write the equation.

The values of the dependent variable h increase by ______ each year.

After multiplying by 2, you add ______ to obtain the correct value for h.

Check each pair of values to determine if the rule $2g + 40$ accurately describes the relationship.

So, the equation to find the height of the cactus h after g years of growth is $h = 2g + 40$.

Think About It!

What are the ordered pairs you can use to write the equation?

Talk About It!

Why is it important to identify the independent and dependent variables before writing the equation?

Check

The graph shows the approximate number of inches of rain r that is equivalent to s inches of snow. Write an equation from the graph that could be used to find the total inches of snow equivalent to any number of inches of rain.

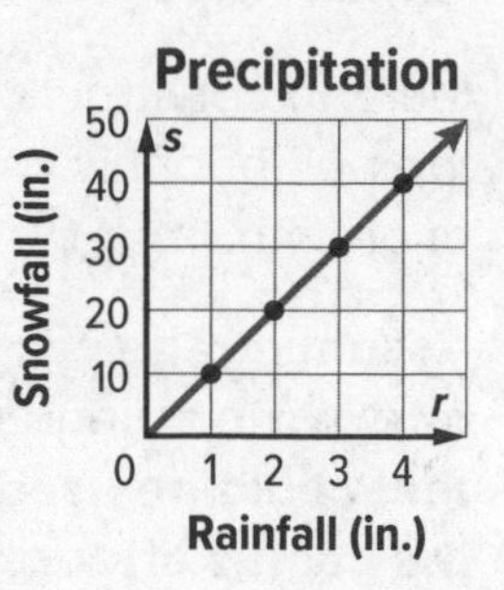

Go Online You can complete an Extra Example online.

Foldables It's time to update your Foldable, located in the Module Review, based on what you learned in this lesson. If you haven't already assembled your Foldable, you can find the instructions on page FL1.

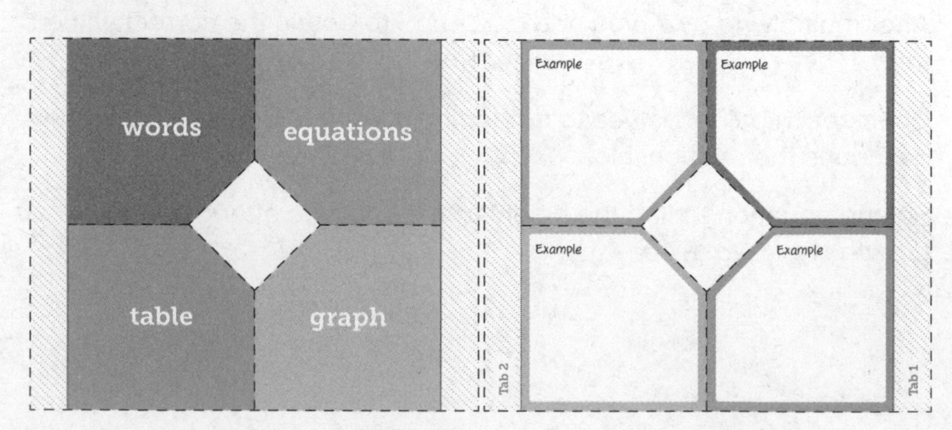

Name ______________________ Period __________ Date __________

Practice

Go Online You can complete your homework online.

1. The equation $p = 144b$ represents the number of pencils p in b boxes. Graph the relationship on the coordinate plane. **(Example 1)**

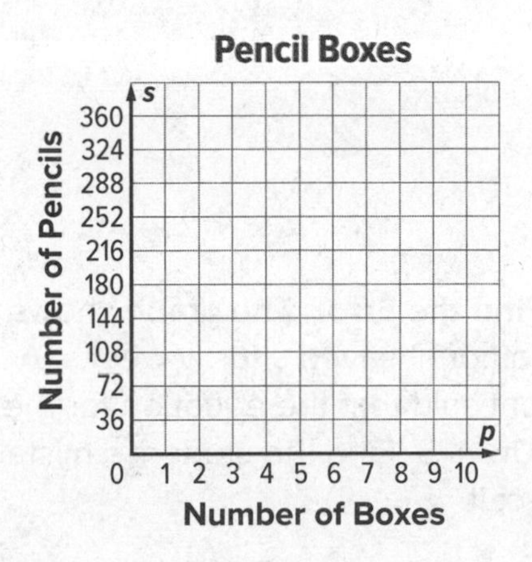

2. The equation $c = 2b + 6$ represents the total cost c of b sets of beads and one necklace string. Graph the relationship on the coordinate plane. **(Example 1)**

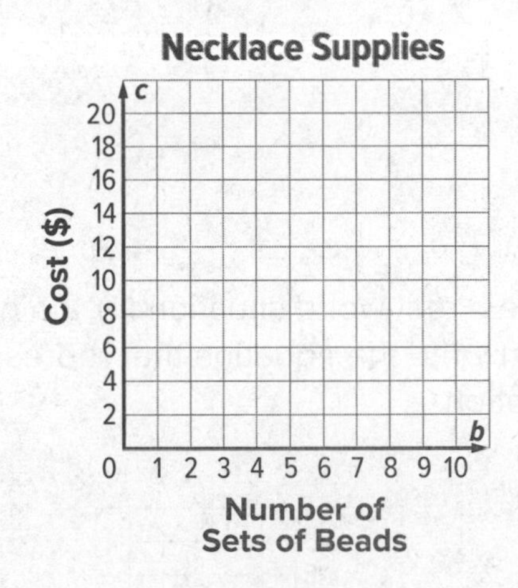

3. The graph shows the total cost c of buying one large bucket of popcorn and d large drinks. Write an equation from the graph that could be used to find the total cost c if you buy one large bucket of popcorn and d large drinks. **(Example 2)**

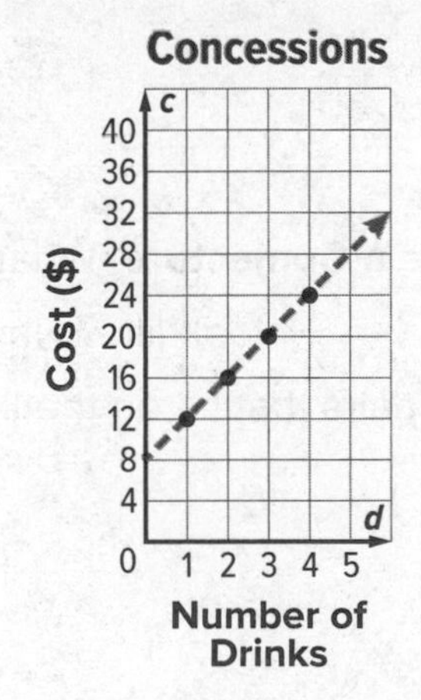

Test Practice

4. **Open Response** The graph shows the total cost c of buying one parking pass and t tickets to a concert. Write an equation from the graph that could be used to find the total cost c if you buy one parking pass and t tickets to a concert. **(Example 2)**

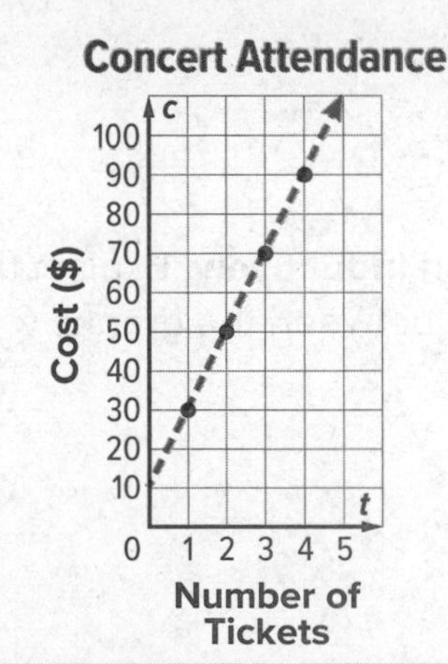

Apply

5. Nancy and Elsa like to ride bikes. The equation $m = 12h$ represents the approximate number of miles m Nancy bikes in h hours. The equation $m = 9h$ represents the approximate number of miles m Elsa bikes in h hours. How much longer will it take Elsa to bike 72 miles than Nancy?

6. Write a real-world situation for the graph. Then write the equation that represents the situation.

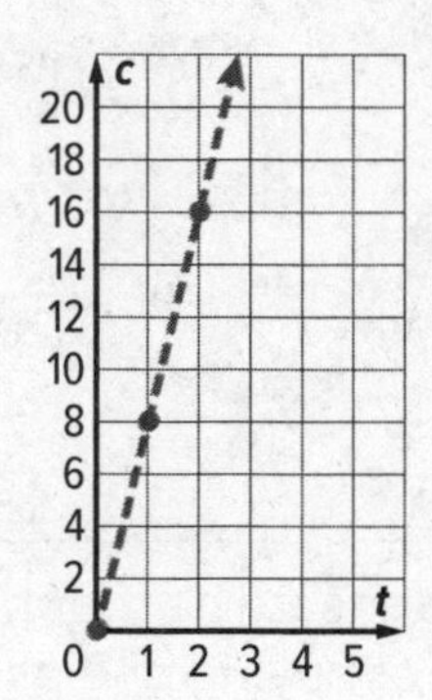

7. MP **Find the Error** The graph shows the total amount saved s for w weeks. A student said that the equation for the line is $s = 10w + 5$. Find the student's mistake and correct it.

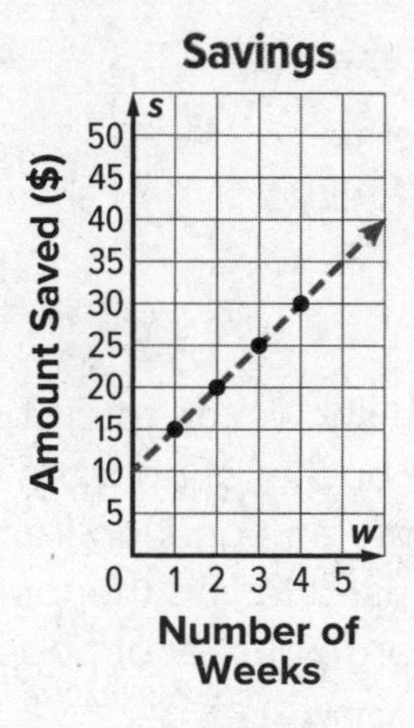

8. MP **Reason Inductively** Explain the difference between the graphs $y = 3x$ and $y = 3x + 2$.

9. MP **Make a Conjecture** What would the graph of $y = \frac{1}{2}x$ look like? Name three ordered pairs that lie on the line.

Multiple Representations

I Can... identify the independent and dependent variables in a given scenario and use that information to create an equation, table, and graph that represent the situation.

Learn Multiple Representations of Relationships

Relationships between two variables can be described using multiple representations, such as words, equations, tables, and graphs. By generating multiple representations of the same relationship, you can identify correspondences between the representations. Each representation describes the same relationship, yet in a different way.

Words

Words help express the relationship, using real-life elements.

On a trip, a cyclist traveled at a constant speed of 14 miles per hour for several hours.

Equation

Equations can be used to readily find other values for the relationship that are not already known.

$d = 14t$

Table

Tables help organize individual pairs of input-output values.

Time (h), t	Distance (mi), d
1	14
2	28
3	42

Graph

Graphs help to show trends in the relationship and can be used to make predictions.

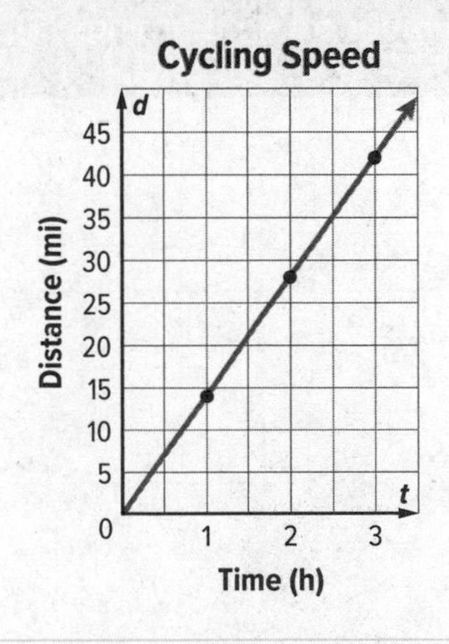

Talk About It!

Give an example of a situation where a table might be a better representation for a relationship than a graph. Explain your reasoning.

Your Notes

Think About It!
What will you do first, write an equation, make a table, or create a graph?

Example 1 Multiple Representations of Relationships

The student council has already earned $150 this year. For the next fundraiser, they are holding a car wash and charging $7 for each car they wash.

Represent the relationship between the number of cars washed *c* and the total earnings *t* with an equation, a table, and a graph.

Part A Represent the relationship with an equation.

Step 1 Determine the independent and dependent variables.

In this relationship, the number of cars washed *c* is the
________________ variable and the total earnings *t* is the
________________ variable.

Step 2 Write the equation.

Before the car wash, the student council had already earned
$__________. For washing cars, they will earn $__________ per car.

To determine the total earned *t*, multiply the number of cars washed *c* by 7 and add 150.

The equation that represents the situation is $t = 7c + 150$.

Part B Represent the relationship with a table.

Number of Cars, *c*	Earnings ($), *t*
1	
2	
3	
4	
5	

(continued on next page)

Part C Represent the relationship with a graph.

Step 1 Write the ordered pairs.

From the table, the ordered pairs are (1, 157), (2, 164), (3, 171), (4, 178), and (5, 185).

Step 2 Graph the ordered pairs and draw the line.

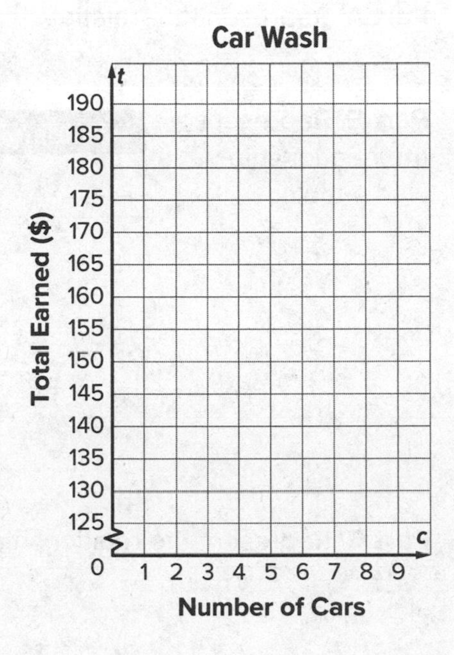

Pause and Reflect

Compare what you learned today with something similar you learned in an earlier module or grade. How are they similar? How are they different?

Record your observations here

Talk About It!

Which representation would be good to use if you wanted to see a trend in the amount of money the student council was earning? Explain your reasoning.

Check

An online stores sells trail mix for $2.75 per pound and charges a shipping fee of $4. Represent the relationship between the pounds of trail mix bought p and the total cost c with an equation, a table, and a graph.

Part A Represent the relationship with an equation.

Part B Represent the relationship with a table.

Pounds of Trail Mix, p	Total Cost ($), c
1	
2	
3	
4	
5	

Part C Represent the relationship with a graph.

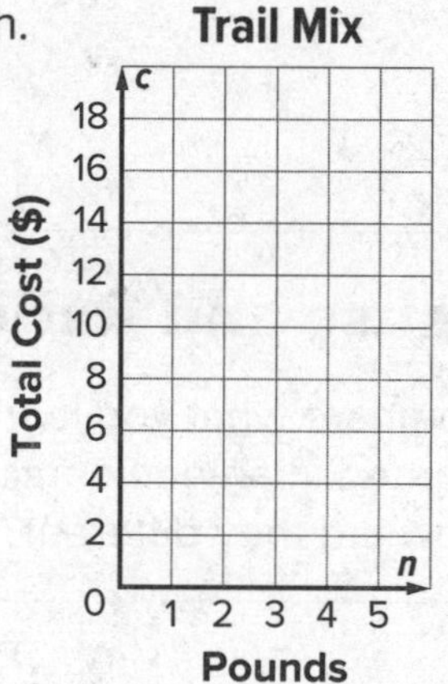

Go Online You can complete an Extra Example online.

Foldables It's time to update your Foldable, located in the Module Review, based on what you learned in this lesson. If you haven't already assembled your Foldable, you can find the instructions on page FL1.

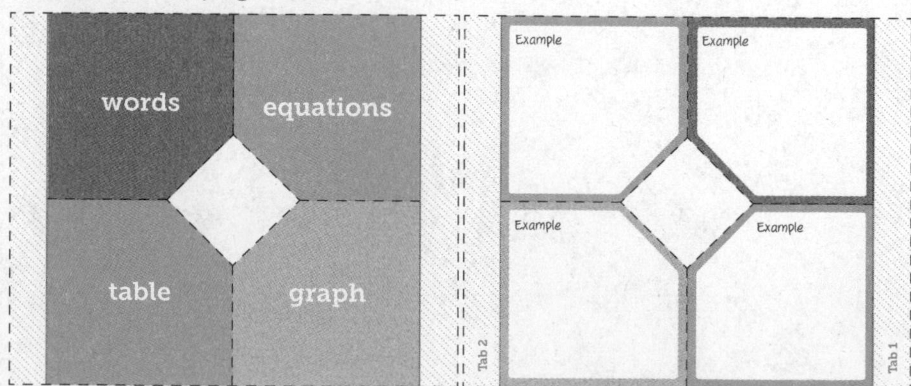

Name ______________________ Period ____________ Date ____________

Practice

Go Online You can complete your homework online.

1. A school sells tickets to their school play through an online ticket company. Each ticket costs $8 and the company charges a $2.50 processing fee per order. Represent the relationship between the number of tickets bought t and the total cost c with an equation, a table, and a graph. **(Example 1)**

a. Represent the relationship with an equation.

b. Represent the relationship with a table.

Number of Tickets, t	Total Cost ($), c
1	
2	
3	
4	

c. Represent the relationship with a graph.

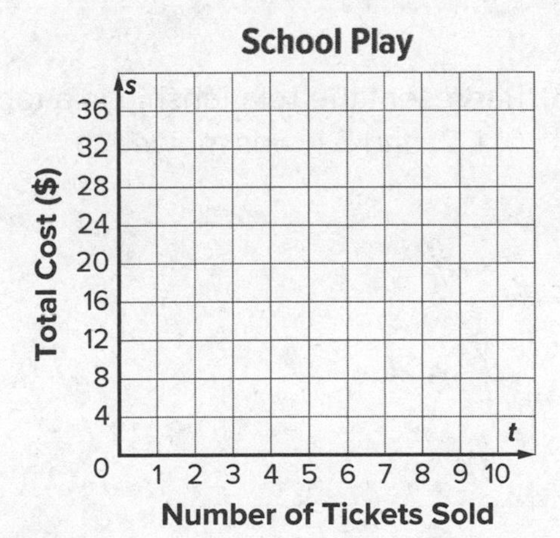

2. Carmelo earns a weekly allowance of $5 plus an additional $0.75 for each chore that he completes. Represent the relationship between the total earned t and the number of chores completed c with an equation, a table, and a graph. **(Example 1)**

a. Represent the relationship with an equation.

b. Represent the relationship with a table.

Number of Chores, c	Total Earned ($), t
1	
2	
3	
4	

c. Represent the relationship with a graph.

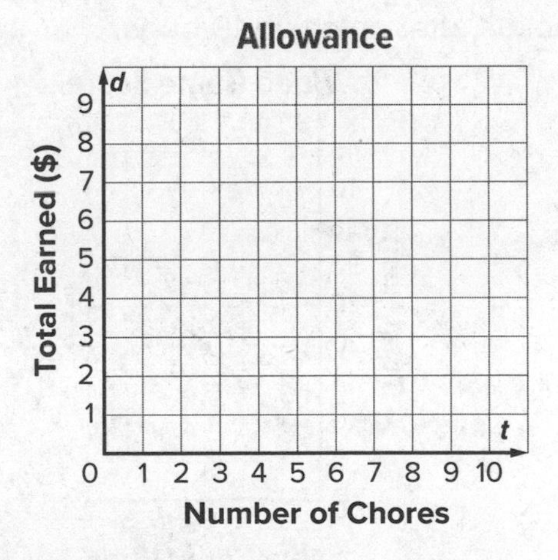

Test Practice

3. Open Response The table shows the earnings for each pie sold at the sixth grade bake sale. Represent the relationship between the number of pies sold p and the total earnings e with an equation.

Number of Pies, p	Total Earnings ($), e
1	6
2	12
3	18

Apply

4. Zari is comparing the costs of having cupcakes delivered from two different bakeries. Betty's Bakery offers free delivery and sells cupcakes by the dozen. The table shows the total cost c of d dozens from Betty's Bakery. The Sweet Shoppe charges $20 for delivery and $18 per dozen. The equation $c = 18d + 20$ represents the total cost c of d dozens of cupcakes and delivery from the Sweet Shoppe. If Zari has $110 to spend, which bakery should she use to order the greatest number of cupcakes? Explain.

Number of Dozens of Cupcakes, d	Total Cost ($), c
1	24
2	48
3	72

5. MP **Persevere with Problems** Ryder plays a video game where each player is given points and players earn more points by catching bugs. Write an equation to represent the total number of points p earned for catching b bugs. Use the equation to find Ryder's points after catching 10 bugs.

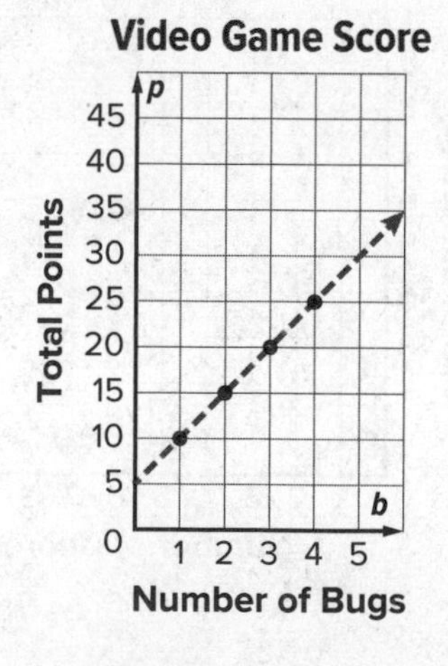

6. **Multiple Representations** Winslow earns $15.50 for each lawn that he mows.

a. Represent the relationship between the number of lawns mowed m and his total earnings e with an equation.

b. Represent the relationship in a table for 0, 1, 2, and 3 lawns mowed.

7. MP **Reason Abstractly** Reese and Tamara both babysit. Reese earns $5 per hour and Tamara earns $10 per hour. Will the amount earned for each girl ever be the same for the same number of hours after zero hours? Explain.

8. Write about a real-world situation that could be represented with an equation, a table, and a graph.

Review

Foldables Use your Foldable to help review the module.

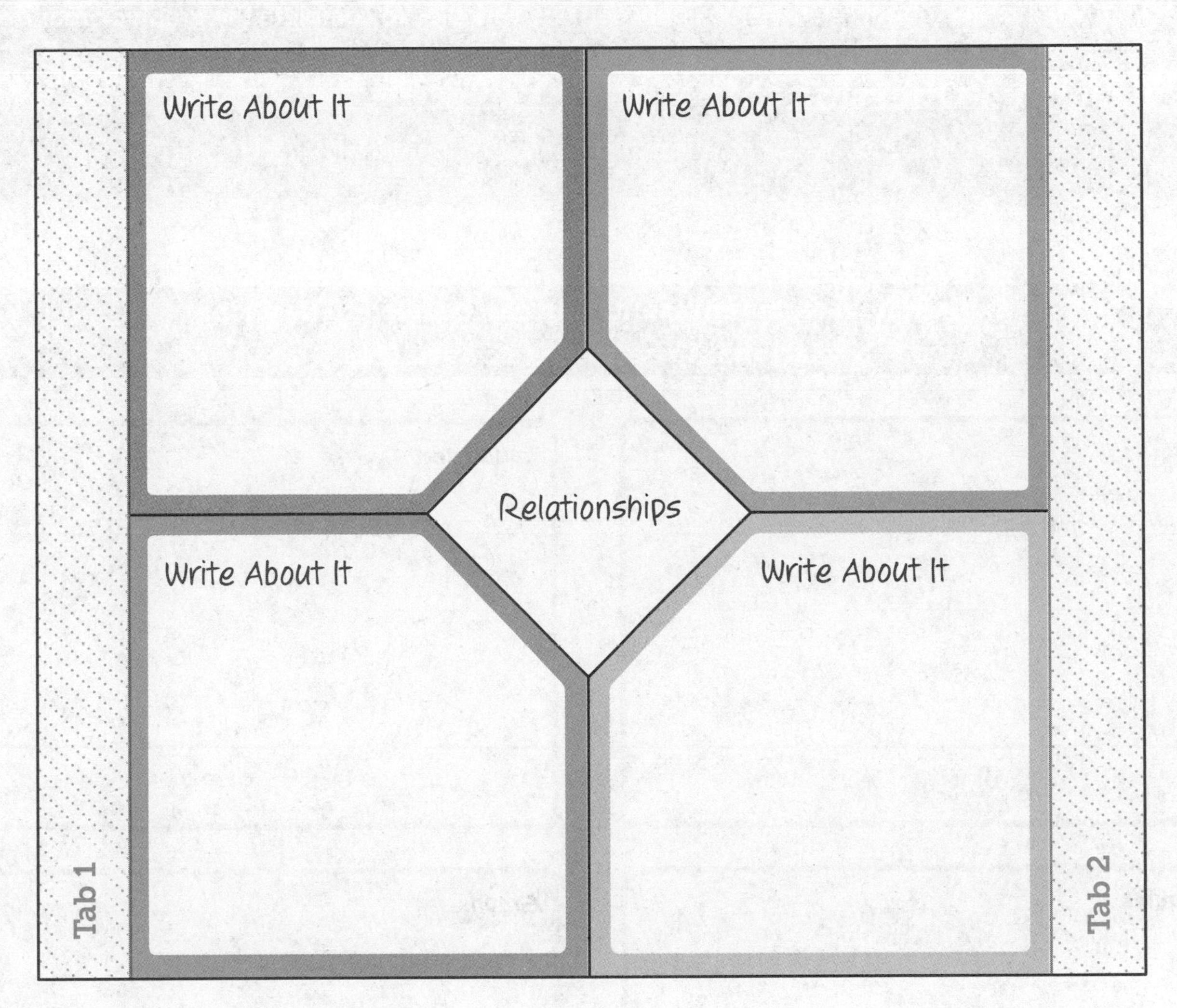

Rate Yourself!   

Complete the chart at the beginning of the module by placing a checkmark in each row that corresponds with how much you know about each topic after completing this module.

Write about one thing you learned.

Write about a question you still have.

Reflect on the Module

Use what you learned about relationships between two variables to complete the graphic organizer.

Essential Question

What are the ways in which a relationship between two variables can be displayed?

Explain how each representation can be used to describe a relationship between two variables.

Words	Equations

Tables	Graphs

Name ______________________ Period __________ Date __________

Test Practice

1. **Equation Editor** A store charges $1.70 for a fountain soft drink. The total cost *c* of *d* soft drinks is equal to 1.7 times *d*. The table below represents this situation. What is the missing value in the output column? **(Lesson 1)**

Input, d	Rule, $1.7d$	Output, c
1	1.7(1)	1.7
2	1.7(2)	3.4
3	1.7(3)	?

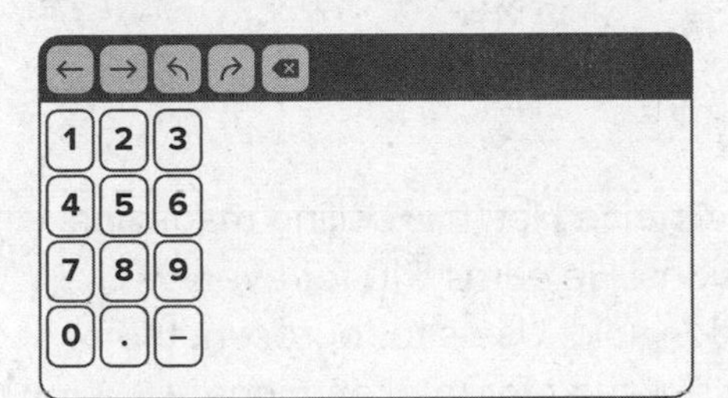

2. **Multiple Choice** Mr. Hamilton has 144 pencils to give to his students. The number of pencils *p* each student will receive is equal to 144 divided by *s*, the number of students. The table below represents this situation. Which of the following numbers should be entered into the input column (top to bottom) in order to complete the table? **(Lesson 1)**

Input, s	Rule, $\frac{144}{s}$	Output, p
?	$\frac{144}{s}$	12
?	$\frac{144}{s}$	9
?	$\frac{144}{s}$	6

Ⓐ 9, 12, 16

Ⓑ 9, 16, 24

Ⓒ 12, 16, 24

Ⓓ 12, 18, 24

3. **Open Response** The table shows the total cost *c* of buying *b* shell bracelets at a souvenir shop. Write an equation to represent the relationship between *c* and *b*. **(Lesson 2)**

Number of Bracelets, b	Total Cost ($), c
1	6
2	12
3	18
4	24
5	30

4. **Equation Editor** The table shows the total number of laps Sue and Kee walked over the past four days. If the pattern continues, how many more laps will Sue have walked than Kee after 7 days? **(Lesson 2)**

Days	Sue	Kee
1	4	2
2	8	4
3	12	6
4	16	8

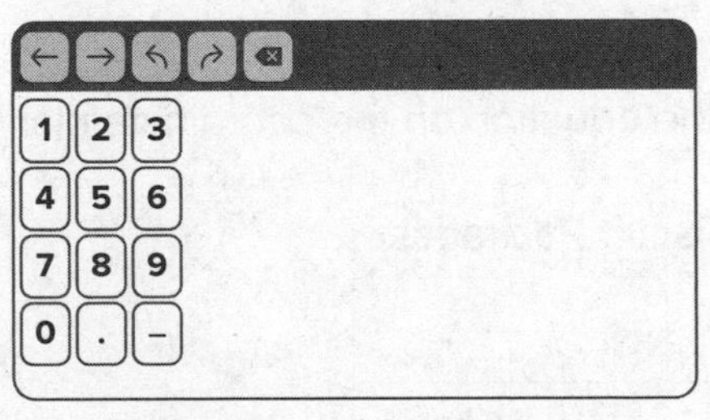

5. **Open Response** The equation $c = 15.25h$ represents the cost *c* for *h* hours of a bicycle rental. What is the cost of a 4-hour bicycle rental? **(Lesson 3)**

6. Multiselect The table shows the total cost for h hours a plumber charges to make a service call to a customer. Which of the following two-step equations represents the total cost for the number of hours of service the plumber provides? Select all that apply. **(Lesson 2)**

Number of Hours, h	Total Cost ($), c
1	70
2	110
3	150
4	190

☐ $c = 30h + 40$

☐ $c = 40h + 30$

☐ $30h + 40 = c$

☐ $20h + 50 = c$

☐ $40h + 30 = c$

7. Grid The equation $b = 8p$ represents the number of biscuits b in p packages of biscuits. **(Lesson 3)**

A. Complete the table of values that represents this situation.

Number of Packages, p	Number of Biscuits, b
0	
1	
2	
3	

B. Graph the equation on the coordinate plane.

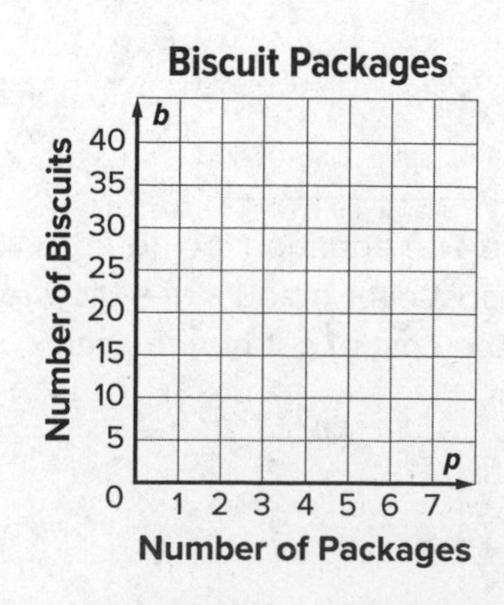

8. Open Response The graph shows the amount of money m, in dollars, Stacey earned for h hours of work. Write an equation that could be used to find the amount of money Stacey earns for any number of hours. **(Lesson 3)**

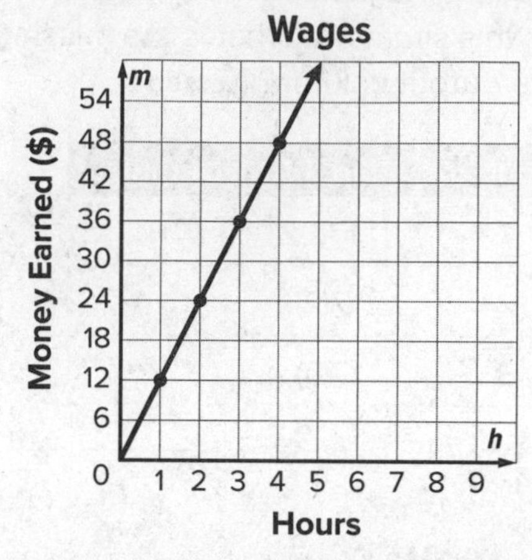

9. Multiple Choice Heath is selling magazine subscriptions. He earns $10 for every subscription sold. Use s to represent the number sold and t for total earnings. **(Lesson 4)**

A. Which of the following equations can be used to find Heath's total earnings t given s subscriptions sold?

Ⓐ $t = 10s$

Ⓑ $t = 10 + s$

Ⓒ $s = 10t$

Ⓓ $s = 10 + t$

B. Graph the ordered pairs and draw the line on the coordinate plane.

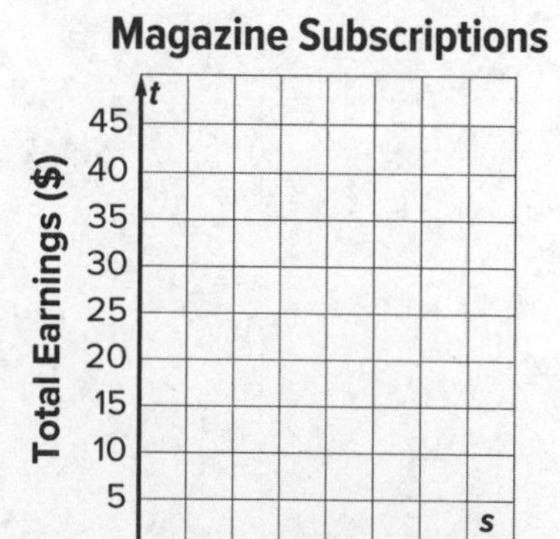

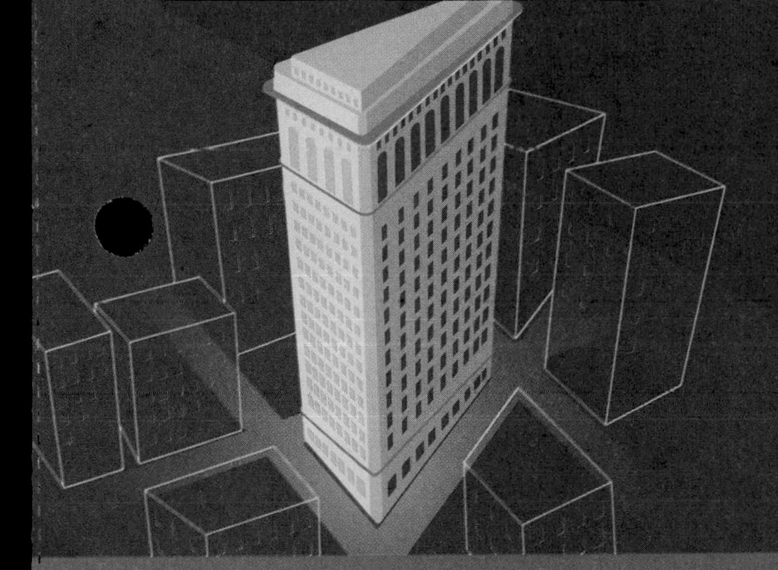

Module 8

Area

Essential Question

How are the areas of triangles and rectangles used to find the areas of other polygons?

What Will You Learn?

Place a checkmark (✓) in each row that corresponds with how much you already know about each topic **before** starting this module.

KEY

□ — I don't know. ◇ — I've heard of it. ☆ — I know it!

	Before			After		
	□	◇	☆	□	◇	☆
finding areas of parallelograms						
finding missing dimensions of parallelograms						
finding areas of triangles						
finding missing dimensions of triangles						
finding areas of trapezoids						
finding missing dimensions of trapezoids						
finding areas of regular polygons						
finding perimeters and areas of polygons on the coordinate plane						

Foldables Cut out the Foldable and tape it to the Module Review at the end of the module. You can use the Foldable throughout the module as you learn about area.

What Vocabulary Will You Learn?

Check the box next to each vocabulary term that you may already know.

- ☐ base
- ☐ congruent figures
- ☐ height
- ☐ parallelogram
- ☐ regular polygon
- ☐ trapezoid

Are You Ready?

Study the Quick Review to see if you are ready to start this module.

Then complete the Quick Check.

Quick Review

Example 1

Find area of rectangles.

Find the area of the rectangle.

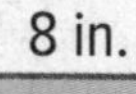

$A = \ell w$ Area of a rectangle

$= 8 \cdot 5$ Replace ℓ with 8 and w with 5.

$= 40$ Multiply.

The area of the rectangle is 40 square inches.

Example 2

Multiply fractions by whole numbers.

Find $\frac{1}{2} \cdot 22$.

$\frac{1}{2} \cdot 22 = \frac{1}{2} \cdot \frac{22}{1}$ Write 22 as $\frac{22}{1}$.

$= \frac{1}{\cancel{2}_1} \cdot \frac{\cancel{22}^{11}}{1}$ Divide the numerator and denominator by their GCF, 2.

$= \frac{11}{1}$ or 11 Simplify.

Quick Check

1. A garden is in the shape of a rectangle. The length of the garden is 12 feet and the width is 7 feet. What is the area of the garden?

2. Find $\frac{1}{2} \cdot 34$.

How Did You Do?

Which exercises did you answer correctly in the Quick Check? Shade those exercise numbers at the right.

Area of Parallelograms

I Can... understand how a parallelogram can be decomposed into a rectangle to find its area, and use the area formula for a parallelogram to find areas or missing dimensions.

What Vocabulary Will You Learn?
base
height
parallelogram

Explore Area of Parallelograms

Online Activity You will use Web Sketchpad to explore the area of a parallelogram.

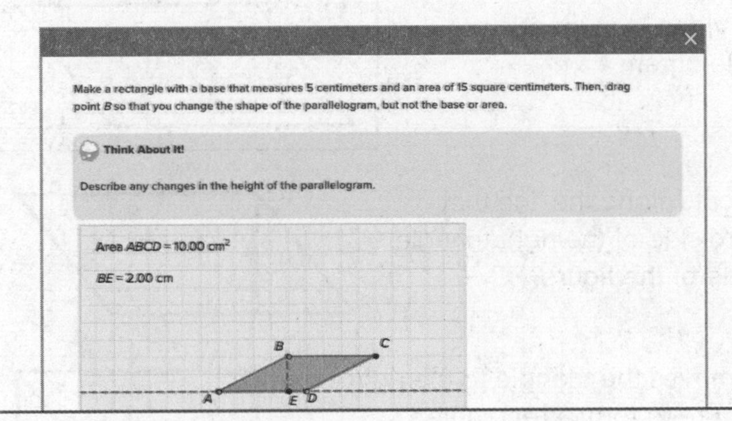

Pause and Reflect

Now that you have completed the Explore activity, what are some concepts you learned in a prior grade that might help you find the area of parallelograms in this lesson?

Your Notes

Learn Area of Parallelograms

A **parallelogram** is a quadrilateral with opposite sides that are parallel and have the same length. Recall that *area* is the measure of the interior surface of a two-dimensional figure and is measured in square units.

Go Online Watch the video to learn how the area of a parallelogram is related to the area of a rectangle.

The video shows how a rectangle can be used to find the area of a parallelogram by following these steps.

A parallelogram is shown on grid paper. In the video, a student cuts out the parallelogram.

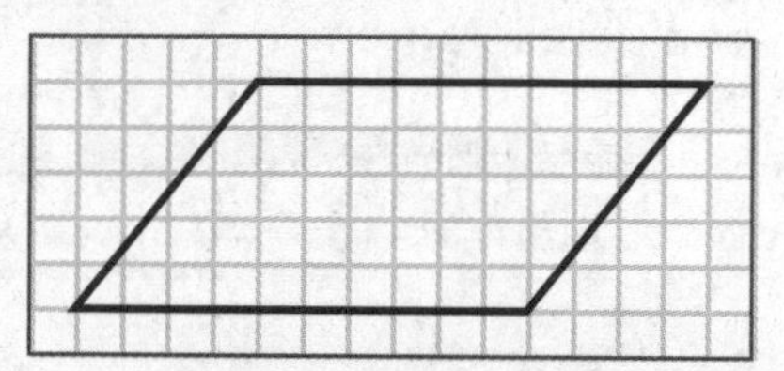

The student cuts along the line that forms the third side of the right triangle on the left side of the figure.

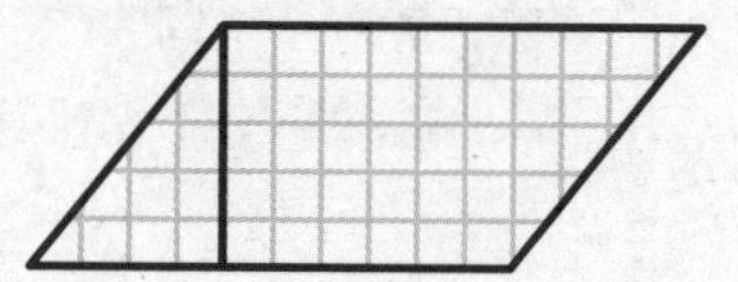

The student moves the triangle to the right side of the figure to form a rectangle.

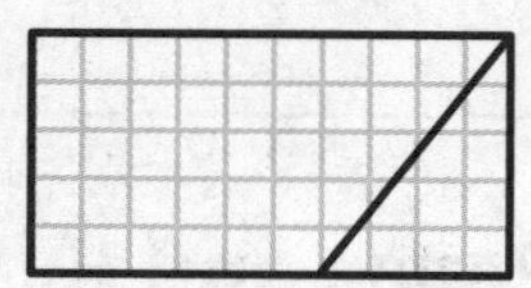

The area of a figure is the number of unit squares needed to cover it. The area of the rectangle formed by moving the right triangle is 50 square units. Because nothing was added or removed, the area of the parallelogram is also 50 square units.

The formula for the area of a parallelogram is similar to the formula for the area of a rectangle, but it uses its **base** and **height** instead of length and width.

The base b of a parallelogram can be any one of its sides.

The height h of a parallelogram is the perpendicular distance from a base to its opposite side.

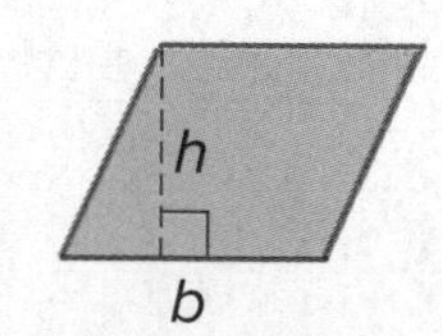

Words	Symbols
The area of a parallelogram is the product of its base b and its height h.	$A = bh$

Talk About It!

How is the formula for the area of a parallelogram, $A = bh$, similar to the area of a rectangle, $A = \ell w$?

Example 1 Find Area of Parallelograms

Romilla is painting a replica of the national flag of Trinidad and Tobago for a research project.

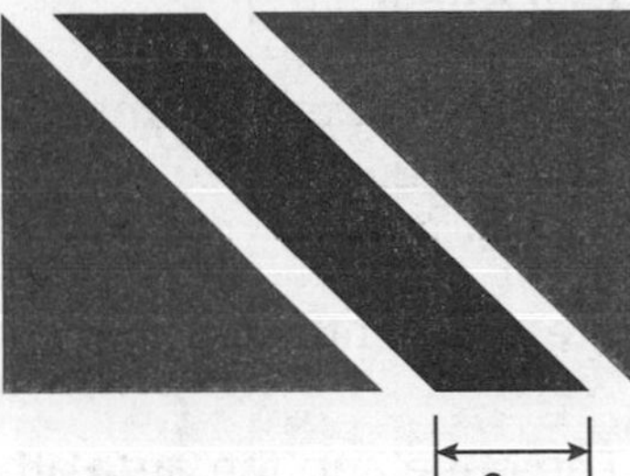

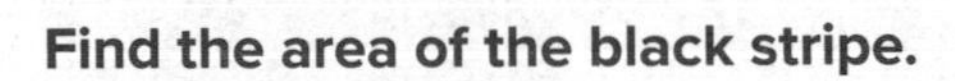

Find the area of the black stripe.

Step 1 Identify the measures of the base and the height of the stripe.

What is the measure of the base? _________ inches

What is the measure of the height? _________ inches

Step 2 Find the area.

$A = bh$	Area of a parallelogram
$A = \left(6\frac{3}{4}\right)(30)$	Replace b and h with the known values.
$A = \boxed{}$	Multiply.

So, the area of the black stripe is $202\frac{1}{2}$ square inches.

Check

Find the area of the parking space shown.

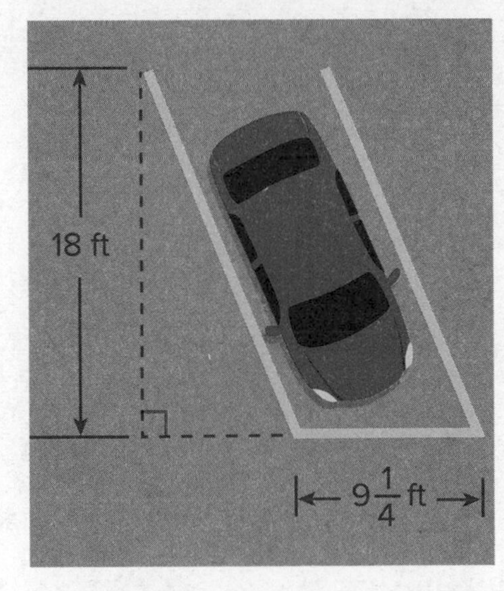

Think About It!

What dimensions do you need to know to find the area of a parallelogram?

Talk About It!

Why are the units that represent the area in square inches, instead of inches?

Go Online You can complete an Extra Example online.

Example 2 Find Missing Dimensions of Parallelograms

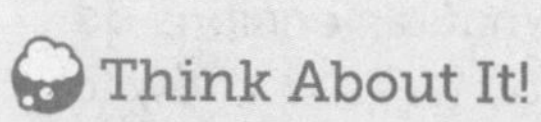

Think About It!
What formula will you use to solve this problem?

Find the missing dimension of the parallelogram.

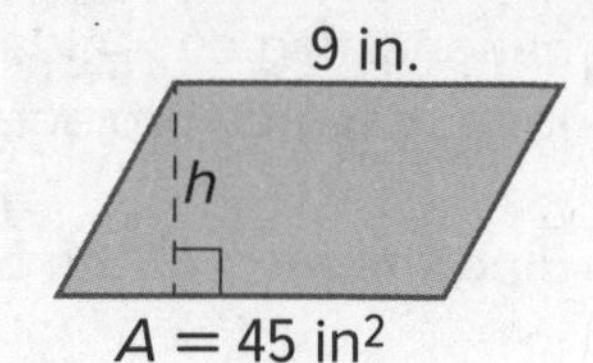

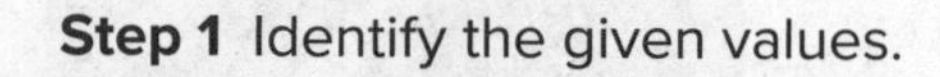

Step 1 Identify the given values.

The base and the area are given.
You need to find the height.

Step 2 Find the missing dimension.

$A = bh$	Area of a parallelogram
$45 = 9h$	Replace A and b with the known values.
$\frac{45}{9} = \frac{9h}{9}$	Divide each side by 9.
$5 = h$	Simplify.

So, the height of the parallelogram is ______ inches.

Talk About It!
Why is the label for height measured as inches, and not square inches?

Check

Find the missing dimension of the parallelogram shown.

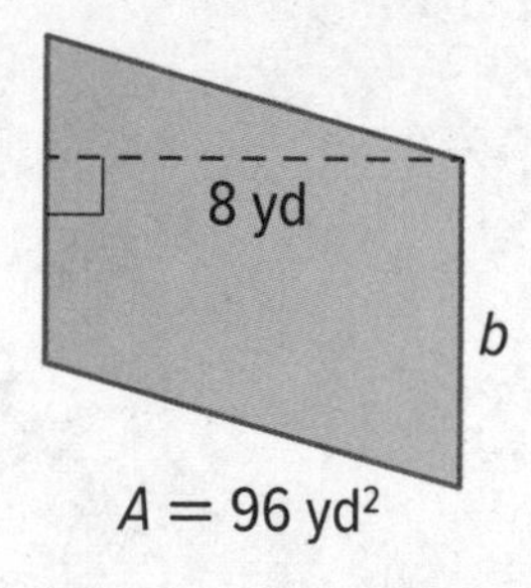

Go Online You can complete an Extra Example online.

Apply Landscaping

Andy, a city horticulturalist, is developing a new park over an old city lot. The center of the park features a koi pond that will cover 1,245 square feet. The remaining space will need to be covered with grass seed. If a 50-pound bag of grass seed covers up to 7,500 square feet, how many bags of grass seed will Andy need to buy to seed the rest of the park?

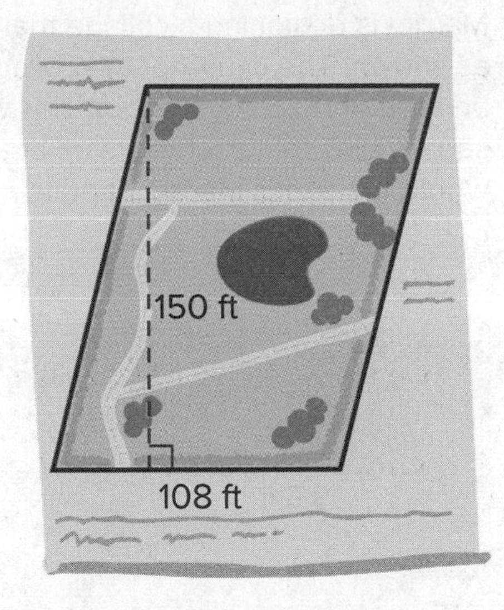

1 What is the task?

Make sure you understand exactly what question to answer or problem to solve. You may want to read the problem three times. Discuss these questions with a partner.

First Time Describe the context of the problem, in your own words.
Second Time What mathematics do you see in the problem?
Third Time What are you wondering about?

2 How can you approach the task? What strategies can you use?

3 What is your solution?

Use your strategy to solve the problem.

4 How can you show your solution is reasonable?

Write About It! Write an argument that can be used to defend your solution.

Talk About It!
Why is the final answer given as a whole number, when the quotient was a decimal?

Check

Margie is designing a collage that will be shaped like a parallelogram as shown. The center of the collage will be a square photo that has an area of 0.25 square foot. This will be surrounded by painted, square tiles that each have an area of 0.0625 square foot. How many whole tiles does Margie need to cover the collage?

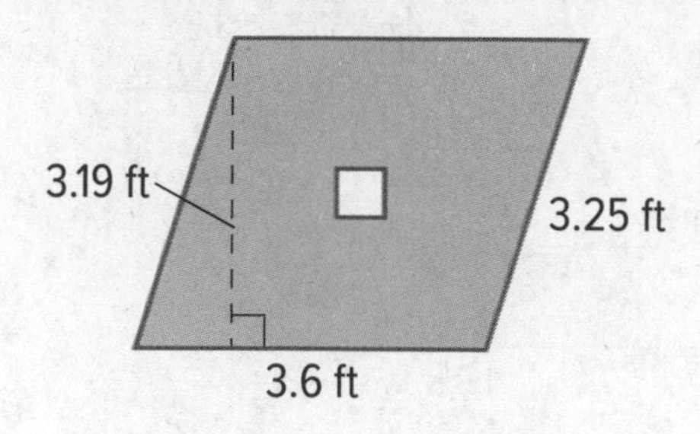

Go Online You can complete an Extra Example online.

Foldables It's time to update your Foldable, located in the Module Review, based on what you learned in this lesson. If you haven't already assembled your Foldable, you can find the instructions on page FL1.

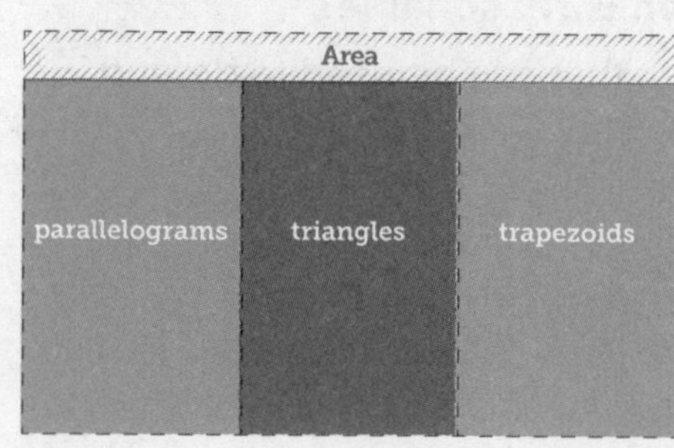

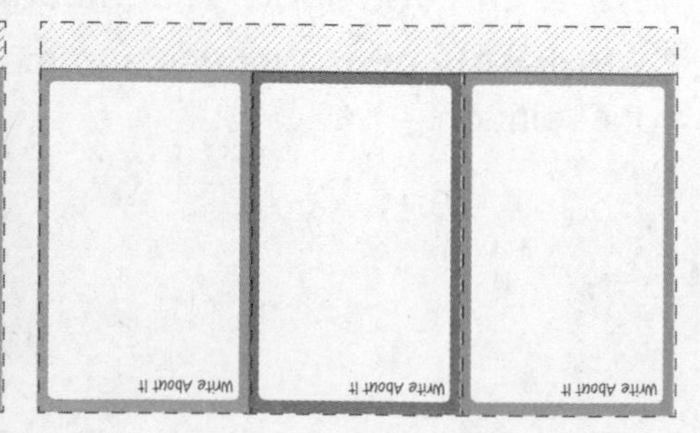

Name ______________________ Period __________ Date __________

Practice

Go Online You can complete your homework online.

1. The pattern shows the dimensions of a quilting square that Nakida will use to make a quilt. How much blue fabric will she need to make one square? **(Example 1)**

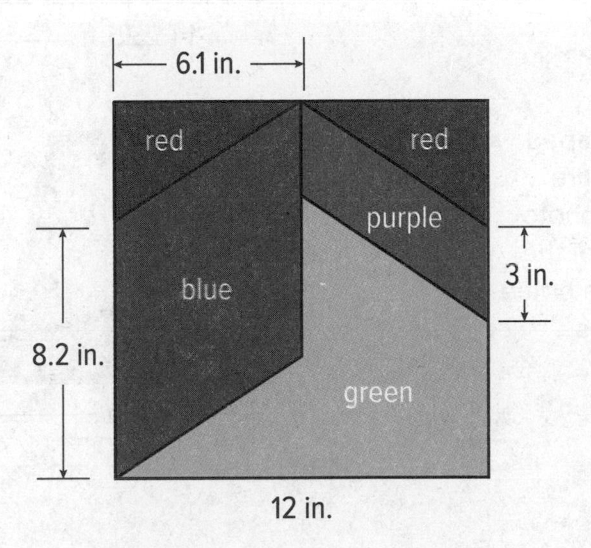

2. A group of students is painting the flag of Brunei for a geography project. Joseph is responsible for painting only the background colors of the flag. How many square inches will he cover with white paint? **(Example 1)**

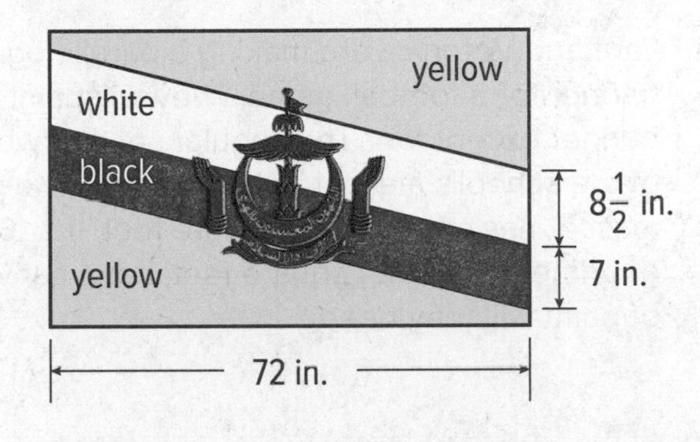

3. Find the missing dimension of the parallelogram. **(Example 2)**

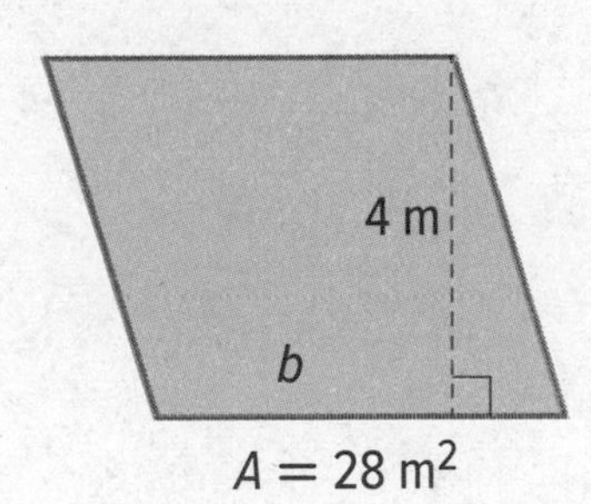

4. Find the missing dimension of the parallelogram. **(Example 2)**

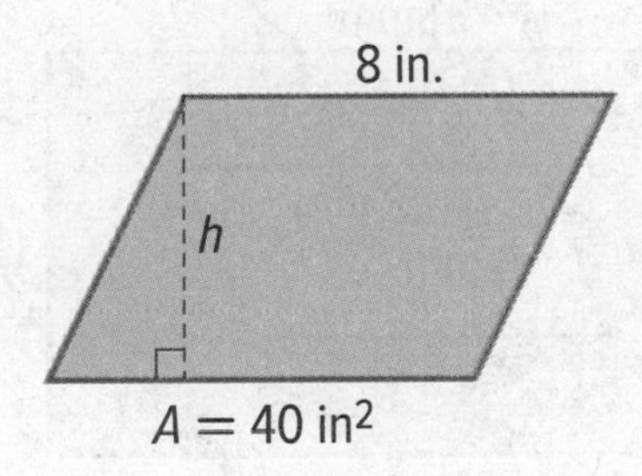

5. Find the area of the yellow striped region of the flag of the Republic of the Congo.

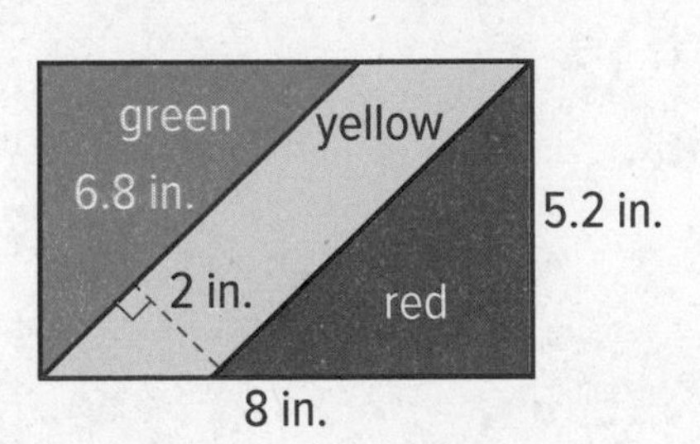

Test Practice

6. Open Response What is the area of the parallelogram?

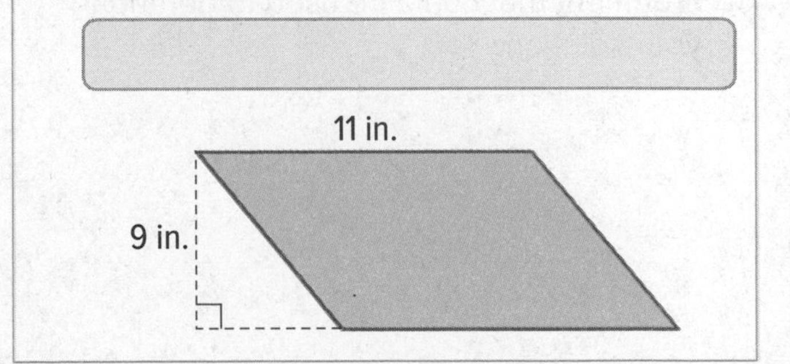

Apply

7. Liam is designing a patio and fountain for his backyard. The fountain will cover 50 square feet. The remaining space will be covered with tiles. If one tile covers 2.25 square feet, how many tiles will Liam need?

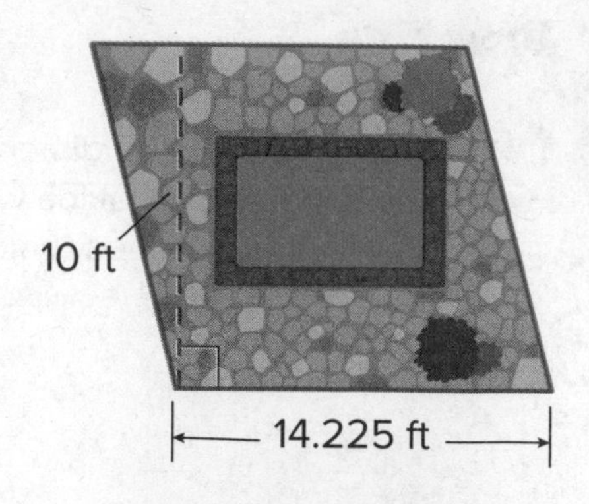

8. Tara and Veronica are making a parallelogram-shaped banner for a football game. They will paint the entire banner except for a rectangular section where a photo of the school's mascot will be placed. The photo of the mascot has an area of 6 square feet. If a 16-ounce bottle of primer covers 24 square feet, how many bottles of paint will they need?

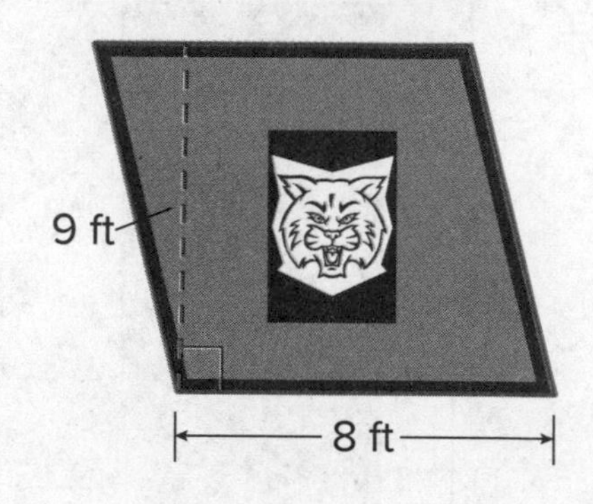

9. MP **Identify Structure** Find the area of the shaded region.

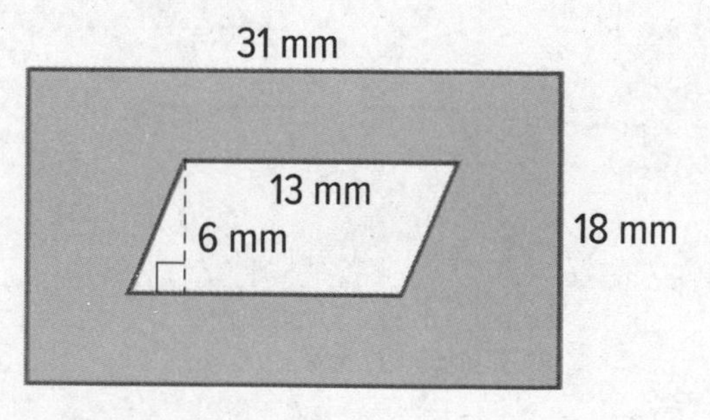

10. **Create** Draw and label a parallelogram with a base that is 2 times its height and has an area that is less than 100 square yards.

11. MP **Reason Abstractly** If you were to draw three different parallelograms each with a base of 5 units and a height of 4 units, how would the areas compare? Write an argument that could be used to defend your solution.

12. MP **Persevere with Problems** A rectangle and a parallelogram have the same area of 24 square inches. Describe the possible dimensions for each figure.

Area of Triangles

I Can... understand how a parallelogram can be decomposed into two congruent triangles to find the area of one triangle, and use the area formula for a triangle to find areas or missing dimensions.

What Vocabulary Will You Learn?
base
congruent figures
height (triangle)

Explore Parallelograms and Area of Triangles

Online Activity You will use Web Sketchpad to explore how the area of a parallelogram is related to the area of triangles.

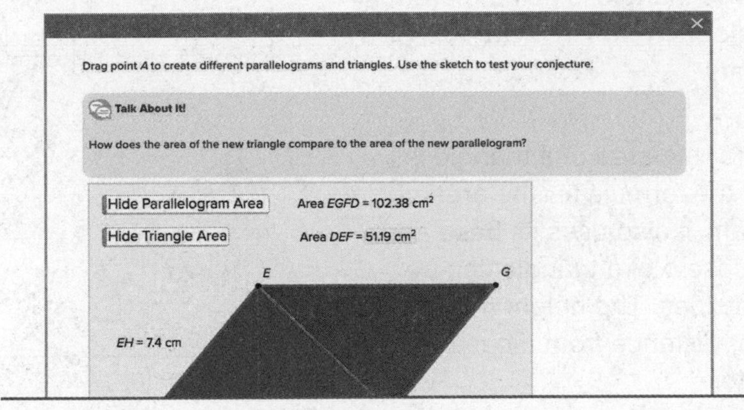

Pause and Reflect

What did you learn in the previous lesson that might help you find the area of triangles in this lesson? What did you learn in the Explore activity that also might help you in this lesson?

Your Notes

Learn Area of Triangles

Congruent figures are figures that have the same shape and size. A diagonal of a parallelogram separates it into two congruent triangles. Since congruent triangles have the same area, the area of a triangle is one-half the area of the parallelogram.

Go Online Watch the video to learn how a parallelogram is used to find the area of a triangle.

The base of the parallelogram is 8 units. The height is 12 units. The area of the parallelogram is 8(12), or 96 square units.

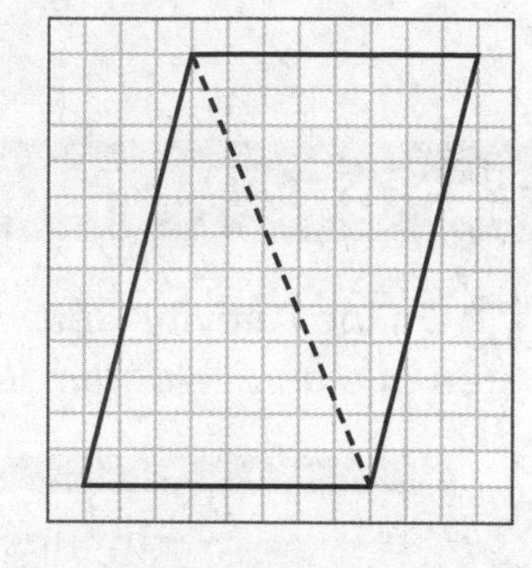

A diagonal line is drawn to form two congruent triangles.

The area of one triangle is half the area of the parallelogram, which is 96 ÷ 2, or 48 square units.

The formula for the area of a triangle is derived from the formula for the area of a parallelogram. It also uses its **base** and **height**. The base b of a triangle can be any one of its sides. The height h is the perpendicular distance from a base to its opposite vertex.

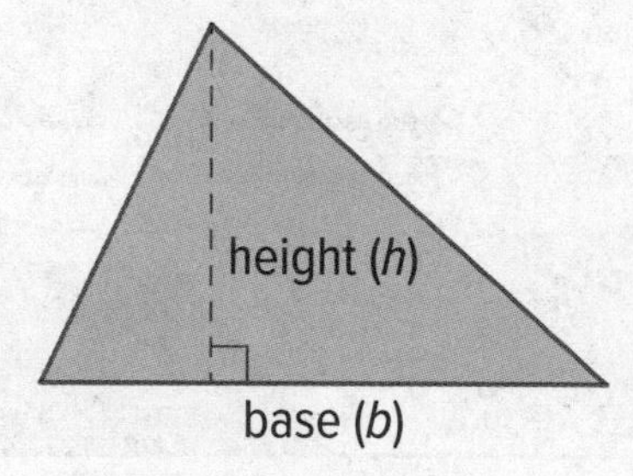

Words	Symbols
The area of a triangle is one half the product of its base b and its height h.	$A = \frac{1}{2}bh$ or $A = \frac{bh}{2}$

Think About It!
What formula will you use to find the area?

Example 1 Find Area of Right Triangles

Find the area of the triangle.

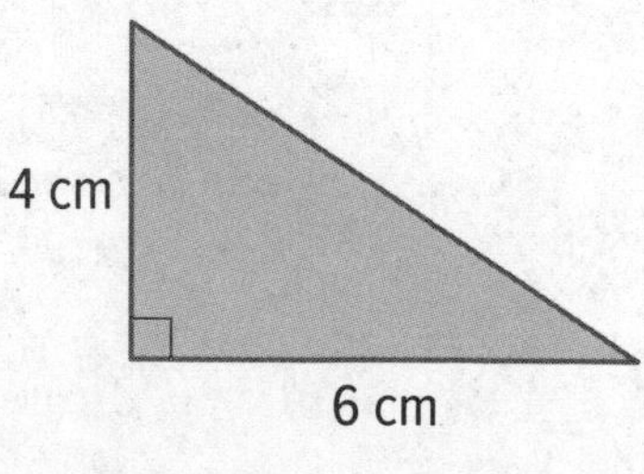

$A = \frac{1}{2}bh$ Area of a triangle

$A = \frac{1}{2}(6)(4)$ Replace b and h with the known values.

$A = \square$ Multiply.

So, the area of the triangle is 12 square centimeters.

Talk About It!
What is the area of a *rectangle* with a base of 6 centimeters and a height of 4 centimeters? How can you use this to check your answer to this example?

Check

Find the area of the triangle.

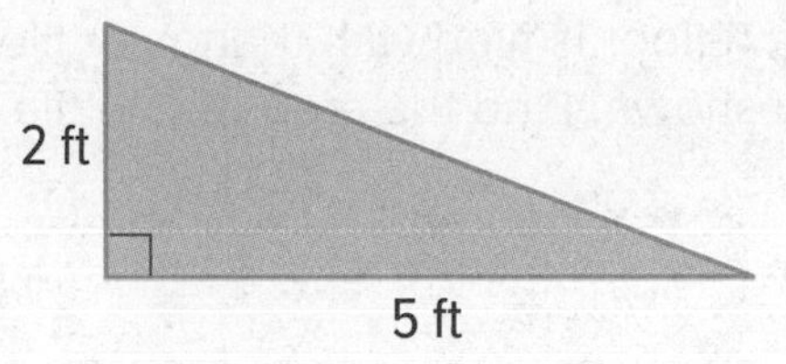

Go Online You can complete an Extra Example online.

Explore Area of Triangles

Online Activity You will use Web Sketchpad to explore the area of a triangle.

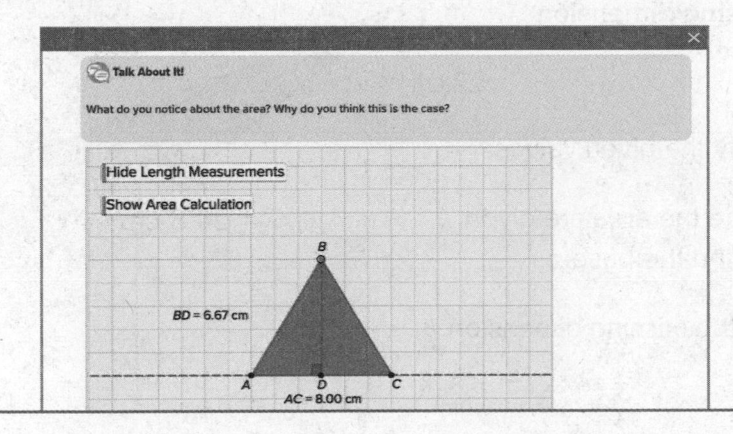

Example 2 Find Area of Triangles

The front of a camping tent has the dimensions shown.

How much material was used to make the front of the tent?

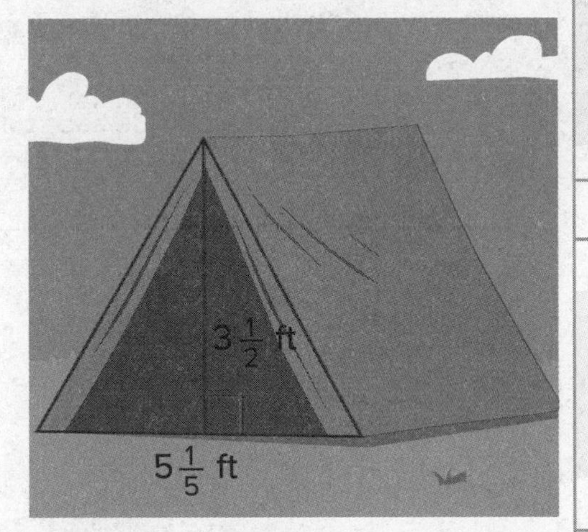

$A = \frac{1}{2}bh$ Area of a triangle

$A = \frac{1}{2}\left(5\frac{1}{5}\right)\left(3\frac{1}{2}\right)$ Replace b and h with the known values.

$A = 9\frac{1}{10}$ Multiply.

So, the amount of fabric used to make the front of the tent is

___________ square feet.

Think About It!

What is a good estimate for the solution?

Talk About It!

Compare your solution to the estimate.

Check

A floor is tiled with triangular tiles as shown. Find the area of one tile.

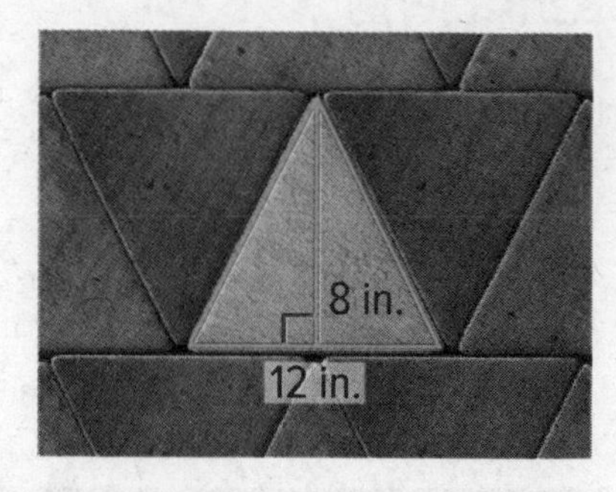

Go Online You can complete an Extra Example online.

Example 3 Find Missing Dimensions of Triangles

Think About It!
What formula will you use to solve the problem?

Find the missing dimension of the triangle.

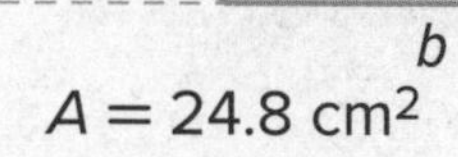

$A = 24.8\text{ cm}^2$

Step 1 Identify the given values.

The height and the area are given. You need to find the base.

Step 2 Find the missing dimension.

$A = \frac{1}{2}bh$	Area of a triangle
$24.8 = \frac{1}{2}b(6.2)$	Replace A and h with the known values.
$49.6 = b(6.2)$	Multiply each side by the reciprocal of $\frac{1}{2}$, 2.
$8 = b$	Divide each side by 6.2.

So, the base of the triangle is ______ centimeters.

Talk About It!
How can you check your answer?

Check

Find the missing dimension of the triangle.

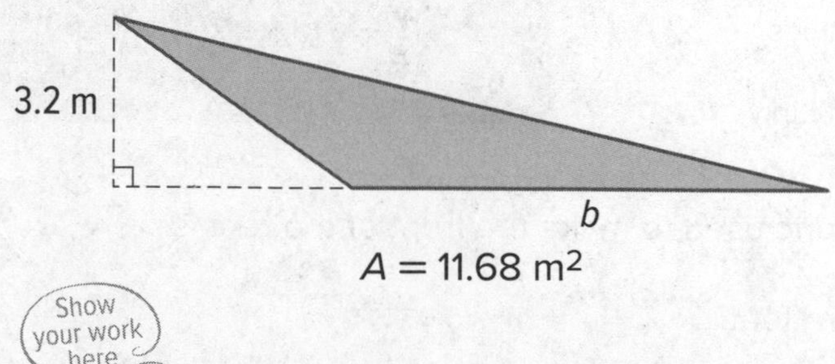

$A = 11.68\text{ m}^2$

Go Online You can complete an Extra Example online.

Apply Home Improvement

Blossom is painting the outlined section of the cabin shown. A gallon of paint costs $24.95 and covers 250 square feet. If the total area of the windows is 0.75 square feet, how much money will Blossom spend on paint?

1 What is the task?

Make sure you understand exactly what question to answer or problem to solve. You may want to read the problem three times. Discuss these questions with a partner.

First Time Describe the context of the problem, in your own words.
Second Time What mathematics do you see in the problem?
Third Time What are you wondering about?

2 How can you approach the task? What strategies can you use?

3 What is your solution?

Use your strategy to solve the problem.

4 How can you show your solution is reasonable?

Write About It! Write an argument that can be used to defend your solution.

Talk About It!

What is the area of a triangle with a base of 35 feet and a height of 25 feet? How can you use this to check your answer to this application problem?

Check

Vladimir is planting wildflowers in the corner of his yard as shown. A packet of wildflower seeds costs $4.95 and covers 50 square feet. How much will Vladimir spend on wildflower seeds?

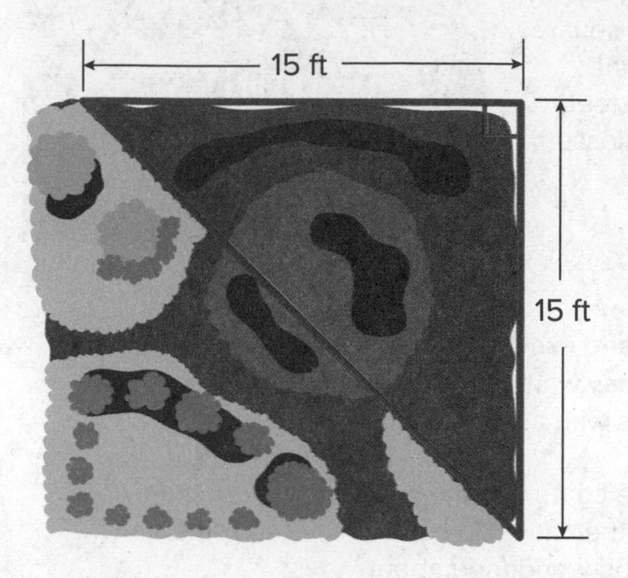

Go Online You can complete an Extra Example online.

Foldables It's time to update your Foldable, located in the Module Review, based on what you learned in this lesson. If you haven't already assembled your Foldable, you can find the instructions on page FL1.

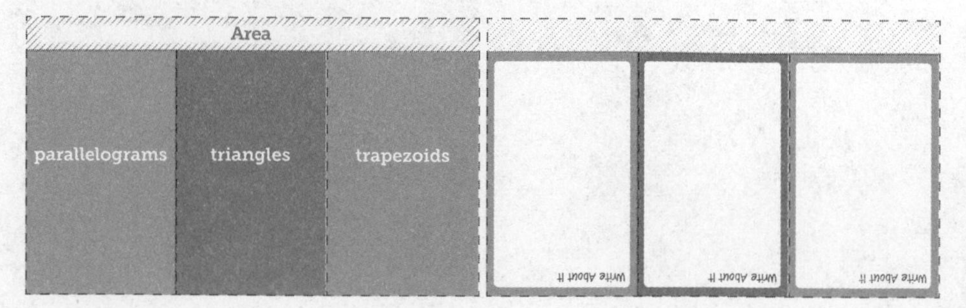

Name ____________________ Period ________ Date ________

Practice

Go Online You can complete your homework online.

Find the area of each triangle. (Example 1)

1.

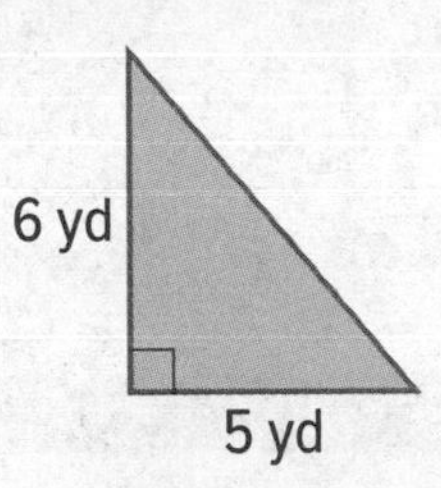

2.

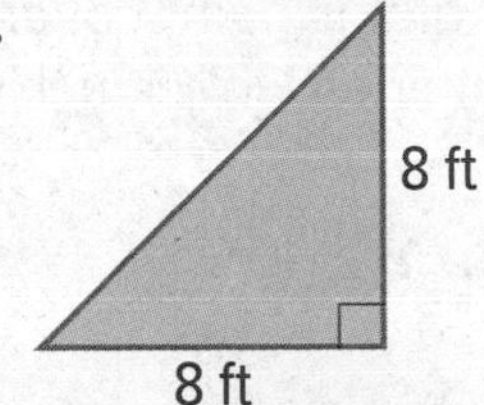

3. Tameeka is in charge of designing a school pennant for spirit week. What is the area of the pennant? **(Example 2)**

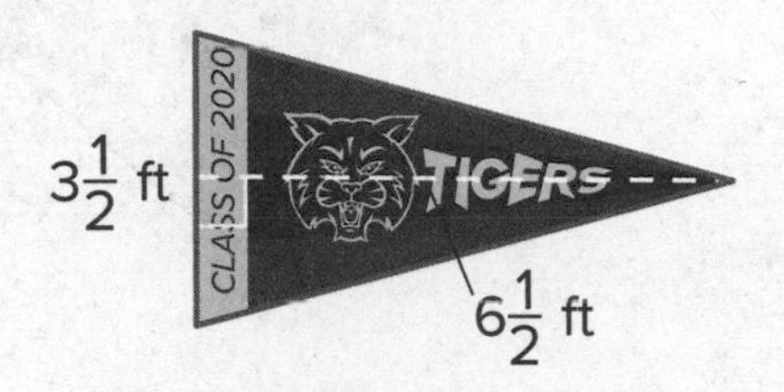

4. Norma has an A-frame cabin. The back is shown below. If the total area of the windows and doors is 3.5 square yards, how many square yards of paint will she need to cover the back of the cabin? **(Example 2)**

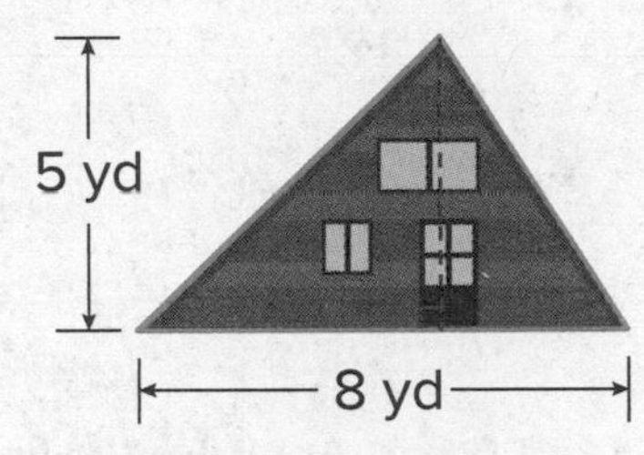

Find the missing dimension in each triangle. (Example 3)

5.

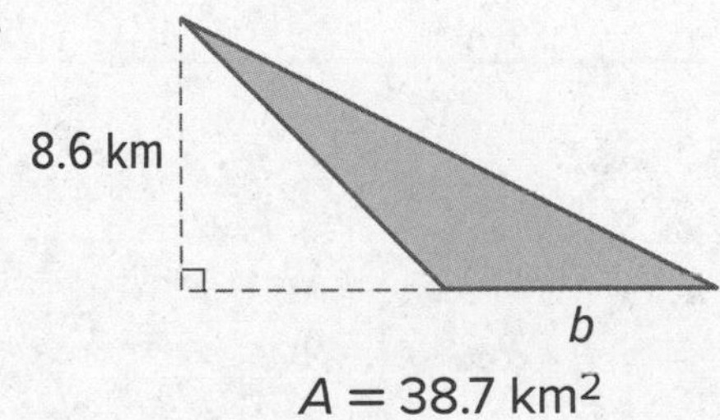

$A = 38.7\text{ km}^2$

6.

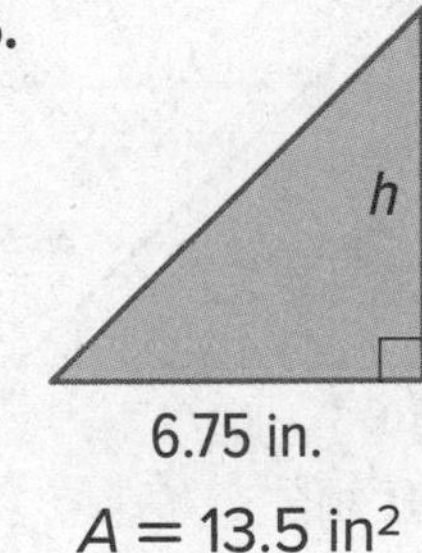

6.75 in.

$A = 13.5\text{ in}^2$

7. The flag of Bosnia and Herzegovina is shown. What is the area of the triangle on the flag?

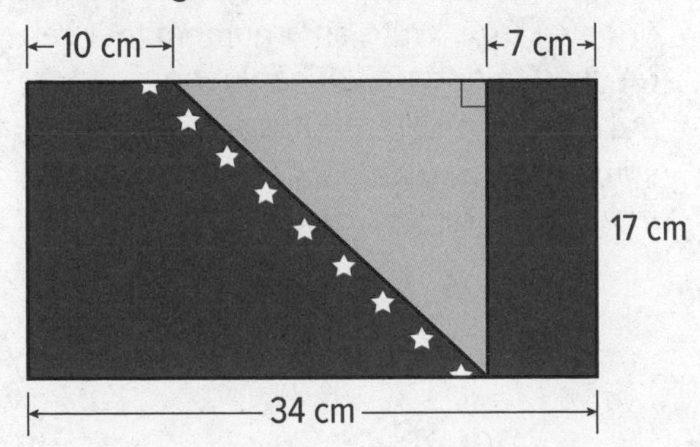

Test Practice

8. Open Response What is the area of the triangle?

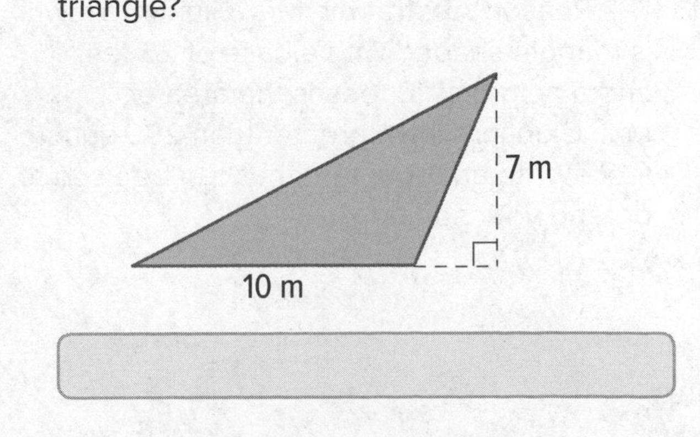

Apply

9. Aubrey is painting a mural of an ocean scene. The triangular sail on a sailboat has a base of 6 feet and a height of 4 feet. Aubrey will paint the sail using a special white paint. A container of this paint covers 10 square feet and costs $6.79 per container. How much will Aubrey spend on the white paint?

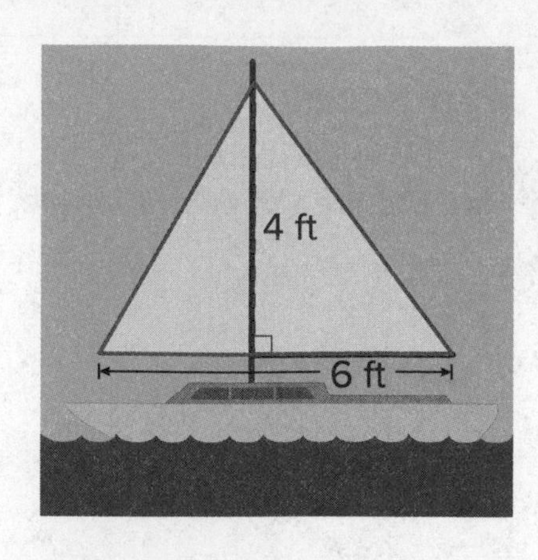

10. Silas is a making a wildflower meadow with the dimensions shown. He plans to cover the entire meadow with a wildflower seed mix. One bag of wildflower seed mix covers 22 square yards and costs $12.79. How much will Silas spend on the wildflower seed mix?

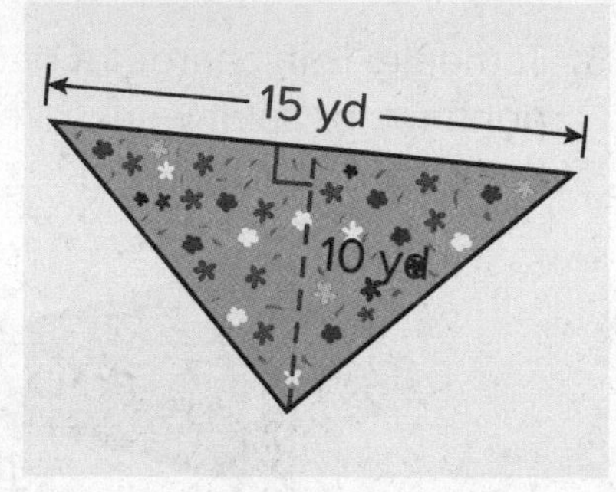

11. MP **Find the Error** A student is finding the height of the triangle. Find the student's mistake and correct it.

$17h = 68$

$h = 4$ meters

h

17 m

$A = 68\ m^2$

12. Create Draw and label a triangle with a base that is 3 times its height and has an area that is less than 50 square inches.

13. MP **Reason Abstractly** Mrs. Giuntini's lawn is triangle-shaped with a base of 25 feet and a height of 10 feet. Is the area of Mrs. Giuntini's lawn greater than 250 square feet? Write an argument that can be used to defend your solution.

14. MP **Justify Conclusions** Determine if the following statement is *always, sometimes,* or *never* true. Write an argument that can be used to defend your solution.

If a triangle and a parallelogram have the same base and height, the area of the triangle will always be greater.

Area of Trapezoids

I Can... understand how to find the area of a trapezoid by decomposing or composing, relate this to the area formula, and find the area of trapezoids or missing dimensions.

What Vocabulary Will You Learn?
base
height (trapezoid)
trapezoid

Learn Find Area of Trapezoids by Decomposing

A **trapezoid** is a quadrilateral with one pair of parallel sides. To find the area of a trapezoid, first decompose, or break down, the trapezoid into triangles and a rectangle. Since you know the formulas for the areas of triangles and rectangles, you can find the area of each smaller section and then add them together to find the area of the trapezoid.

Go Online Watch the animation to see how to find the area of a trapezoid by decomposing.

The animation shows how to find the area of the trapezoid, by first finding the areas of the shapes that make up the trapezoid.

The trapezoid shown is made up of one rectangle and two congruent triangles.

9 in.
6 in.
4 in.
4 in.

Step 1 Find the area of the rectangle.

$A = \ell w$ Area of a rectangle

$= 9(6)$ Replace ℓ and w with the known values.

$= 54$ Multiply.

Step 2 Find the areas of the triangles.

The two triangles are congruent, so the areas are the same. You only need to find the area of one triangle.

$A = \frac{1}{2}bh$ Area of a triangle

$= \frac{1}{2}(4)(6)$ Replace b and h with the known values.

$= 12$ Multiply.

Step 3 Add the areas of the rectangle and the two congruent triangles.

$54 + 12 + 12 = 78$

So, the area of the trapezoid is 78 square inches.

Talk About It!
How does decomposing the trapezoid help determine the area?

Your Notes

Think About It!
How can you divide the trapezoid into shapes with which you are familiar?

Talk About It!
Is there another way you can decompose the trapezoid? Will this result in the same area measurement? Explain your reasoning.

Example 1 Find Area of Trapezoids by Decomposing

Decompose the trapezoid to find its area.

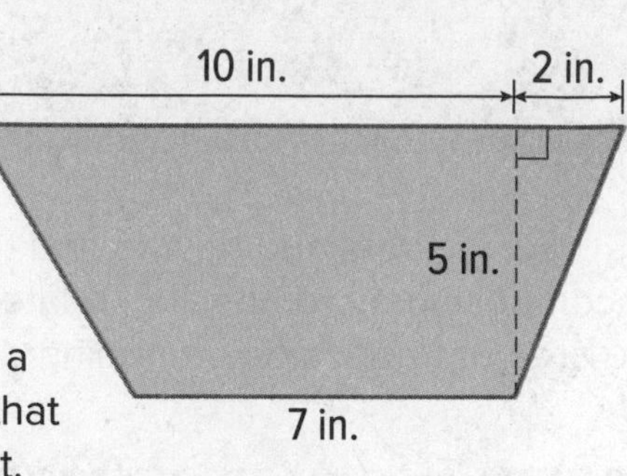

Step 1 Decompose the trapezoid.

The trapezoid is decomposed into a rectangle and two triangles. Note that the two triangles are not congruent.

3 in.
7 in.
2 in.
5 in.
Triangle 1
Rectangle
Triangle 2
7 in.

Step 2 Find the area of each shape.

Triangle 1	**Rectangle**	**Triangle 2**
$A = \frac{1}{2}bh$	$A = \ell w$	$A = \frac{1}{2}bh$
$= \frac{1}{2}(3)(5)$	$= 7(5)$	$= \frac{1}{2}(2)(5)$
$= 7.5$	$= 35$	$= 5$

Step 3 Find the total area.

$A = 7.5 + 35 + 5$

So, the area of the trapezoid is __________ square inches.

Check

Decompose the trapezoid to find its area.

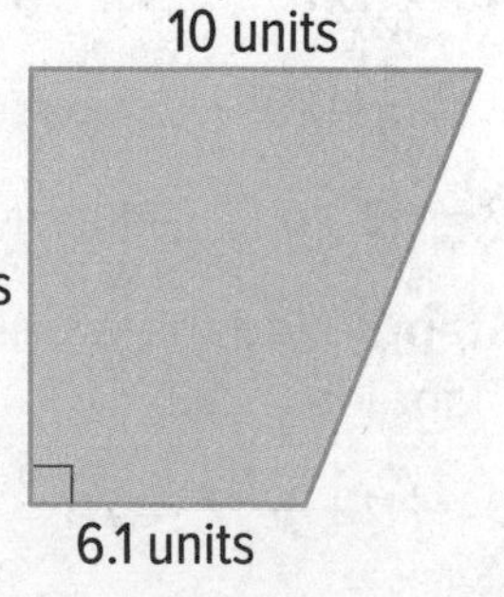

Show your work here

Go Online You can complete an Extra Example online.

Learn Find Area of Trapezoids by Composing

Two congruent trapezoids can be composed, or combined, to form a parallelogram. Since you know the formula for the area of a parallelogram, you can use that formula to help you find the area of a trapezoid.

Go Online Watch the video to see how to find the area of a trapezoid by composing.

The video shows that a parallelogram can be used to find the area of a trapezoid.

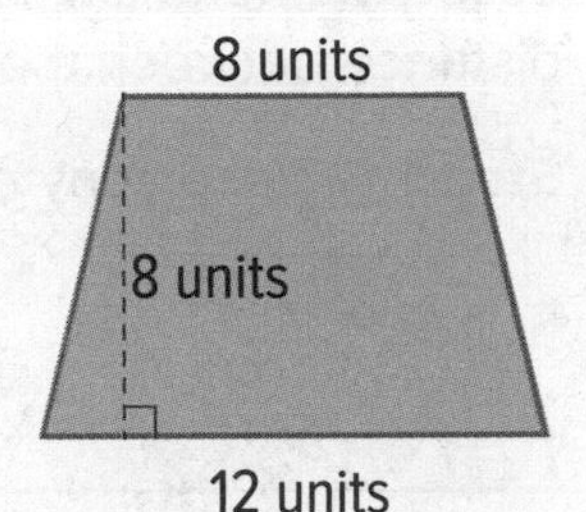

To find the area of the trapezoid shown, draw the trapezoid on grid paper.

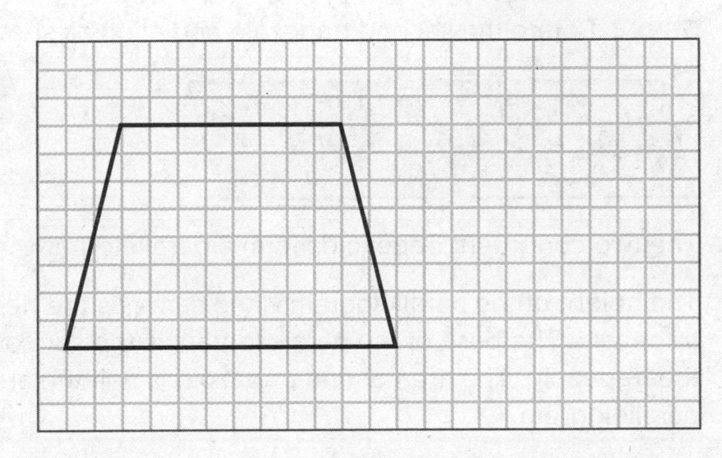

Flip the trapezoid and align it as shown. Draw the second trapezoid.

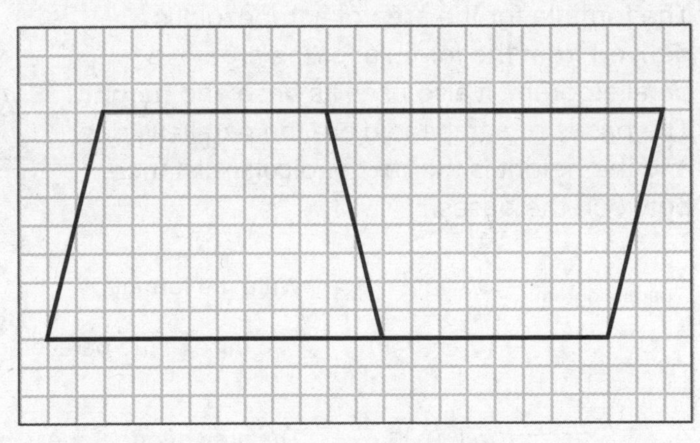

The two congruent trapezoids form a parallelogram. Find the area of the parallelogram.

$A_{\text{parallelogram}} = 12(8)$ or 96 units^2

The parallelogram has a base of 12 units and a height of 8 units.

Because the parallelogram is composed of two congruent trapezoids, the area of one trapezoid is half the area of the parallelogram.

$A_{\text{trapezoid}} = 96 \div 2$ or 48 units^2

Talk About It!

How can you use the concept of composing to find area if you do not know the formula for the area of a trapezoid?

Math History Minute

Júlio César de Mello e Souza (1895–1974) was a Brazilian mathematician, professor, and writer. His writings weave mathematics into entertaining word problems and puzzles. His most famous book, *The Man Who Counted,* tells of the adventures of Beremiz Samir who uses mathematics as a superpower. In Rio de Janeiro, Brazil, his birthday, May 6, is declared as Mathematician's Day.

Learn Find Area of Trapezoids by Using the Formula

Go Online Watch the animation to see how the formula for the area of a trapezoid is derived by composing it into a parallelogram.

In the trapezoid shown, base one, b_1, is the shorter base and base two, b_2, is the longer base. The height, h, is the perpendicular distance between the bases.

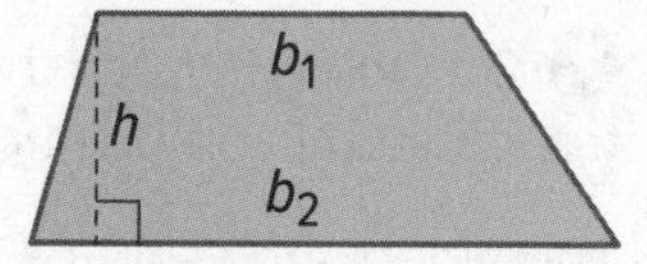

Step 1 Make a copy of the trapezoid.

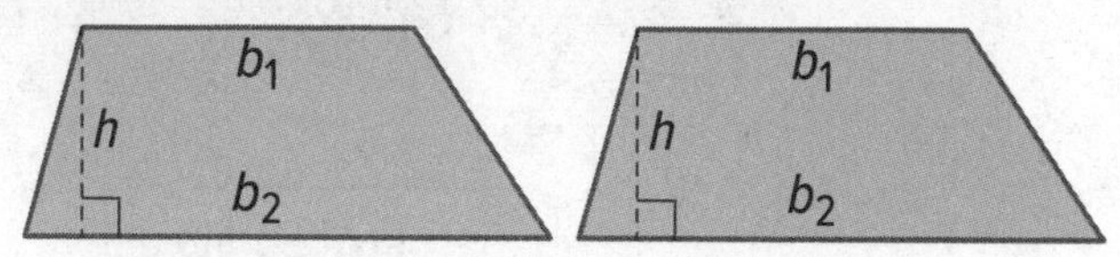

Step 2 Rotate the second trapezoid and align as shown.

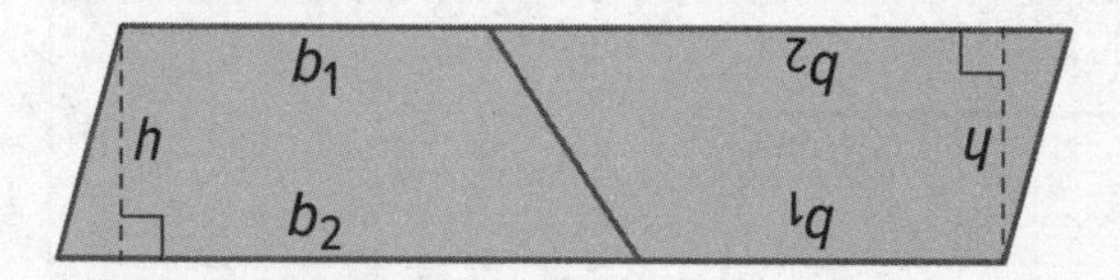

The two congruent trapezoids form a parallelogram.

The height of the parallelogram is the same as the height of the trapezoid. The base of the parallelogram is the sum of b_1 and b_2 of the trapezoid. The area of one trapezoid is half the area of the parallelogram.

The formula for the area of a trapezoid is derived from the formula for the area of a parallelogram. It also uses its **base** and **height**. The bases of a trapezoid are the parallel sides, and the height is the perpendicular distance between the bases.

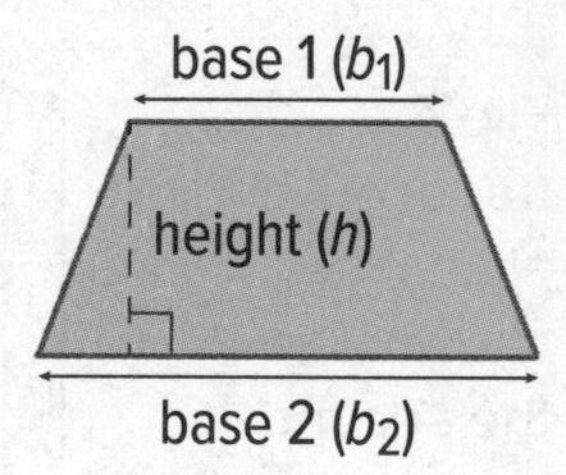

$A_{parallelogram} = bh$	Write the formula.
$= (b_1 + b_2)h$	The base of the parallelogram is $b_1 + b_2$.
$A_{trapezoid} = \frac{1}{2}(b_1 + b_2)h$	The area of one trapezoid is half the area of the parallelogram.
$= \frac{1}{2}h(b_1 + b_2)$	Commutative Property

Words	Symbols
The area of a trapezoid is one half the product of the height, h, and the sum of its bases, b_1 and b_2.	$A = \frac{1}{2}h(b_1 + b_2)$

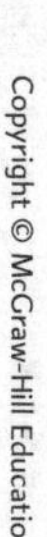

Example 2 Find Area of Trapezoids

Find the area of the trapezoid.

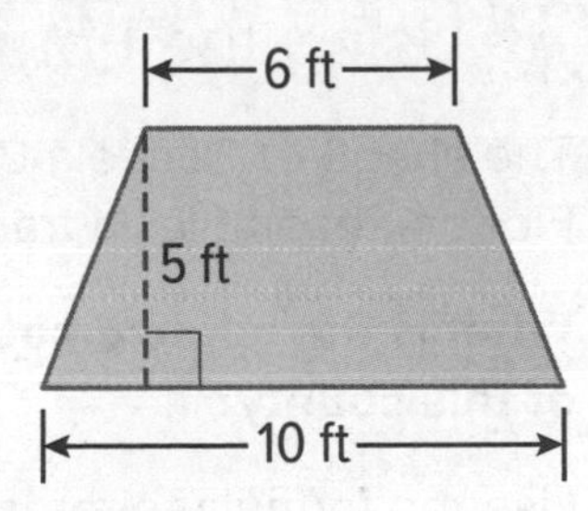

Method 1 Find the area by composing.

Step 1 Compose the trapezoid into a parallelogram.

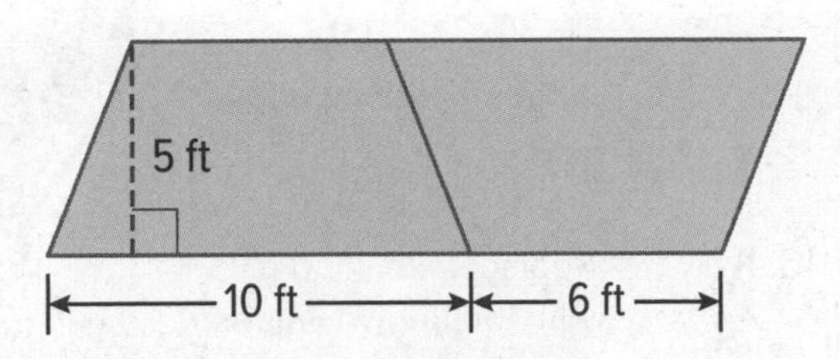

Step 2 Find the area of the parallelogram.

$A = (16)(5)$ The base is $10 + 6$, or 16 feet. The height is 5 feet.

$= 80$ The area of the parallelogram is 80 square feet.

Step 3 Find the area of the trapezoid.

$80 \div 2 =$ ☐ Divide the area of the parallelogram by 2. The area of the trapezoid is 40 square feet.

Method 2 Find the area using the formula.

$A = \frac{1}{2}h(b_1 + b_2)$ Area of a trapezoid

$A = \frac{1}{2}(5)(10 + 6)$ Replace h, b_1, and b_2 with the known values.

$A = \frac{1}{2}(5)(16)$ Add.

$A = 40$ Multiply.

So, using either method, the area of the trapezoid is

________ square feet.

Check

Find the area of the trapezoid.

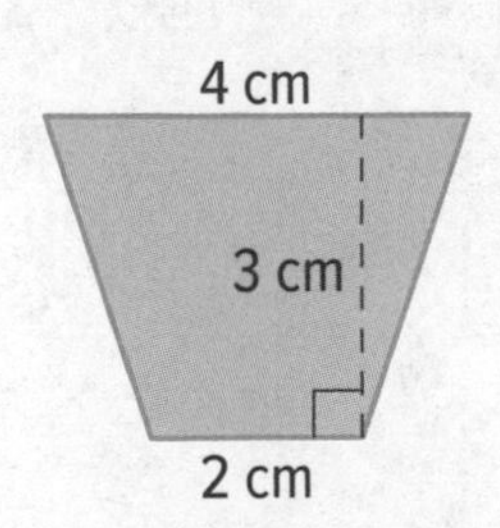

Go Online You can complete an Extra Example online.

Think About It!
How can you find the area of the trapezoid by composing?

Talk About It!
Compare the two methods.

Think About It!

What measurements do you need to find the area of the trapezoid?

Talk About It!

Why is one of the sides of the trapezoid the height?

Example 3 Find Area of Right Trapezoids by Using the Formula

The shape of Osceola County, Florida, resembles a trapezoid.

48 mi
OSCEOLA COUNTY
51 mi
16.4 mi

What is the approximate area of this county?

Use the formula for area of a trapezoid.

$A = \frac{1}{2}h(b_1 + b_2)$	Area of a trapezoid
$A = \frac{1}{2}(51)(48 + 16.4)$	Replace h, b_1, and b_2 with the known values.
$A = \frac{1}{2}(51)(64.4)$	Add.
$A = 1{,}642.2$	Multiply.

So, the approximate area of the county is ________ square miles.

Check

The shape of the driveway resembles a trapezoid. Find the area of the driveway.

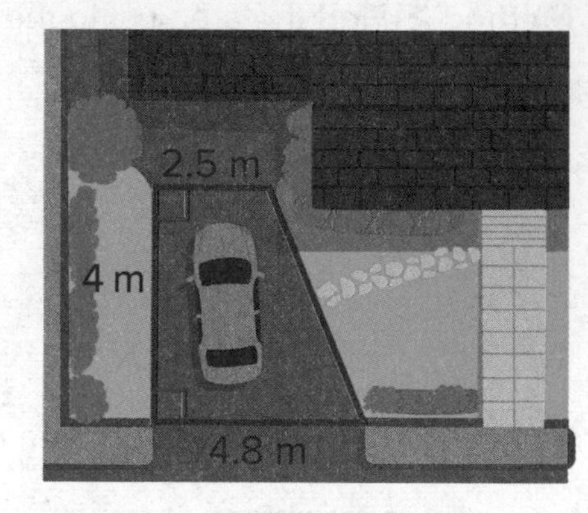

Go Online You can complete an Extra Example online.

Example 4 Find Area of Trapezoids

Each of the airplane's wings in the drawing is in the shape of a trapezoid.

Find the area of one wing.

Use the area formula for a trapezoid.

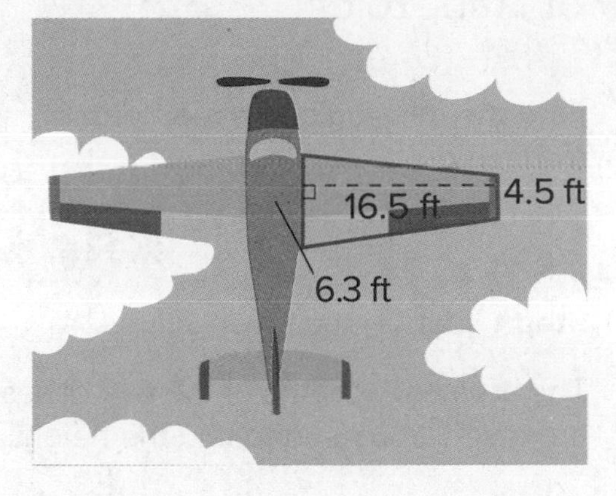

$A = \frac{1}{2}h(b_1 + b_2)$ Area of a trapezoid

$A = \frac{1}{2}(16.5)(4.5 + 6.3)$ Replace h, b_1, and b_2 with the known values.

$A = \frac{1}{2}(16.5)(10.8)$ Add.

$A = 89.1$ Multiply.

So, the area of one wing is ________ square feet.

Check

A teacher's small-group table is in the shape of a trapezoid. Find the area of the table.

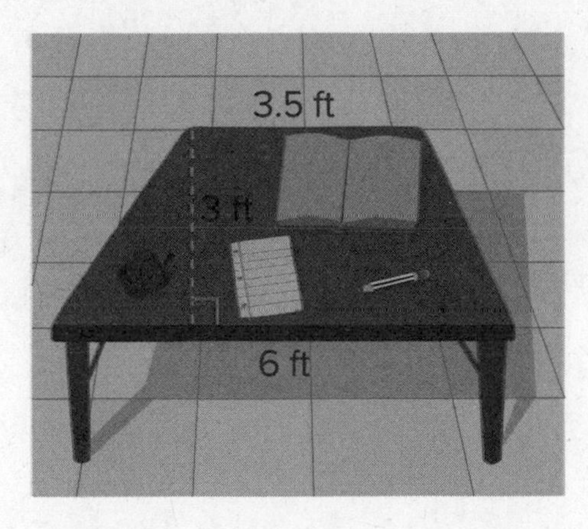

Go Online You can complete an Extra Example online.

Think About It!
What formula will you use to solve the problem?

Talk About It!
Does the solution change depending on which value you choose for b_1 and which value you choose for b_2? Explain.

Example 5 Find Missing Dimensions of Trapezoids

Find the missing dimension of the trapezoid.

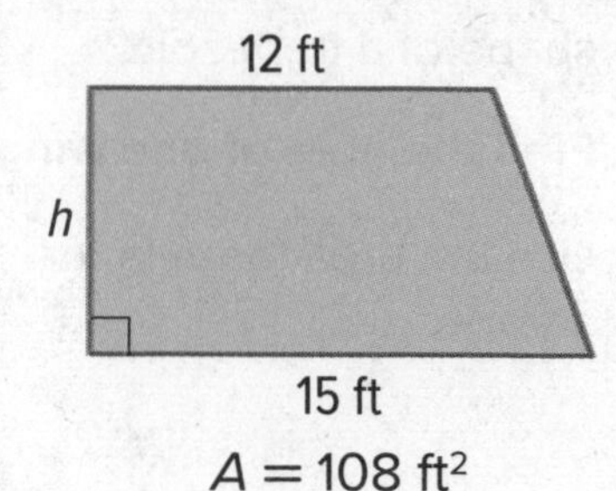

$A = 108\ ft^2$

Step 1 Identify the given values.

The area and lengths of the two bases are given. You need to find the height.

Step 2 Find the missing dimension.

To find a missing dimension of a trapezoid, use the formula for the area of a trapezoid. First replace the variables with the known measurements. Then solve the equation for the remaining variable.

$A = \frac{1}{2}h(b_1 + b_2)$	Area of a trapezoid
$108 = \frac{1}{2}h(15 + 12)$	Replace A, b_1, and b_2 with the known values.
$108 = \frac{1}{2}h(27)$	Add.
$108 = h(13.5)$	Multiply.
$8 = h$	Divide each side by 13.5.

So, the height of the trapezoid is ______ feet.

Check

Find the missing dimension of the trapezoid.

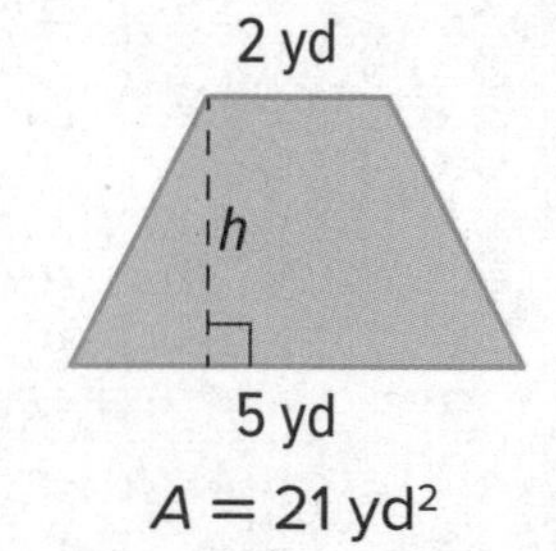

$A = 21\ yd^2$

Go Online You can complete an Extra Example online.

Apply Budgets

The parking lot shown is being repaved. The office manager budgeted $10,000 for the repaving project. Asphalt for the parking lot costs $8.95 per square foot. Find the cost of the asphalt to determine if the office manager budgeted enough money to complete the project.

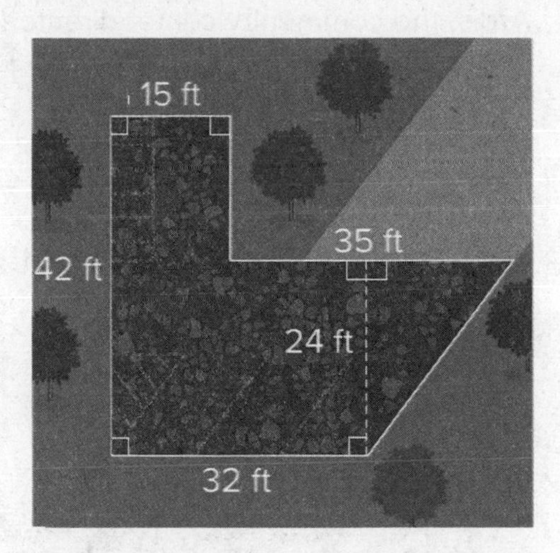

Go Online
Watch the animation.

1 What is the task?

Make sure you understand exactly what question to answer or problem to solve. You may want to read the problem three times. Discuss these questions with a partner.

First Time Describe the context of the problem, in your own words.
Second Time What mathematics do you see in the problem?
Third Time What are you wondering about?

2 How can you approach the task? What strategies can you use?

Talk About It!
How can you solve the problem another way?

3 What is your solution?

Use your strategy to solve the problem.

4 How can you show your solution is reasonable?

Write About It! Write an argument that can be used to defend your solution.

Check

Ardis, the community center director, is having the swimming pool floor resurfaced and has budgeted $20,000.00. The new pebble-based cement material costs $8.45 per square foot. Find the cost to resurface the pool and determine if Ardis has budgeted enough money to complete the project.

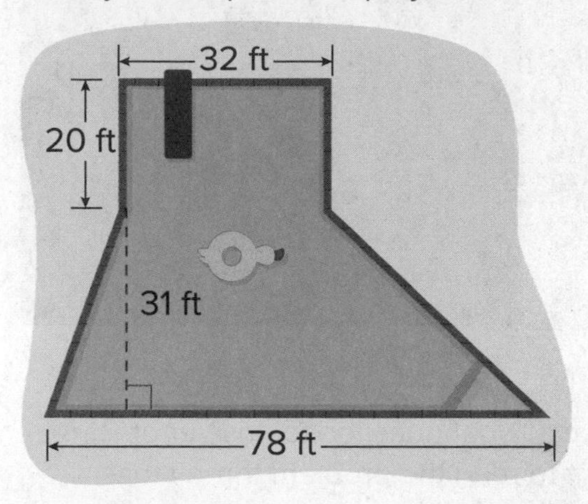

Go Online You can complete an Extra Example online.

Foldables It's time to update your Foldable, located in the Module Review, based on what you learned in this lesson. If you haven't already assembled your Foldable, you can find the instructions on page FL1.

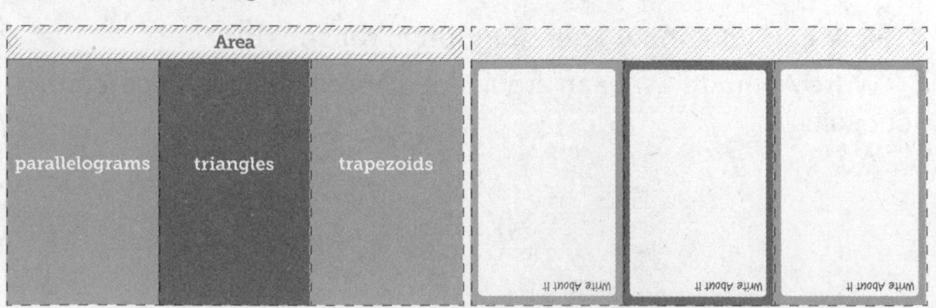

Name ______________________ Period __________ Date __________

Practice

Go Online You can complete your homework online.

Decompose each trapezoid to find its area. **(Example 1)**

1.

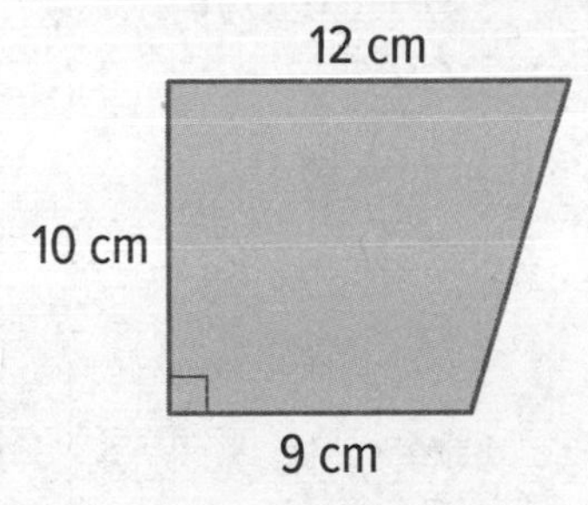

2.

2 ft 7 ft 2 ft

7 ft

7 ft

Find the area of each trapezoid. **(Example 2)**

3.

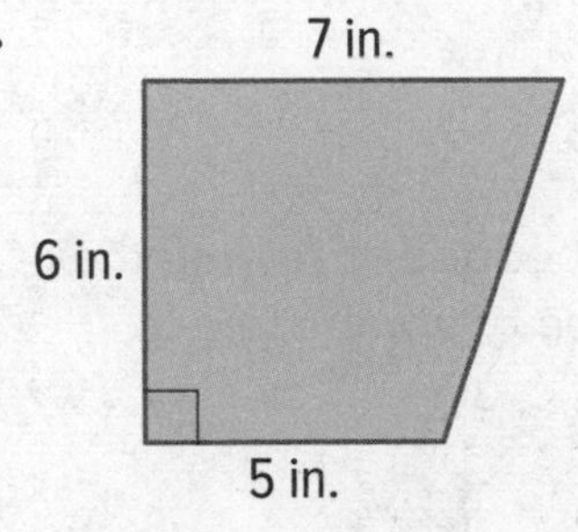

4.

4 cm

12 cm

7 cm

5. The shape of Arkansas resembles a trapezoid. What is the approximate area of Arkansas? **(Example 3)**

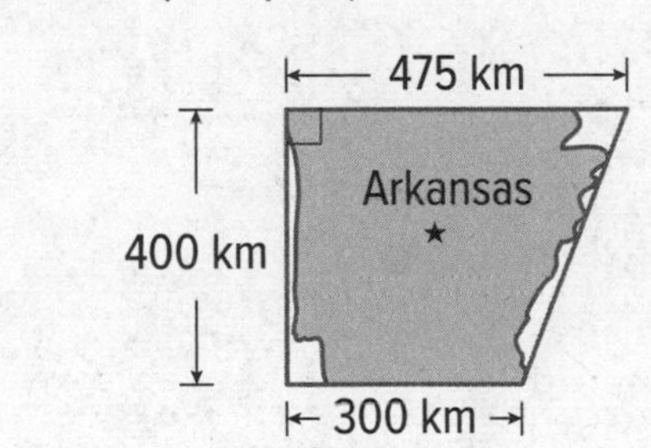

6. The top of the desk shown is in the shape of a trapezoid. What is the area of the top of the desk? **(Example 4)**

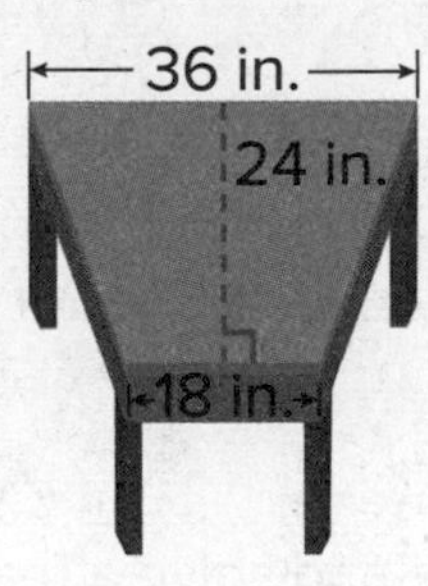

7. Find the missing dimension of the trapezoid. **(Example 5)**

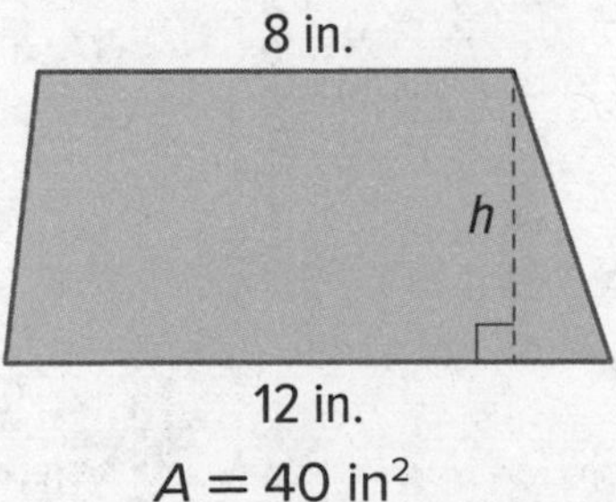

$A = 40$ in^2

Test Practice

8. Open Response Ciro made a sign in the shape of a trapezoid. What was the area of Ciro's sign?

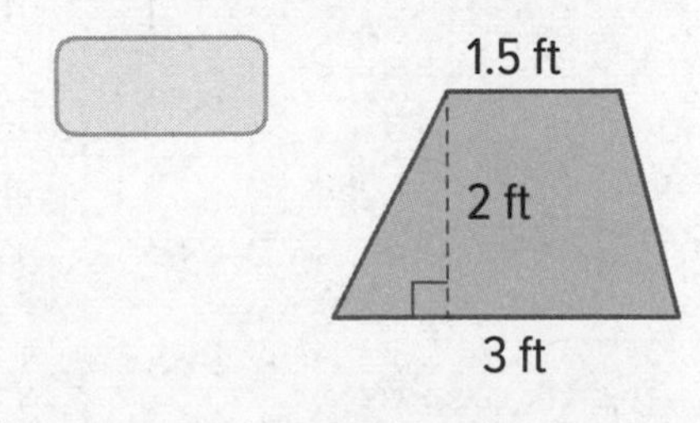

Apply

9. Greta has budgeted $1,500 to have a concrete patio poured in her backyard like the one shown. The cost per square foot of the concrete is $5.50. Find the cost of the patio to determine if Greta has budgeted enough money to complete the project.

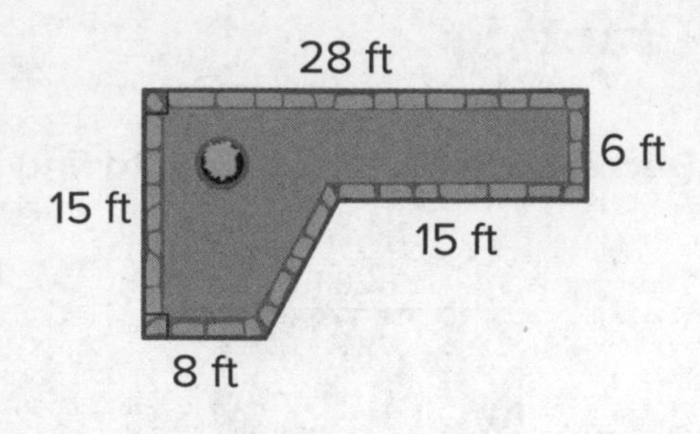

10. Create Draw and label a trapezoid that has no right angles and an area greater than 75 square meters.

11. Explain the steps needed to rewrite the formula for the area of a trapezoid to find b_2.

12. Create Write and solve a real-world problem where you need to find the area of a trapezoid.

13. MP **Reason Inductively** The area of a trapezoid is 48 square centimeters. The height is 6 centimeters and one base is 3 times the length of the other base. What are the lengths of the bases?

Area of Regular Polygons

I Can... decompose a polygon into triangles, parallelograms, and trapezoids, find the areas of the decomposed figures, and then add or multiply to find the area of the polygon.

What Vocabulary Will You Learn?

regular polygon

Explore Area of Regular Polygons

Online Activity You will use Web Sketchpad to explore how the area of triangles, parallelograms, and trapezoids can be used to find the area of regular polygons.

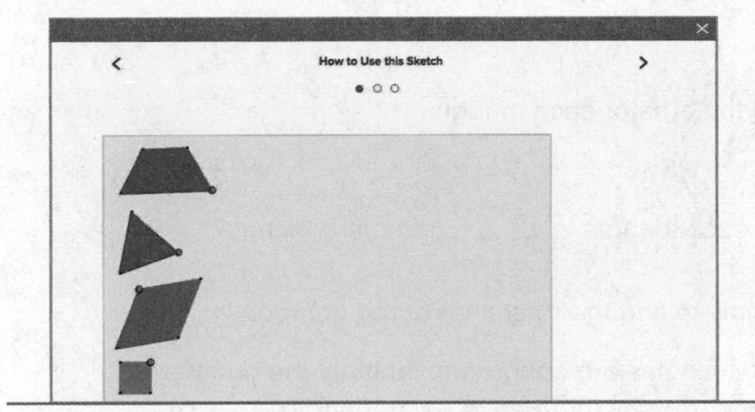

Learn Area of Regular Polygons

To find the area of a **regular polygon**, a polygon in which all sides and all angles are congruent, you can decompose the figure into triangles, parallelograms, or trapezoids. Find the area of each smaller figure, and then add or multiply to find the total area.

Go Online Watch the animation to learn how to decompose a regular polygon to find its area.

The animation shows how to find the area of the hexagon shown.

Step 1 Decompose the figure into two congruent trapezoids.

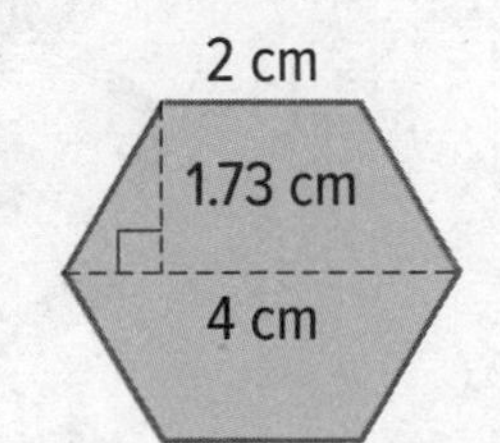

Step 2 Find the area of one trapezoid.

$A = \frac{1}{2}h(b_1 + b_2)$	Area of a trapezoid
$A = \frac{1}{2}(1.73)(4 + 2)$	$h = 1.73$; $b_1 = 4$; $b_2 = 2$
$A = 5.19$	Simplify. $A = 5.19$ cm^2

The trapezoids are congruent, so the areas are the same.

Step 3 Add the areas. $5.19 + 5.19 = 10.38$ cm^2

Talk About It!

Is there another way to decompose the figure in the animation?

Think About It!

Is the area less than, greater than, or equal to 36^2, or 1,296 square inches? How do you know?

Talk About It!

With the given information, can you decompose the octagon into different shapes to find the area? Why or why not?

Example 1 Find Area of Regular Polygons

A stop sign is shaped like a regular octagon. Each side of the sign is 15 inches long and measures 36 inches between parallel sides.

15 in.
18 in.
STOP

Find the area of the octagon.

Step 1 Decompose the octagon into congruent triangles.

The octagon decomposes into 8 congruent triangles.

Step 2 Find the area of each triangle.

$A = \frac{1}{2}(15)(18) = 135$

The area of each triangle is ________ square inches.

Step 3 Multiply to find the total area of the octagon.

Because the triangles are congruent, multiply the number of triangles, ________, by the area of each triangle.

$8(135) = 1{,}080$

So, the area of the stop sign is ________ square inches.

Check

The white section of the soccer ball is a regular hexagon. Each side of the hexagon is 1.8 inches. Find the area of the hexagon. Round to the nearest hundredth.

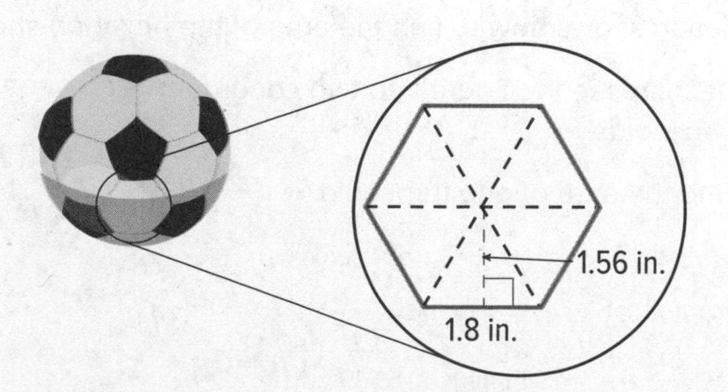

Go Online You can complete an Extra Example online.

Apply Home Improvement

Keilani designed a gazebo and wants to cover the floor with tiles. The gazebo is shaped like a decagon with 3.75 foot sides. If floor tiles cost $2.89 per square foot, what is the least amount she will spend on the tiles?

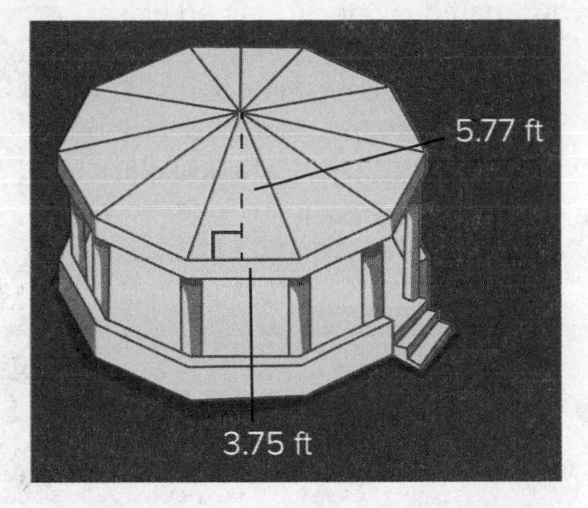

1 What is the task?

Make sure you understand exactly what question to answer or problem to solve. You may want to read the problem three times. Discuss these questions with a partner.

First Time Describe the context of the problem, in your own words.
Second Time What mathematics do you see in the problem?
Third Time What are you wondering about?

2 How can you approach the task? What strategies can you use?

Talk About It!
How can you solve the problem another way?

3 What is your solution?

Use your strategy to solve the problem.

4 How can you show your solution is reasonable?

Write About It! Write an argument that can be used to defend your solution.

Check

Morgan designed a stained glass window to be added above the door at the community center. The window is shaped like an octagon with 15-inch sides. If stained glass costs \$0.70 per square inch, how much will she spend on the window?

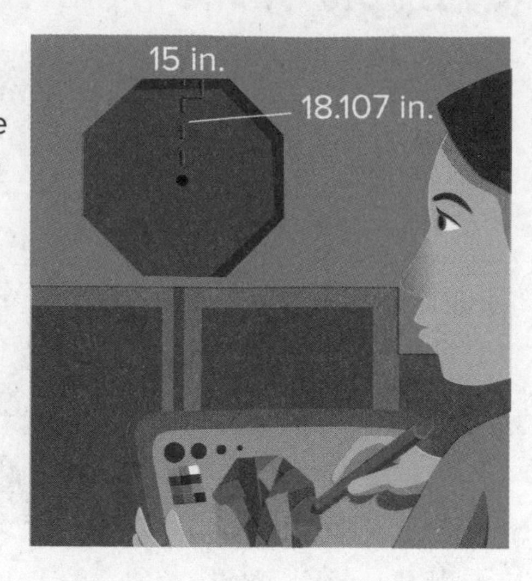

Go Online You can complete an Extra Example online.

Pause and Reflect

Compare finding the area of regular polygons with finding the area of irregular polygons. What are some similarities? What are some differences?

Name ____________________ Date ____________

Practice

Go Online You can complete your homework online.

1. Kendra knitted the coaster shown as a present for her grandmother. The coaster is shaped like a regular hexagon. Each side of the hexagon is 3.5 inches. Find the area of the coaster. Round to the nearest hundredth. **(Example 1)**

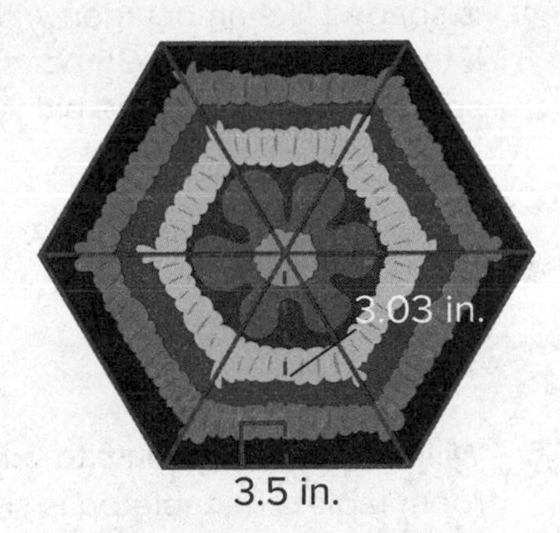

2. Paul bought a new rug in the shape of a regular decagon. Each side of the decagon is 4.25 feet. Find the area of the rug. Round to the nearest hundredth. **(Example 1)**

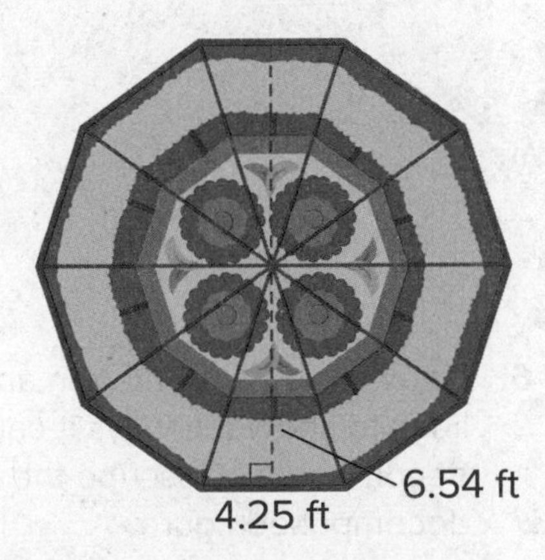

Test Practice

3. Open Response A regular pentagon is shown. What is the area of the pentagon?

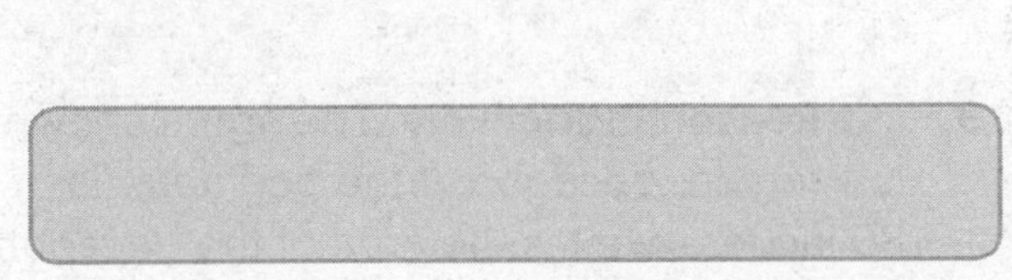

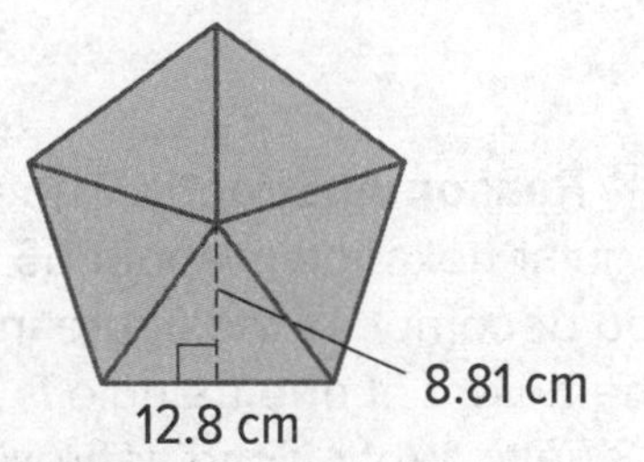

Apply

4. Julian is going to build a picnic table. The top of the picnic table is shaped like an octagon with sides measuring 2.5 feet. If the wood costs $3.95 per square foot, what is the least he will spend on the top of the picnic table?

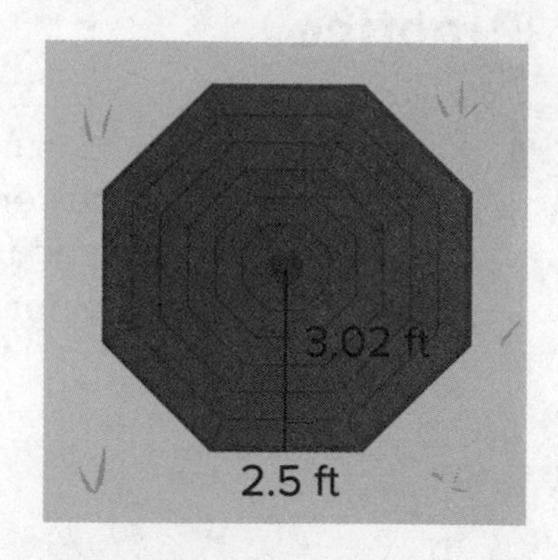

5. Williana's mother wants to buy a glass tabletop for their dining room table. The tabletop is shaped like a hexagon with sides measuring 27.75 inches. If the glass costs $0.06 per square inch, how much will she spend on the glass table top?

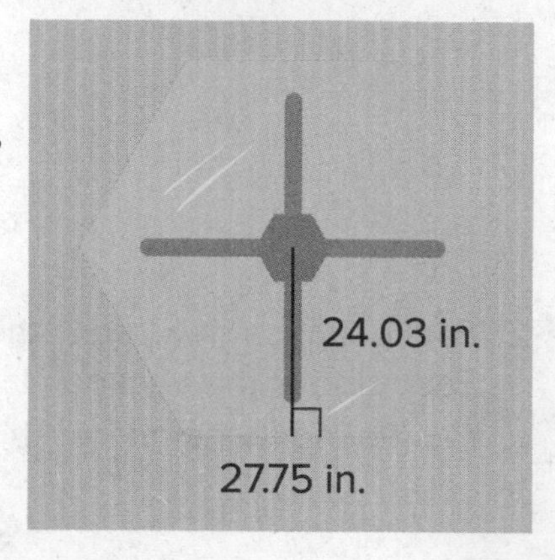

6. Draw a regular pentagon and use dashed lines to show the ways it can be decomposed. Describe the shapes in the decomposed figure.

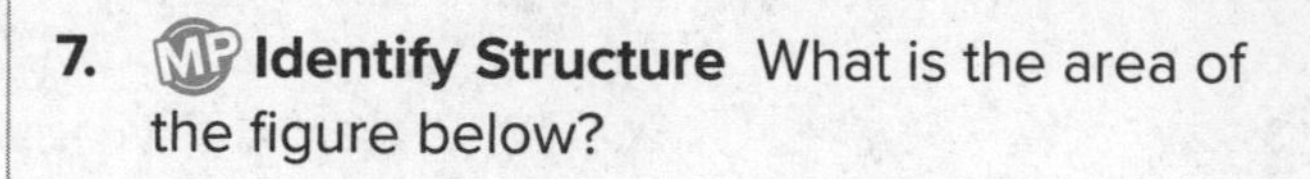

7. **Identify Structure** What is the area of the figure below?

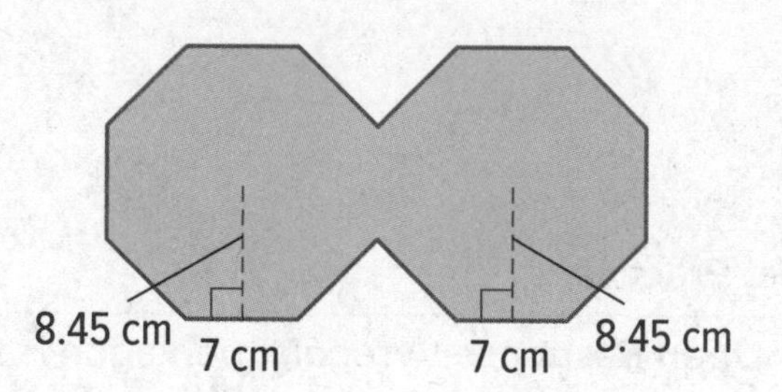

8. **Reason Abstractly** The area of a regular hexagon is about 65 square units. You decompose the figure into 6 triangles. The height of one triangle is about 4.3 units. What is the approximate length of the base of the triangle?

9. **Reason Inductively** The figure shown is a regular decagon. If the perimeter is 80 inches, what is the area of the decagon? Write an argument that can be used to defend your solution.

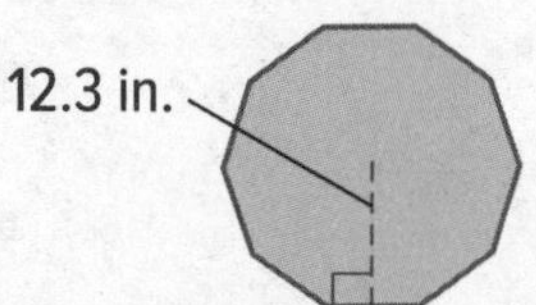

Polygons on the Coordinate Plane

I Can... graph the vertices of a polygon, draw the shape represented by the points, and then use the graphed polygon to find its area and perimeter.

Explore Explore the Coordinate Plane

Online Activity You will use Web Sketchpad to explore finding perimeter and area of polygons graphed on the coordinate plane.

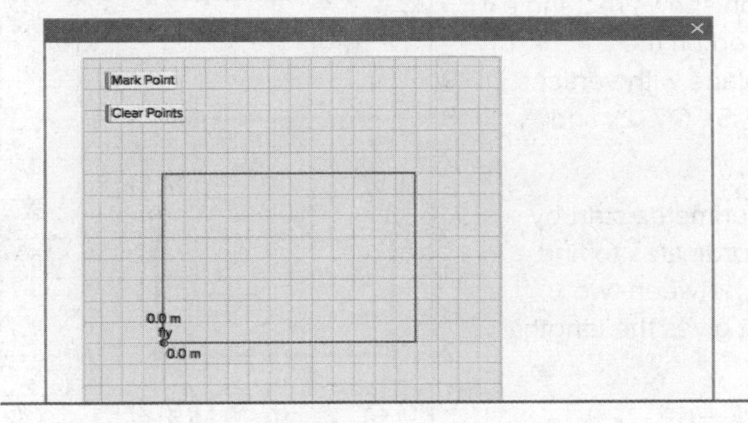

Learn Draw Polygons on the Coordinate Plane

You already know how to graph points on the coordinate plane. You can also graph polygons on the coordinate plane.

Go Online Use Web Sketchpad to complete the activity.

The sketch shows points *A*, *B*, *C*, and *D* graphed on a coordinate plane.

The points *A*(0, 0), *B*(3, 4), *C*(5, 4), and *D*(7, 0) form a polygon.

What polygon was created?

____

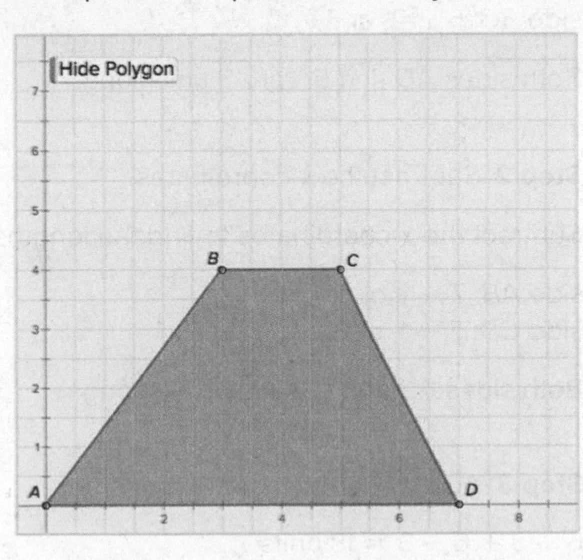

Talk About It!

What does the number of coordinate points given tell you about the polygon?

Your Notes

Talk About It!

How can the coordinates of vertices be used to find the area of a polygon?

Learn Find Perimeter and Area on the Coordinate Plane

You can use the coordinates of a polygon to find its dimensions by finding the distance between two points.

To find the distance between two points with the same x-coordinates, subtract their y-coordinates. To find the distance between two points with the same y-coordinates, subtract their x-coordinates. You can use those dimensions to find the perimeter and area of a polygon.

Go Online Watch the animation to learn about finding perimeter on the coordinate plane.

The animation shows rectangle $ABCD$ graphed on the coordinate plane with vertices at $A(1, 5)$, $B(7, 5)$, $C(7, 2)$, and $D(1, 2)$.

To find the perimeter, start by using the coordinates to find the distance between two points, which gives the length of the side.

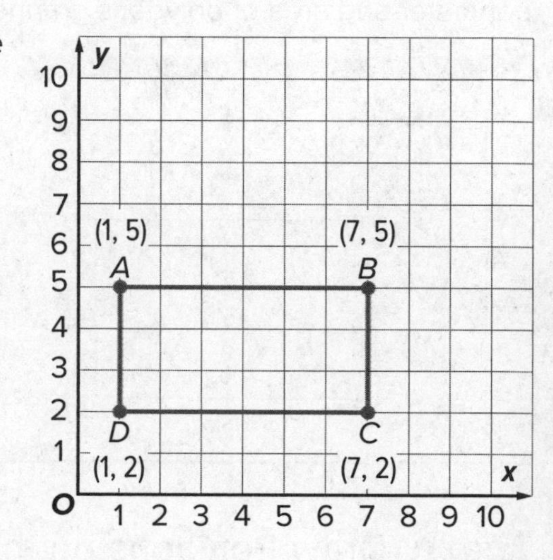

Step 1 Subtract the y-coordinates.

Subtract the y-coordinates to find the lengths of sides AD and BC.

side AD: $5 - 2$, or 3
side BC: $5 - 2$, or 3

Both sides AD and BC are 3 units long.

Step 2 Subtract the x-coordinates.

Subtract the x-coordinates to find the lengths of sides AB and CD.

side AB: $7 - 1$, or 6
side CD: $7 - 1$, or 6

Both sides AB and CD are 6 units long.

Step 3 Add the lengths of the four sides.

$6 + 3 + 6 + 3 = 18$ units

So, rectangle $ABCD$ has a perimeter of ________ units.

Example 1 Find Perimeter of an Irregular Figure

Find the perimeter of the exhibit shown on the coordinate plane.

Method 1 Count the units.

Count the units as you move along the perimeter of the exhibit.

Start at the entrance, or (0, 0). How many units do you need to travel along the *y*-axis to reach the monkeys? ________ units

How many units do you need to travel along the *x*-axis from the monkeys to reach the gorillas?

________ units

Continue counting along the perimeter until you return to the entrance.

Add to find the perimeter.

$10 + 7 + 3 + 4 + 4 + 4 + 3 + 7 = \square$ units

Method 2 Use the coordinates to find the distances.

Find the lengths of the horizontal line segments by subtracting the *x*-coordinates.

tigers to elephants:
$11 - 7 = 4$

aquarium to rhinoceros:
$11 - 7 = 4$

reptiles to entrance:
$7 - 0 = 7$

gorillas to monkeys:
$7 - 0 = 7$

Find the lengths of the vertical line segments by subtracting the *y*-coordinates.

gorillas to elephants:
$10 - 7 = 3$

tigers to aquarium:
$7 - 3 = 4$

rhinoceros to reptiles:
$3 - 0 = 3$

monkeys to entrance:
$10 - 0 = 10$

Find the sum of the sides.

$4 + 4 + 7 + 7 + 3 + 4 + 3 + 10 =$ ________

So, using either method, the perimeter of the exhibit is 42 units.

Think About It!

How can you find the distance between two points on the coordinate plane?

Talk About It!

Compare the two methods.

Check

Find the perimeter of the outlined section of the park shown on the coordinate plane.

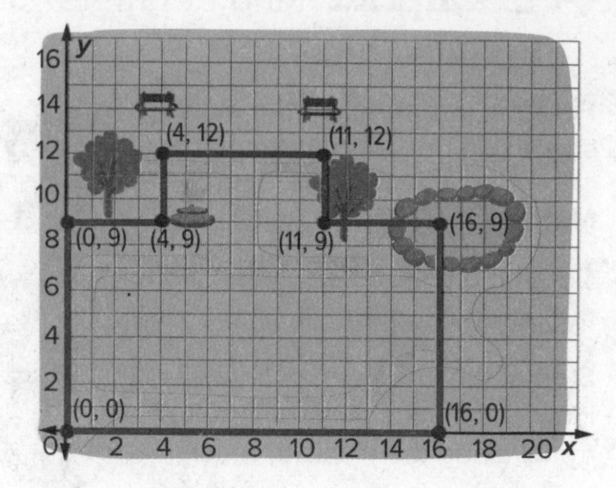

Go Online You can complete an Extra Example online.

Example 2 Find Perimeter Using Coordinates

A rectangle has vertices $A(2, 8)$, $B(7, 8)$, $C(7, 5)$, and $D(2, 5)$.

Use the coordinates to find the perimeter of the rectangle.

Step 1 Identify the sides of the rectangle.

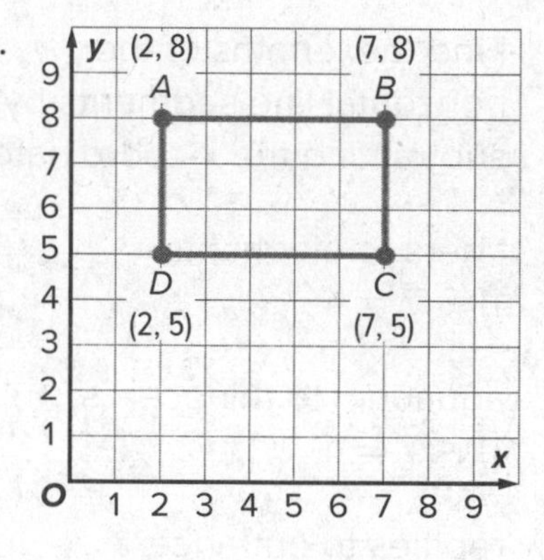

Graph the vertices on the coordinate plane. Then draw line segments to connect them to form a rectangle.

The horizontal sides are $\overline{AB}$ and $\overline{CD}$. You can also determine this from studying the coordinates.

Points A and B have the same y-coordinate, so they are endpoints of horizontal side $\overline{AB}$.

Points C and D have the same y-coordinate, so they are endpoints of horizontal side $\overline{CD}$.

The vertical sides are $\overline{AD}$ and $\overline{BC}$. You can also determine this from studying the coordinates.

Points A and D have the same x-coordinate, so they are endpoints of vertical side $\overline{AD}$.

Points B and C have the same x-coordinate, so they are endpoints of vertical side $\overline{BC}$.

(continued on next page)

Think About It!

How can you find the length or horizontal distance? How can you find the width or vertical distance?

Step 2 Find the perimeter of the rectangle.

Find the length of each side. You can count the units along each side of the rectangle's graph, or you can use the coordinates of the vertices and subtract to find the length of each side.

Length of $\overline{AB}$: ☐ units   Length of $\overline{CD}$: ☐ units

Length of $\overline{AD}$: ☐ units   Length of $\overline{BC}$: ☐ units

So, rectangle *ABCD* has a perimeter of 5 + 5 + 3 + 3, or ______ units.

Check

A rectangle has vertices *A*(1, 4), *B*(1, 9), *C*(8, 9), and *D*(8, 4). Find the perimeter of the rectangle. Use the coordinate plane if needed.

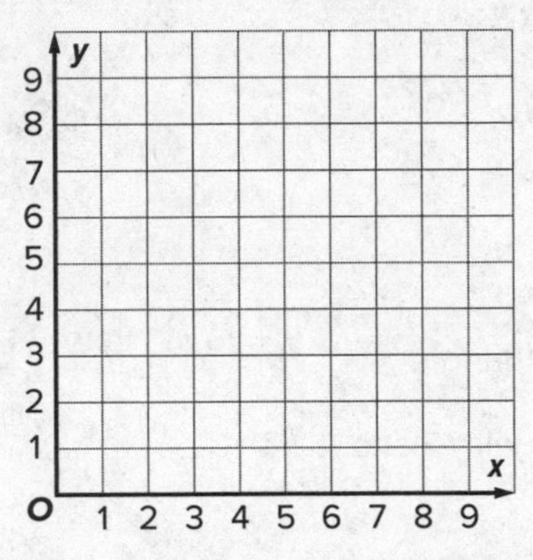

Go Online You can complete an Extra Example online.

Pause and Reflect

Suppose a classmate was having difficulty finding perimeter on the coordinate plane. How can you explain how to use the different methods to help the classmate understand?

Talk About It!

Why do the vertical sides share *x*-coordinates and horizontal sides share *y*-coordinates? Explain your reasoning.

Example 3 Find Area Using Coordinates

A polygon has vertices $A(2, 5)$, $B(2, 8)$, and $C(5, 8)$.

Find the area of the polygon.

Step 1 Identify the polygon.

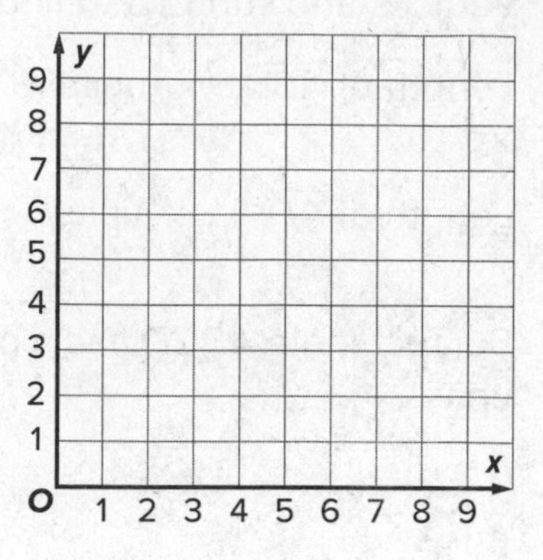

Graph the vertices. Draw line segments to connect them to form the polygon.

What polygon is formed? ________

Step 2 Find the area of the polygon.

The base is side $\overline{AB}$, and the height is side $\overline{BC}$.

Length of $\overline{AB}$: ☐ units

Length of $\overline{BC}$: ☐ units

Find the area.

$A = \frac{1}{2}bh$ Area of a triangle

$= \frac{1}{2}(3)(3)$ Replace b with 3 and h with 3.

$= 4\frac{1}{2}$ Simplify.

So, the area of the polygon is ☐ square units.

Check

A polygon has vertices $A(2, 4)$, $B(2, 9)$, and $C(9, 9)$. Find the area of the polygon. Use the coordinate plane if needed.

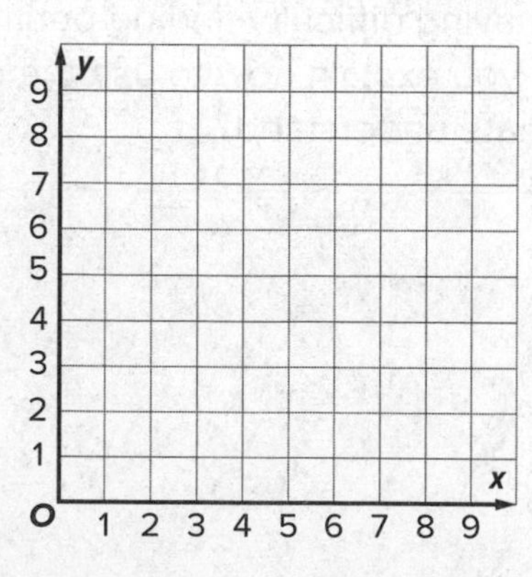

Go Online You can complete an Extra Example online.

Apply Business Finance

Miyu, a craft store owner, plans to rent a location in the mall and is considering the two spaces shown. On the map, one unit is equal to one foot. Space A has a monthly rental cost of $13.89 per square foot. Space B has a monthly rental cost of $13.49 per square foot. Miyu wants to pay the lower total monthly rental price. Which location should she choose to rent?

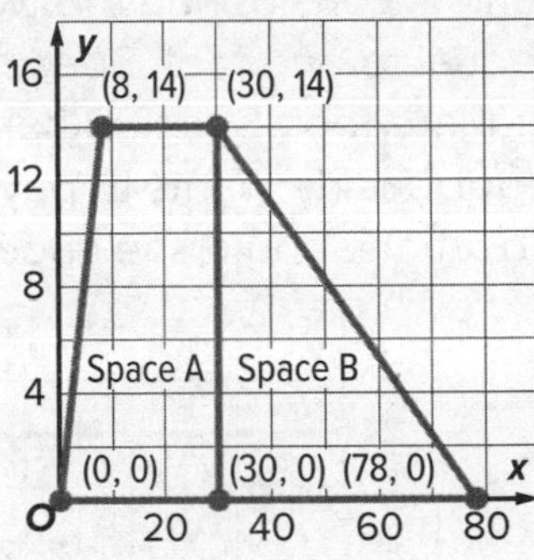

Go Online Watch the animation.

1 What is the task?

Make sure you understand exactly what question to answer or problem to solve. You may want to read the problem three times. Discuss these questions with a partner.

First Time Describe the context of the problem, in your own words.
Second Time What mathematics do you see in the problem?
Third Time What are you wondering about?

2 How can you approach the task? What strategies can you use?

Talk About It! In this problem, why is it possible to determine which space she should rent, without figuring out the monthly rental cost?

3 What is your solution?

Use your strategy to solve the problem.

4 How can you show your solution is reasonable?

Write About It! Write an argument that can be used to defend your solution.

Check

Jackie, a sports store owner, plans to rent a location in a strip mall and is considering the two spaces shown. On the map, one unit is equal to one foot. Space A has a monthly rental cost of $14.59 per square foot. Space B has a monthly rental cost of $15.15 per square foot. Jackie wants to pay the lower total monthly rental price. Which location should she choose to rent?

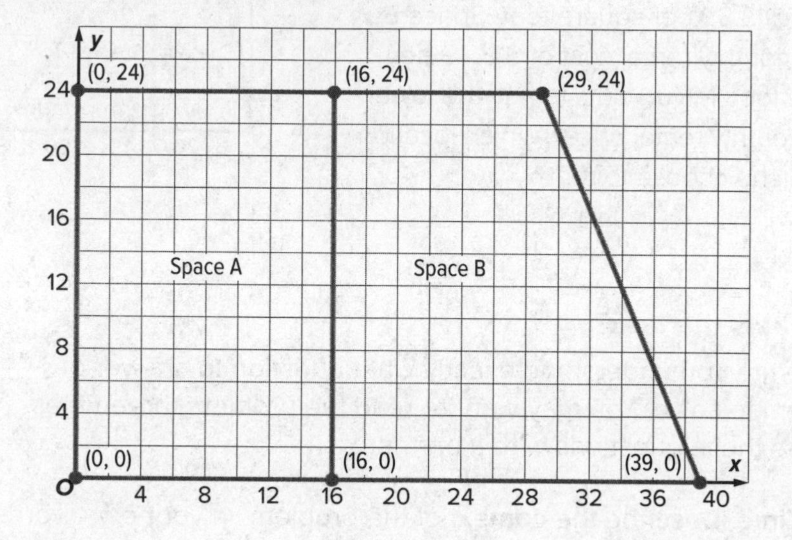

Go Online You can complete an Extra Example online.

Name ______________________ Period __________ Date __________

Practice

Go Online You can complete your homework online.

1. Find the perimeter of the summer camp shown on the coordinate plane. **(Example 1)**

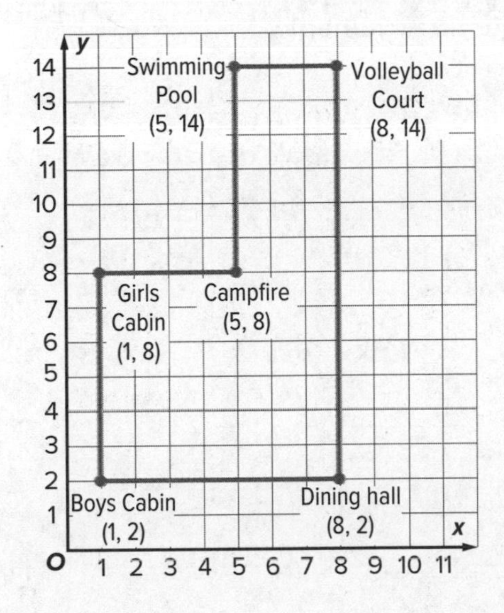

2. Find the perimeter of the science center shown on the coordinate plane. **(Example 1)**

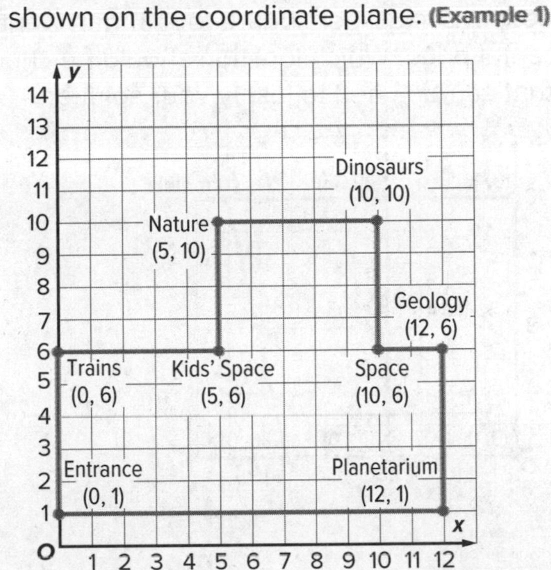

3. A rectangle has vertices $W(2, 7)$, $X(2, 0)$, $Y(6, 0)$, and $Z(6, 7)$. Use the coordinates to find the perimeter of the rectangle. **(Example 2)**

4. A rectangle has vertices $H(3, 0)$, $I(3, 7)$, $J(6, 7)$, and $K(6, 0)$. Use the coordinates to find the perimeter of the rectangle. **(Example 2)**

5. A polygon has vertices $A(3, 3)$, $B(3, 6)$, and $C(9, 3)$. Find the area of the polygon. **(Example 3)**

Test Practice

6. Multiple Choice A polygon has vertices $J(2, 3)$, $K(4, 3)$, $L(4, 7)$, and $M(2, 7)$. What is the area of the polygon? **(Example 3)**

Ⓐ 8 square units

Ⓑ 10 square units

Ⓒ 12 square units

Ⓓ 16 square units

Apply

7. Ethan wants to open a pet store in a town mall and is considering the two spaces shown. On the map, one unit is equal to one foot. Space A has a monthly rental cost of $14.75 per square foot. Space B has a monthly rental cost of $14.50 per square foot. Ethan wants to pay the lower total monthly rental price. Which location should he choose to rent? Write an argument that can be used to justify your solution.

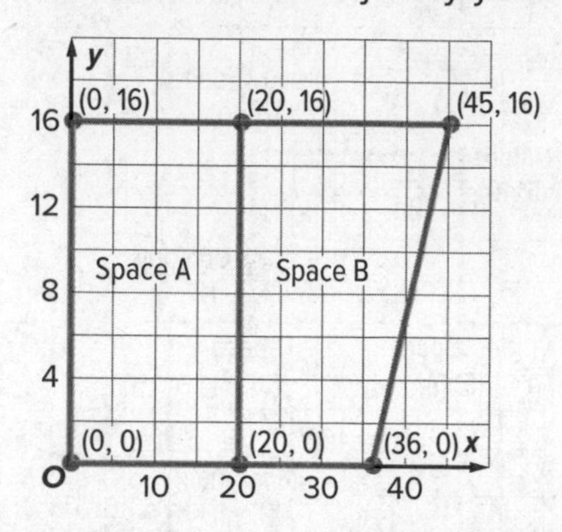

8. Draw and label a triangle on the coordinate plane that has an area of 20 square units.

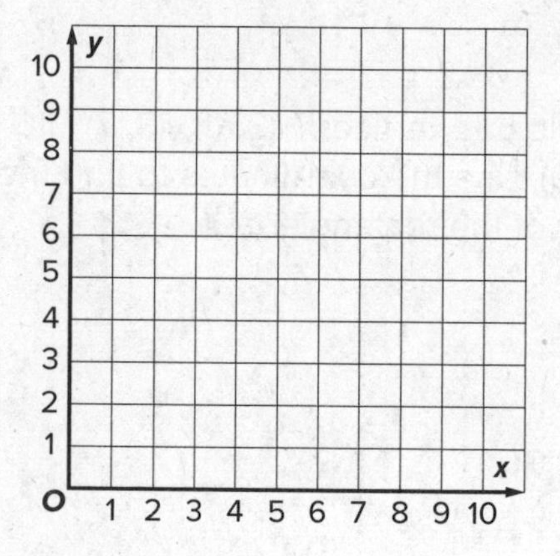

9. MP **Reason Inductively** A certain rectangle has a perimeter of 10 units and an area of 6 units. Two of the vertices have coordinates (1, 7) and (1, 4). Find the two missing coordinates.

10. MP **Persevere with Problems** Mrs. Palmer is placing a retaining wall around a garden. The coordinates of the vertices of the wall are (1, 1), (1, 5), (6, 5), and (6, 1). If each grid square has a length of 2 feet, what is the perimeter of the area? Write an argument that can be used to justify your solution.

11. MP **Find the Error** Rectangle *ABCD* has vertices *A*(2, 1), *B*(2, 7), *C*(10, 7), and *D*(10, 1). A classmate states that the perimeter of the rectangle is 16 units. Find the student's mistake and correct it.

Review

Foldables Use your Foldable to help review the module.

Area		
Real-World Examples	Real-World Examples	Real-World Examples

Rate Yourself!   

Complete the chart at the beginning of the module by placing a checkmark in each row that corresponds with how much you know about each topic after completing this module.

Write about one thing you learned.

Write about a question you still have.

Reflect on the Module

Use what you learned about area to complete the graphic organizer.

Essential Question

How are the areas of triangles and rectangles used to find the areas of other polygons?

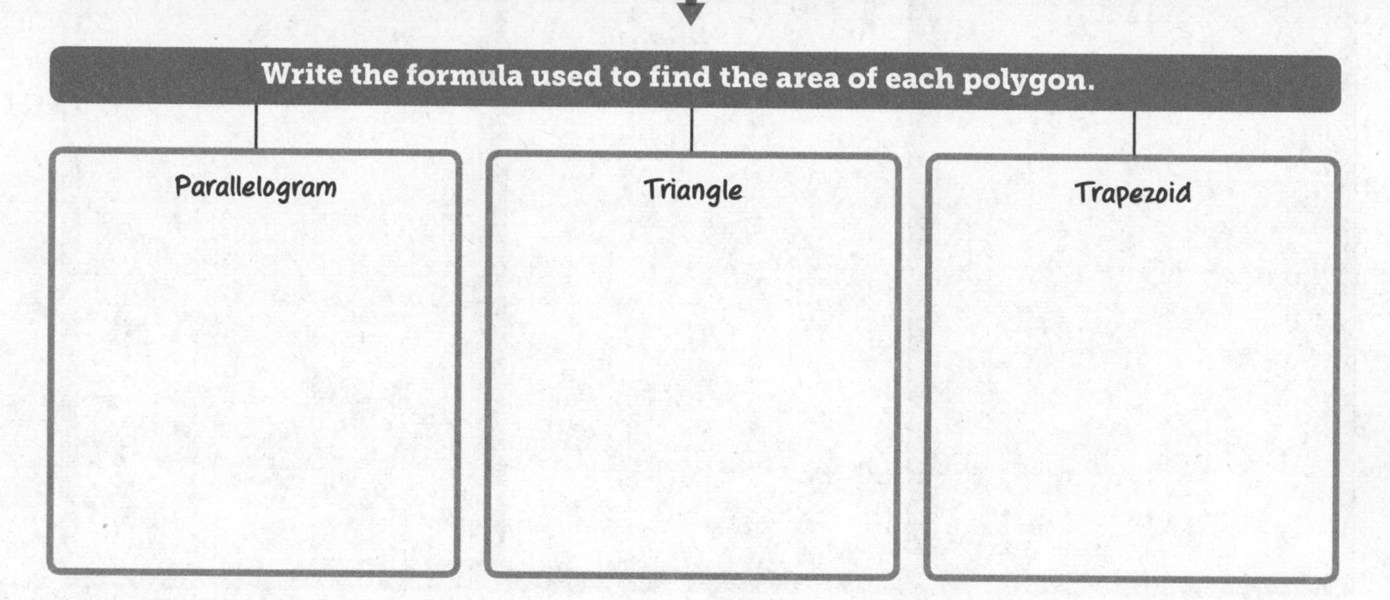

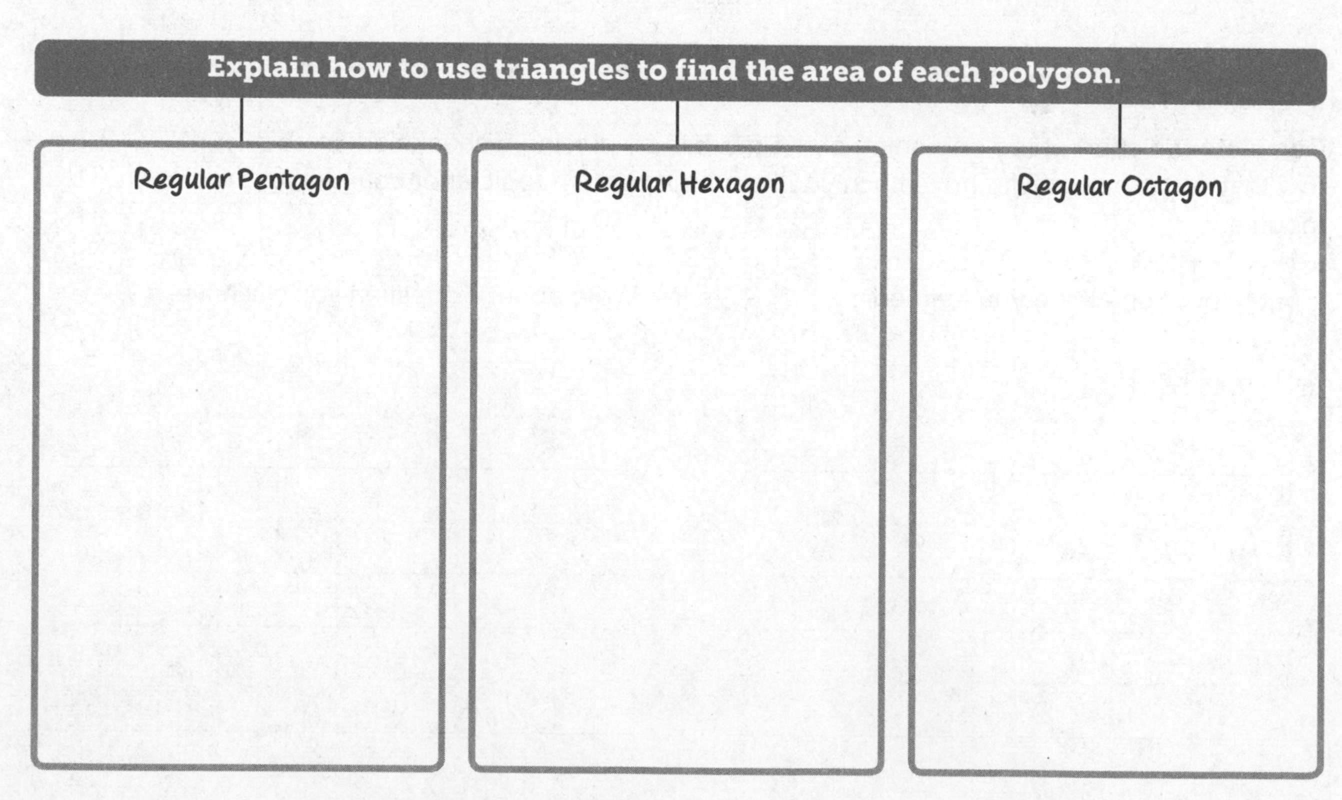

Name ______________________ Period __________ Date __________

Test Practice

1. Open Response Find the area of one of the parallelograms in the quilt pattern shown. **(Lesson 1)**

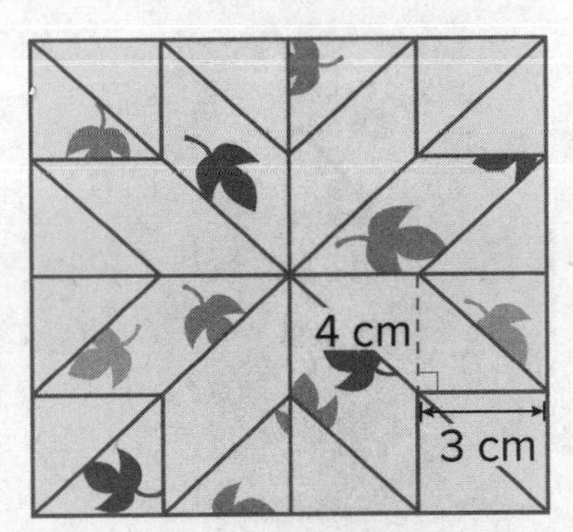

2. Multiple Choice What is the height of the parallelogram? **(Lesson 1)**

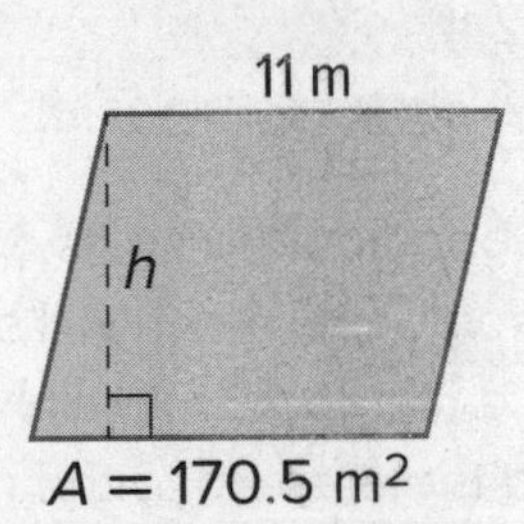

Ⓐ 14 meters

Ⓑ 14.5 meters

Ⓒ 15 meters

Ⓓ 15.5 meters

3. Open Response Find the area of the triangle. **(Lesson 2)**

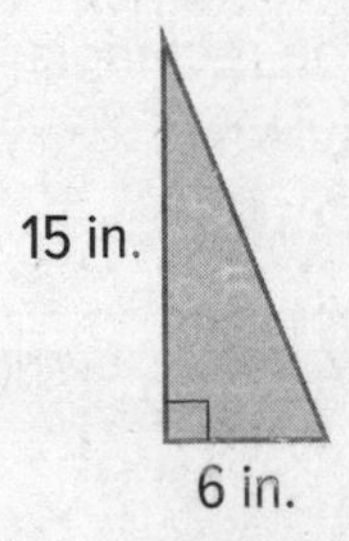

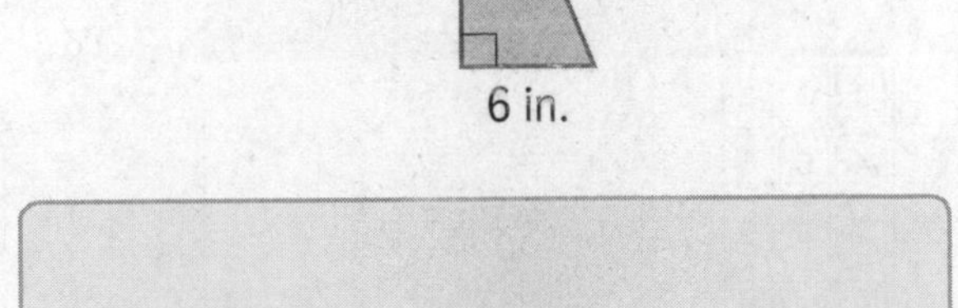

4. Multiselect Yolanda wants to replace the grass in this triangular section of her yard with mulch. A bag of mulch costs $4.85 and covers 3 square feet. Which of the following statements accurately describe this situation? Select all that apply. **(Lesson 2)**

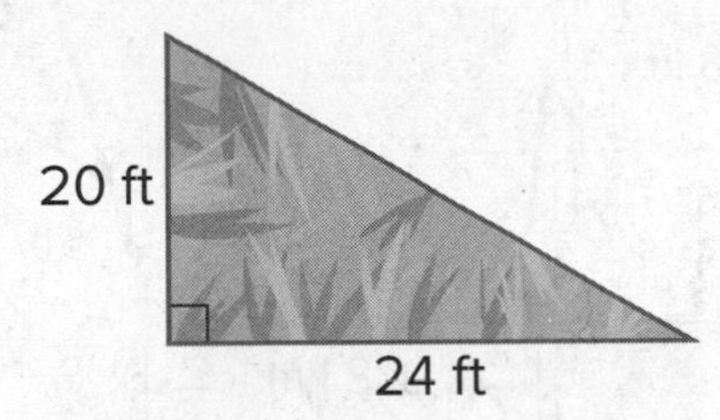

- ☐ The area of the triangle is 480 square feet.
- ☐ The area of the triangle is 240 square feet.
- ☐ Yolanda will need 80 bags of mulch.
- ☐ Yolanda will need 120 bags of mulch.
- ☐ Yolanda will spend $363.75 on mulch.
- ☐ Yolanda will spend $388 on mulch.

5. Equation Editor What is the height of the trapezoid in yards? **(Lesson 3)**

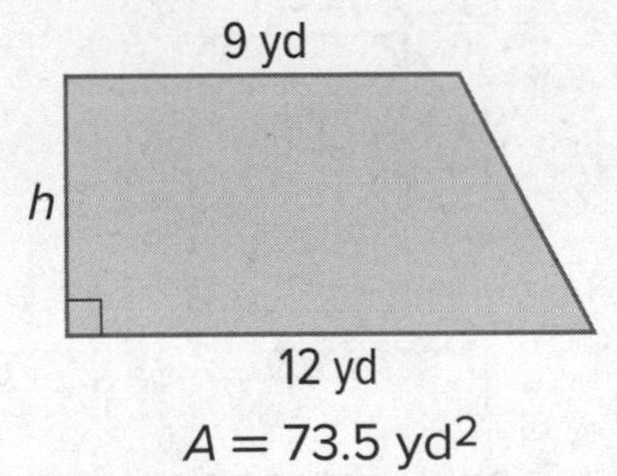

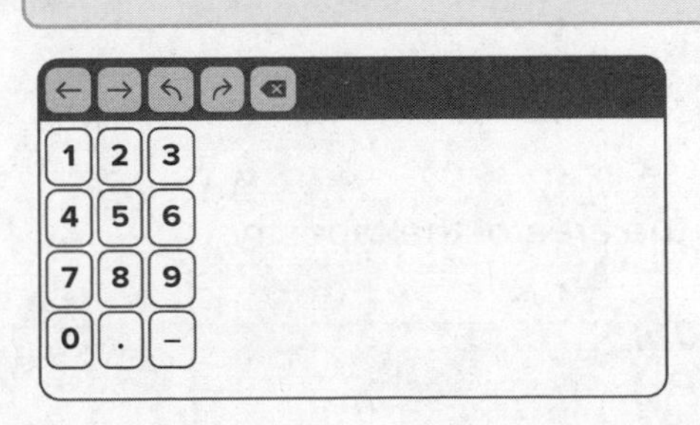

6. Multiple Choice The shape of the park resembles a trapezoid. Which of the following is the approximate area of the park? **(Lesson 3)**

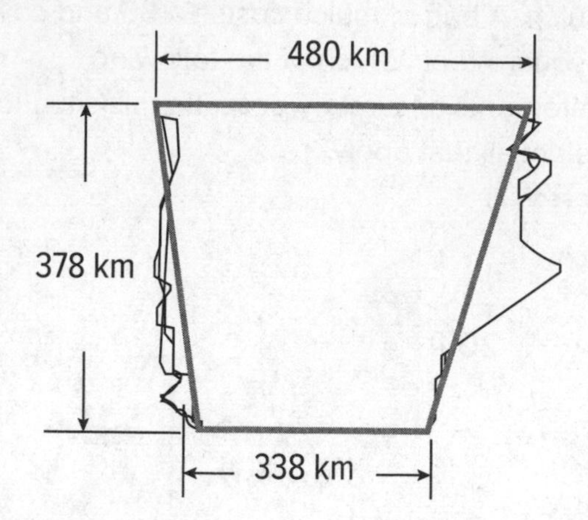

Ⓐ 109,242 km^2

Ⓑ 154,602 km^2

Ⓒ 231,903 km^2

Ⓓ 309,204 km^2

7. Open Response A tapestry is shaped like a regular hexagon. **(Lesson 4)**

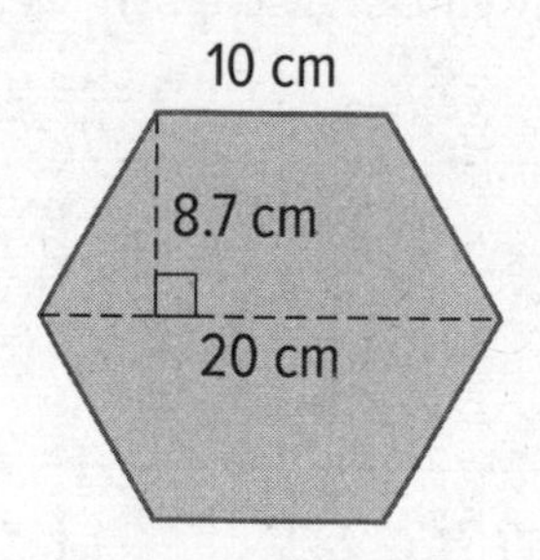

A. Explain how you can decompose the hexagon in order to find its area.

B. Find the area of the tapestry.

8. Open Response Kim wants to replace the area covered by this rug with hardwood flooring. The rug is shaped like a regular octagon with 3-foot sides. **(Lesson 4)**

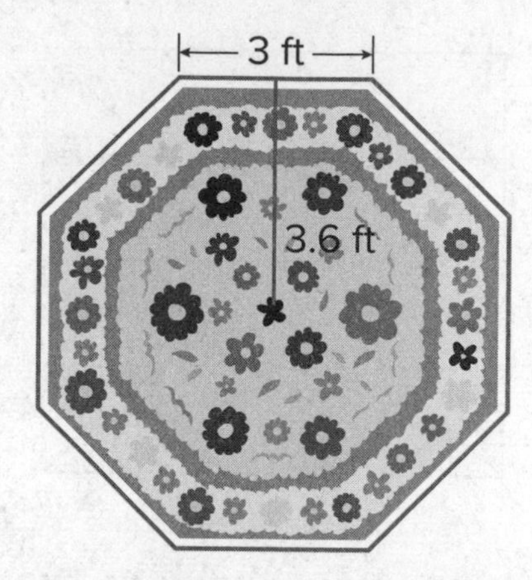

A. What is the area of the floor?

Ⓐ 38.2 ft^2

Ⓑ 40.1 ft^2

Ⓒ 43.2 ft^2

Ⓓ 45.0 ft^2

B. If hardwood flooring costs $9.50 per square foot, how much will she spend to resurface the floor? Explain why you need to round the area up to the nearest whole square foot in order to calculate the cost.

9. Equation Editor A rectangle has vertices $A(1,2)$, $B(1, 9)$, $C(7, 9)$, and $D(7, 2)$. Find the perimeter of the rectangle in units. **(Lesson 5)**

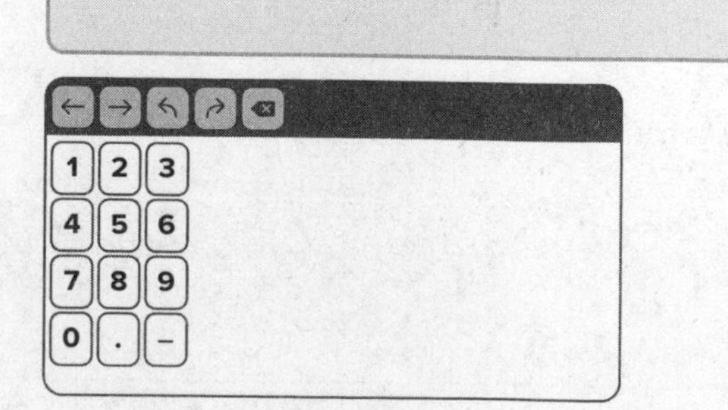

Module 9

Volume and Surface Area

Essential Question

How can you describe the size of a three-dimensional figure?

What Will You Learn?

Place a checkmark (✓) in each row that corresponds with how much you already know about each topic **before** starting this module.

KEY

□ — I don't know. ◇ — I've heard of it. ☆ — I know it!

	Before			After		
	□	◇	☆	□	◇	☆
finding volume of rectangular prisms						
finding missing dimensions of rectangular prisms						
making nets to represent rectangular prisms						
finding surface areas of rectangular prisms						
making nets to represent triangular prisms						
finding surface areas of triangular prisms						
making nets to represent pyramids						
finding surface areas of pyramids						

Foldables Cut out the Foldable and tape it to the Module Review at the end of the module. You can use the Foldable throughout the module as you learn about volume and surface area.

What Vocabulary Will You Learn?

Check the box next to each vocabulary term that you may already know.

- ☐ cubic units
- ☐ lateral face
- ☐ net
- ☐ prism
- ☐ pyramid
- ☐ rectangular prism
- ☐ slant height
- ☐ surface area
- ☐ three-dimensional figure
- ☐ triangular prism
- ☐ volume

Are You Ready?

Study the Quick Review to see if you are ready to start this module.

Then complete the Quick Check.

Quick Review

Example 1	Example 2
Multiply rational numbers.	**Evaluate numerical expressions.**
Find $12 \times 3.5 \times 18$.	Evaluate $(8 \times 6) + (3 \times 9)$.
$12 \times 3.5 \times 18 = 42 \times 18$ Multiply 12 and 3.5.	$(8 \times 6) + (3 \times 9) = 48 + 27$ Multiply.
$= 756$ Multiply by 18.	$= 75$ Add.

Quick Check

1. Find $12 \times 2.2 \times 17.5$.	**2.** Evaluate $(12.5 \times 40) + (16.25 \times 6)$.

How Did You Do?

Which exercises did you answer correctly in the Quick Check? Shade those exercise numbers at the right.

Volume of Rectangular Prisms

I Can... find the volume of a rectangular prism by using unit cubes and by using the volume formula when given the length, width, and height of the prism.

What Vocabulary Will You Learn?
cubic units
prism
rectangular prism
three-dimensional figure
volume

Learn Volume

A **three-dimensional figure** has length, width, and height. A **prism** is a three-dimensional figure with two parallel bases that are congruent polygons. In a **rectangular prism**, the bases are congruent rectangles.

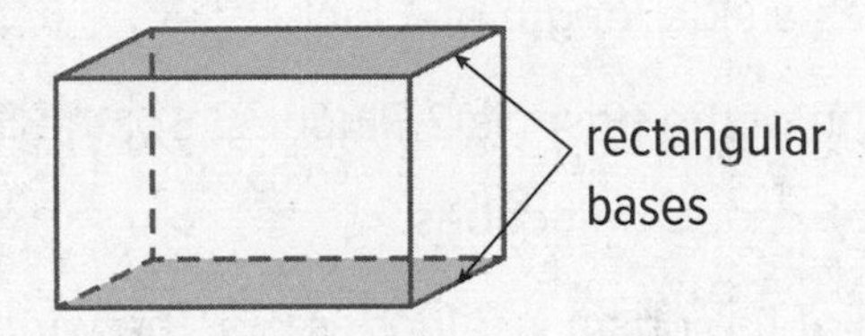

Volume is the amount of space inside a three-dimensional figure. It is measured in **cubic units**, which can be written using abbreviations and an exponent of 3, such as units^3 or in^3.

You can find the volume of a rectangular prism with whole number measurements by packing the prism with unit cubes. Decomposing the prism tells you the number of cubes of a given size it will take to fill the prism. The volume of a rectangular prism is related to its dimensions: length, width, and height.

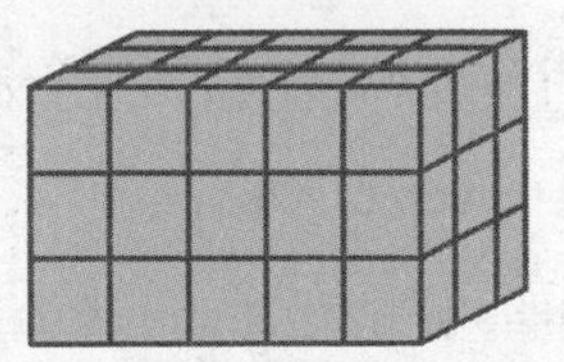

The rectangular prism shown has a length of 5 units, a width of 3 units, and a height of 3 units. There is a total of 15 unit cubes in the base layer of the prism. The prism has 3 layers. So, the volume of the prism is 15 + 15 + 15, or 3(15), or 45 cubic units.

Recall that you learned how to find the volume of a rectangular prism in an earlier grade, by using the volume formula, $V = \ell wh$, where V represents the volume, ℓ represents the length, w represents the width, and h represents the height. Using this method, the volume of the prism shown is 5(3)(3), or 45 cubic units.

Your Notes

Math History Minute

Benjamin Banneker (1731–1806) was an African-American mathematician, astronomer, inventor, and writer. When he was 22, he used his own drawings and calculations to construct a working clock that was made almost entirely out of wood.

Learn Volume of a Rectangular Prism

You can find the volume of a rectangular prism with fractional measurements using different methods.

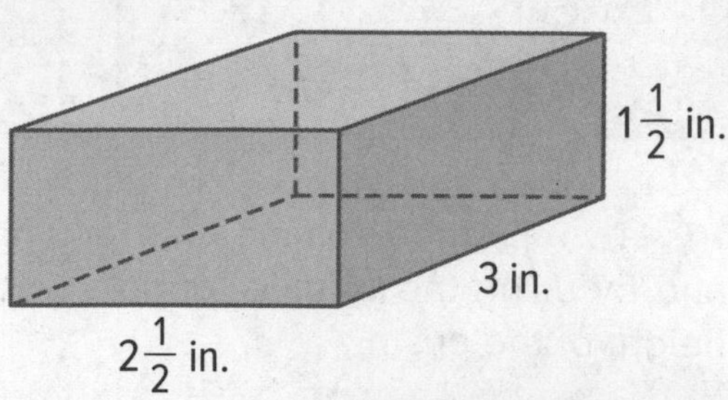

Method 1 Use unit cubes.

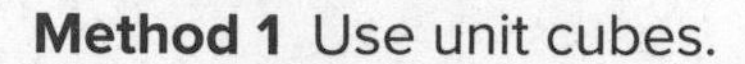

You can pack a rectangular prism with unit cubes. A cube is a special rectangular prism with all sides congruent. The volume of a cube is found by cubing the side length.

Step 1 Find the number of unit cubes needed to fill the prism. Each unit cube has a side length of $\frac{1}{2}$ inch.

Length The length of the prism is $2\frac{1}{2}$ inches. So, the length is composed of $2\frac{1}{2} \div \frac{1}{2}$ or 5 unit cubes.

Width The width of the prism is 3 inches. So, the width is composed of $3 \div \frac{1}{2}$ or 6 unit cubes.

Height The height of the prism is $1\frac{1}{2}$ inches. So, the height is composed of $1\frac{1}{2} \div \frac{1}{2}$ or 3 unit cubes.

The base layer of the prism contains 5×6, or 30 unit cubes. There are three total layers in the prism. So, the rectangular prism contains 30×3, or 90 unit cubes.

Step 2 Find the volume of one unit cube.

$V = s^3$	Volume of a cube with side length s.
$= \left(\frac{1}{2}\right)^3$	Replace s with $\frac{1}{2}$.
$= \left(\frac{1}{2}\right)\left(\frac{1}{2}\right)\left(\frac{1}{2}\right)$	Definition of exponent
$= \frac{1}{8}$	Multiply. The volume of each cube is $\frac{1}{8}$ in^3.

Step 3 Multiply the volume of each unit cube by the total number of unit cubes, 90.

$V = 90\left(\frac{1}{8}\right)$	There are 90 unit cubes, each with a volume of $\frac{1}{8}$ in^3.
$= 11\frac{1}{4}$	Multiply. The volume of the prism is $11\frac{1}{4}$ in^3.

(continued on next page)

Method 2 Use the formula.

The formula for the volume of a right prism is $V = Bh$ where B represents the area of the base of the prism and h represents the height of the prism. In a rectangular prism the base is a rectangle, so $B = \ell w$. So, the volume of a right rectangular prism can also be found using the formula $V = \ell wh$.

$V = \ell wh$	Volume formula
$V = 2\frac{1}{2} \cdot 3 \cdot 1\frac{1}{2}$	$\ell = 2\frac{1}{2}, w = 3, h = 1\frac{1}{2}$
$V = 11\frac{1}{4}$	Multiply.

So, using either method, the volume of the rectangular prism is $11\frac{1}{4}$ cubic inches.

Example 1 Find the Volume of a Rectangular Prism

Mini sugar cubes measure $\frac{1}{4}$ inch on each side. The box shown is packed full of sugar cubes.

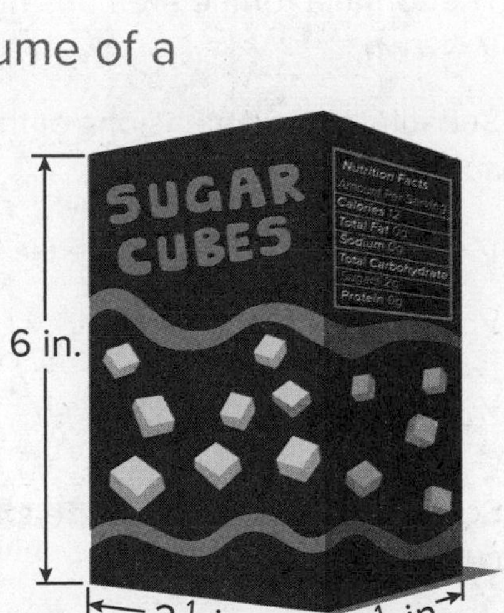

What is the volume of the box?

Method 1 Use unit cubes.

Step 1 Find the number of mini sugar cubes.

Each sugar cube has a side length of $\frac{1}{4}$ inch.

Length of Prism: $3\frac{1}{2}$ inches

Because $3\frac{1}{2} \div \frac{1}{4} = 14$, there are 14 mini sugar cubes that fit along the length of the prism.

Width of Prism: $1\frac{1}{2}$ inches

Because $1\frac{1}{2} \div \frac{1}{4} = 6$, there are 6 mini sugar cubes that fit along the width of the prism.

Height of Prism: 6 inches

Because $6 \div \frac{1}{4} = 24$, there are 24 mini sugar cubes that fit along the height of the prism.

The base layer of the prism contains 14×6, or ________ mini sugar cubes. There are 24 total layers of unit cubes in the prism. So, the rectangular prism contains ________ $\times$ 24, or ________ total mini sugar cubes.

(continued on next page)

Talk About It!

The formula $V = Bh$ can be used to find the volume of any right prism. You know that for a right rectangular prism, the area of the base, B, is represented by the expression ℓw. Think of a prism that doesn't have a rectangular base, such as a triangular prism. What expression can you use to represent the area of the base?

Think About It!

Estimate the volume of the box of mini sugar cubes.

Step 2 Find the volume of one mini sugar cube.

$V = s^3$	Volume of a cube with side length s.
$= \left(\frac{1}{4}\right)^3$	Replace s with $\frac{1}{4}$.
$= \left(\frac{1}{4}\right)\left(\frac{1}{4}\right)\left(\frac{1}{4}\right)$	Definition of exponent
$= \frac{1}{64}$	Multiply. The volume of each cube is $\frac{1}{64}$ in^3.

Step 3 Multiply the volume of each cube by the total number of unit cubes, 2,016.

$V = \square \left(\frac{1}{64}\right)$	There are 2,016 unit cubes, each with a volume of $\frac{1}{64}$ in^3.
$= \square$	Multiply. The volume of the prism is $31\frac{1}{2}$ in^3.

Method 2 Use the volume formula.

The formula for the area of a right rectangular prism is $V = Bh$ or $V = \ell wh$.

Substitute the dimensions of the box for the variables in the formula and multiply.

$V = \ell wh$	Write the volume formula.
$V = 3\frac{1}{2} \cdot 1\frac{1}{2} \cdot 6$	Replace ℓ with $3\frac{1}{2}$, w with $1\frac{1}{2}$, and h with 6.
$V = 31\frac{1}{2}$ in^3	Multiply.

So, using either method, the total volume of the box is ________ cubic inches.

Talk About It!

Use what you know about the formula for the volume of a prism to explain why the volume of a cube can be found by cubing the side length.

Check

Find the volume of the prism.

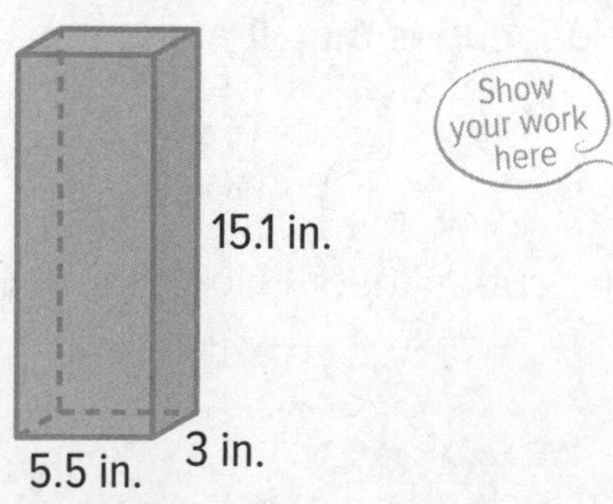

Go Online You can complete an Extra Example online.

Learn Find Missing Dimensions

When you know the volume of a rectangular prism and 2 out of 3 dimensions, you can write and solve an equation to find the missing dimension. Using the volume formula, replace the variables with the known values. Then solve the equation to find the unknown value.

Go Online Watch the animation to learn how to find the missing dimension for the rectangular prism shown.

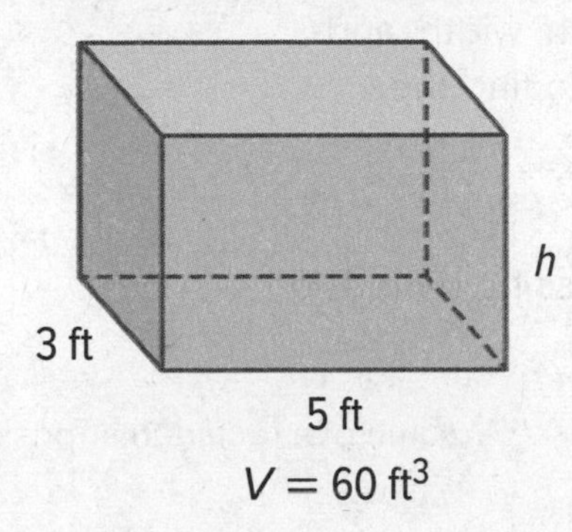

$V = 60 \text{ ft}^3$

The rectangular prism shown has a volume of 60 cubic feet. The width of the prism is 3 feet and the length is 5 feet. The height of the prism is unknown.

To find the unknown height, you can use the formula for volume of a rectangular prism.

$V = \ell wh$	Write the volume formula.
$60 = (5)(3)h$	Replace V with 60, ℓ with 5, and w with 3.
$60 = 15h$	Multiply.
$\frac{60}{15} = \frac{15h}{15}$	Division Property of Equality
$4 = h$	Simplify.

So, the height of the rectangular prism is ________ feet.

Talk About It!

How can understanding variables and equations help you solve geometry problems?

Think About It!

What formula can you use to solve this problem?

Talk About It!

Why is the unit for the height of the prism inches and not cubic inches? Explain your reasoning.

Example 2 Find Missing Dimensions

The rectangular prism shown has a volume of $94\frac{1}{2}$ cubic inches.

What is the height of the prism?

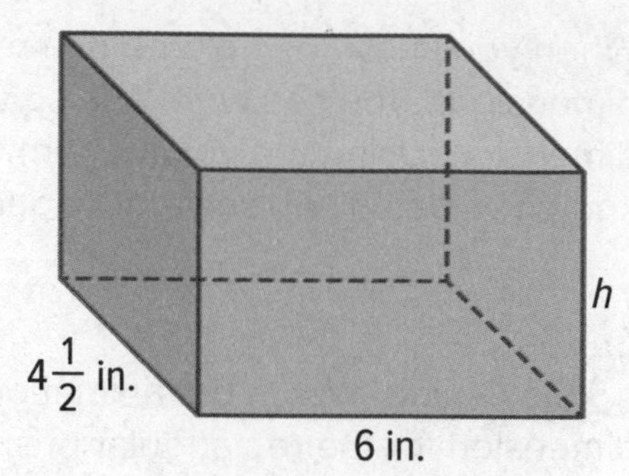

Step 1 Identify the known dimensions.

You know the length, width, and volume. You need to find the height.

Step 2 Find the missing dimension.

$V = \ell wh$	Volume of a rectangular prism
$94\frac{1}{2} = 6 \cdot 4\frac{1}{2} \cdot h$	Substitute the known quantities.
$94\frac{1}{2} = 27h$	Multiply.
$\frac{94\frac{1}{2}}{27} = \frac{27h}{27}$	Divide.
$3\frac{1}{2} = h$	Simplify.

So, the height of the prism is ________ inches.

Check

Find the height of a rectangular prism with a volume of 126 cubic inches, a width of $7\frac{7}{8}$ inches, and a length of 2 inches. ________

Go Online You can complete an Extra Example online.

Apply Comparisons

A movie theater sells three different-sized boxes of popcorn. If the boxes are rectangular prisms, which size of popcorn is the best buy?

Size	Length (in.)	Width (in.)	Height (in.)	Price ($)
Small	$5\frac{1}{2}$	4	$8\frac{1}{4}$	4.50
Medium	$6\frac{3}{4}$	5	$10\frac{1}{2}$	5.75
Large	$10\frac{1}{4}$	6	$11\frac{1}{2}$	7.00

1 What is the task?

Make sure you understand exactly what question to answer or problem to solve. You may want to read the problem three times. Discuss these questions with a partner.

First Time Describe the context of the problem, in your own words.
Second Time What mathematics do you see in the problem?
Third Time What are you wondering about?

2 How can you approach the task? What strategies can you use?

3 What is your solution?

Use your strategy to solve the problem.

4 How can you show your solution is reasonable?

Write About It! Write an argument that can be used to defend your solution.

Talk About It!

Suppose the dimensions of each box doubled. Would the answer remain the same? Explain your reasoning.

Check

A storage cube that has an edge length of 16 centimeters is being packed in a cardboard box with a length of 28 centimeters, a width of 18 centimeters, and a height of 22 centimeters. The extra space is being filled with packing peanuts. The packing peanuts cost $0.002 per cubic centimeter. How much will it cost to fill the extra space with packing peanuts?

Go Online You can complete an Extra Example online.

Foldables It's time to update your Foldable, located in the Module Review, based on what you learned in this lesson. If you haven't already assembled your Foldable, you can find the instructions on page FL1.

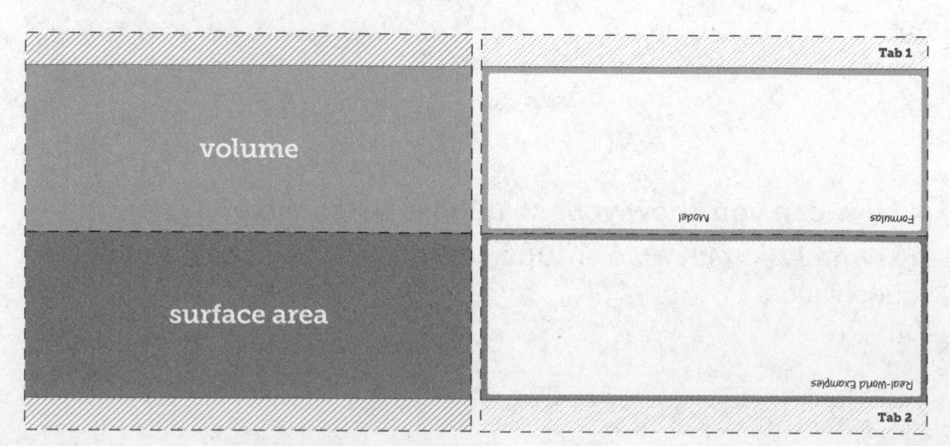

Name ______________________ Period ____________ Date ____________

Practice

Go Online You can complete your homework online.

1. Geneva's younger brother has a toy box that is shaped like a rectangular prism with the dimensions shown. What is the volume of the toy box? **(Example 1)**

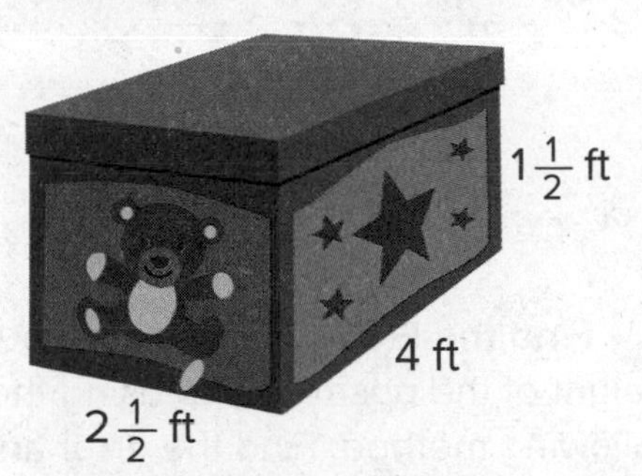

2. Roy made a jewelry box in the shape of a rectangular prism with the dimensions shown. What is the volume of the jewelry box? **(Example 1)**

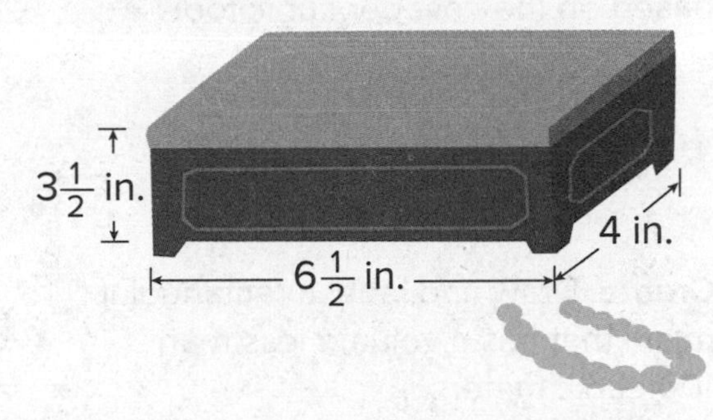

3. The rectangular prism shown has a volume of 52 cubic meters. What is the width of the prism? **(Example 2)**

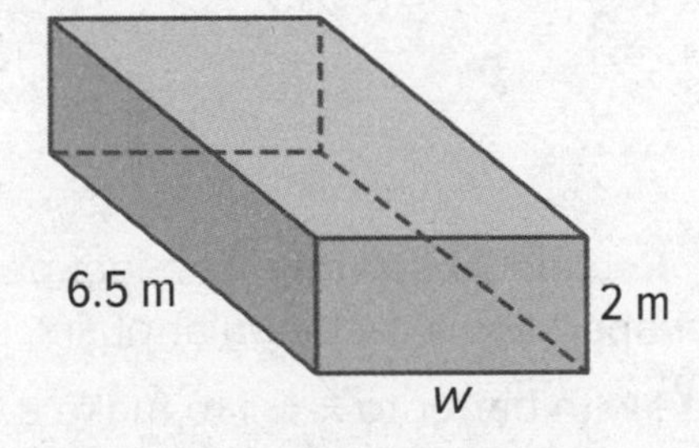

4. The rectangular prism shown has a volume of 115 cubic yards. What is the length of the prism? **(Example 2)**

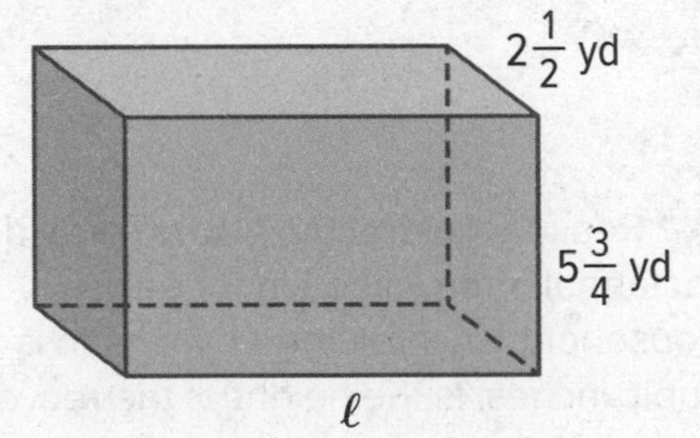

5. Raphael drives a standard-sized dump truck with a rectangular prism shaped bed. The volume of the bed of the truck is 720 cubic feet. If the length of the bed is 15 feet and the width is 8 feet, what is the height of the bed of the dump truck?

Test Practice

6. Open Response A rectangular prism has a length of 8 inches, a width of $7\frac{1}{2}$ inches, and a height of $6\frac{1}{4}$ inches. What is the volume of the prism?

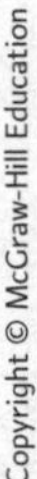

Apply

7. The Lagusch family needs to rent a dumpster. The dumpsters they can choose from are shaped like rectangular prisms and have the dimensions shown. Which size dumpster is the best value to rent based on the cost per cubic foot?

Size	Length (ft)	Width (ft)	Height (ft)	Cost ($)
Small	16	8	2	204.80
Medium	20	8	3.5	420.00
Large	22	8	5	677.60

8. **Create** Draw and label a rectangular prism that has a volume less than 100 cubic meters.

9. **MP Find the Error** A classmate found the height of the prism shown using the following method. Find the error and correct it.

$h = 1.5(1.2)(2.5)$

$= 4.5$ cm

2.5 cm

h

1.2 cm

$V = 1.5\text{ cm}^3$

10. **MP Reason Abstractly** A town provides a rectangular recycling bin for each household. The volume of each bin is 3,840 cubic inches. Is the height of the recycling bin greater than one foot? Write an argument that can be used to defend your solution.

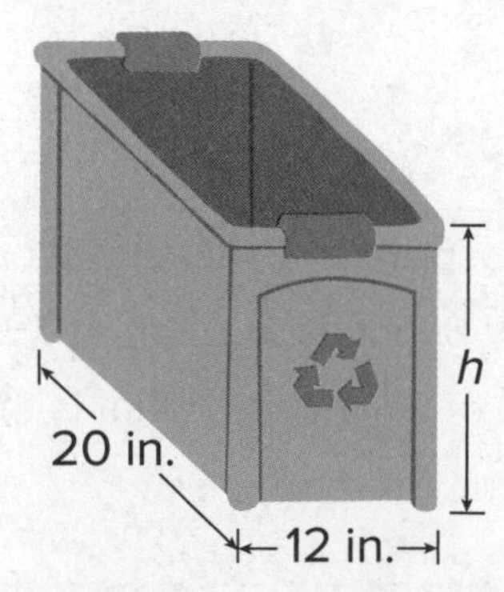

11. **MP Reason Abstractly** The loaf pan shown is shaped like a rectangular prism. It will be filled with batter to $\frac{2}{3}$ full to make a loaf of bread without overflowing while baking. How much batter would it take to fill the pan $\frac{2}{3}$ of the way? Write an argument that can be used to defend your solution.

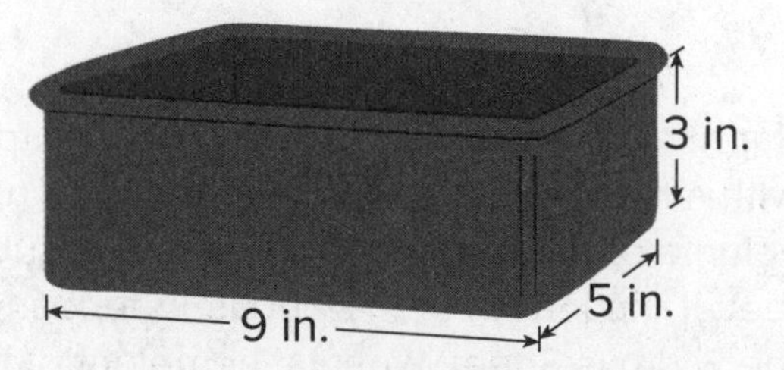

Surface Area of Rectangular Prisms

I Can... represent a rectangular prism with its net to find the surface area in mathematical and real-world contexts.

What Vocabulary Will You Learn?
net
surface area

Explore Cube Nets

Online Activity You will use models to explore nets of prisms.

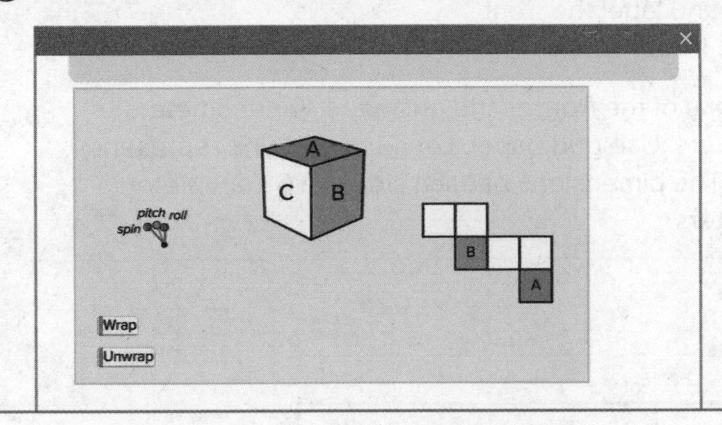

Learn Make a Net to Represent a Rectangular Prism

A **net** is a two-dimensional representation of a three-dimensional figure. When you construct a net, you are deconstructing the three-dimensional figure using its two-dimensional faces. A rectangular prism has six rectangular faces. The top and bottom faces are congruent. The front and back faces are congruent. The two side faces are congruent.

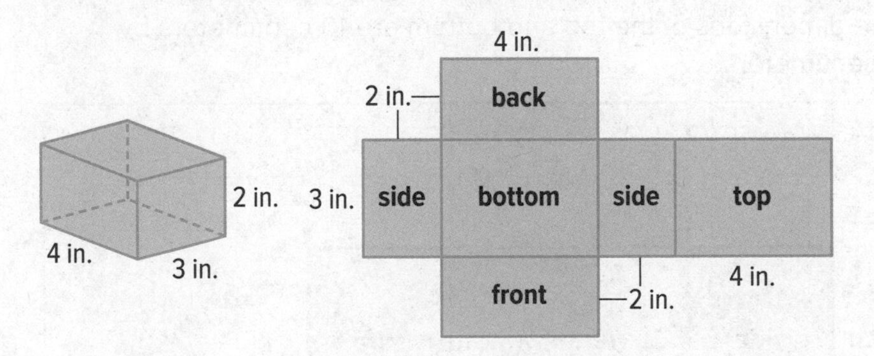

Talk About It!
What similarities do you notice between the length, width, and height of the prism, and the dimensions given in the net?

Your Notes

Think About It!
How can the prism be unfolded to make a two-dimensional net?

Example 1 Make a Net to Represent a Rectangular Prism

A rectangular prism has a length of 10 centimeters, a width of 6 centimeters, and a height of 8 centimeters.

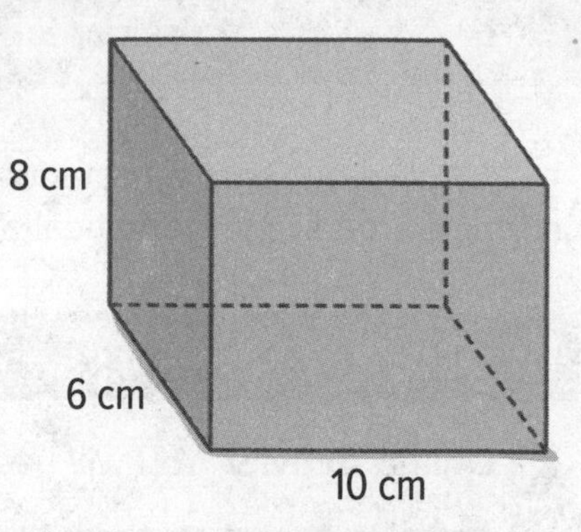

Draw and label a net to represent the rectangular prism.

Step 1 Draw and label the front face and side faces.

The dimensions of the front of the prism are 10 centimeters by 8 centimeters. Use grid paper. Let each grid unit represent 1 centimeter. The dimensions of each side are 6 centimeters by 8 centimeters.

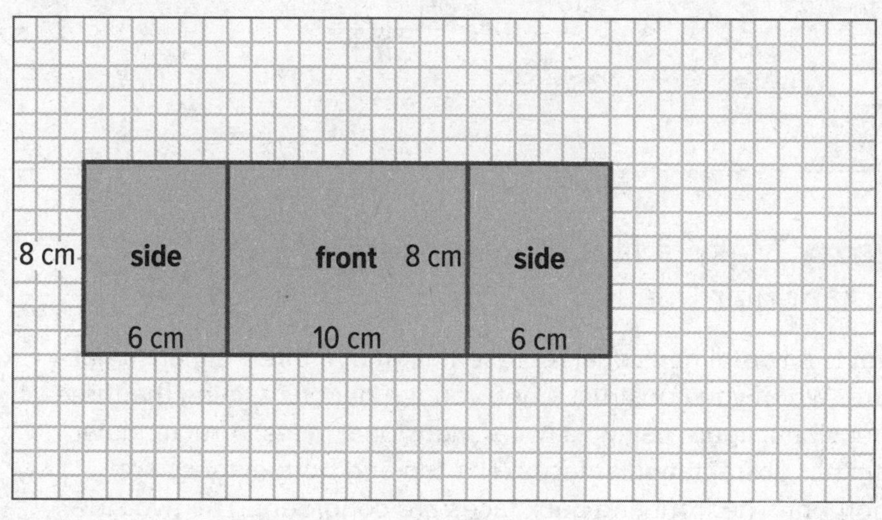

Step 2 Draw and label the top and bottom faces.

The dimensions of the top and bottom are 10 centimeters by 6 centimeters.

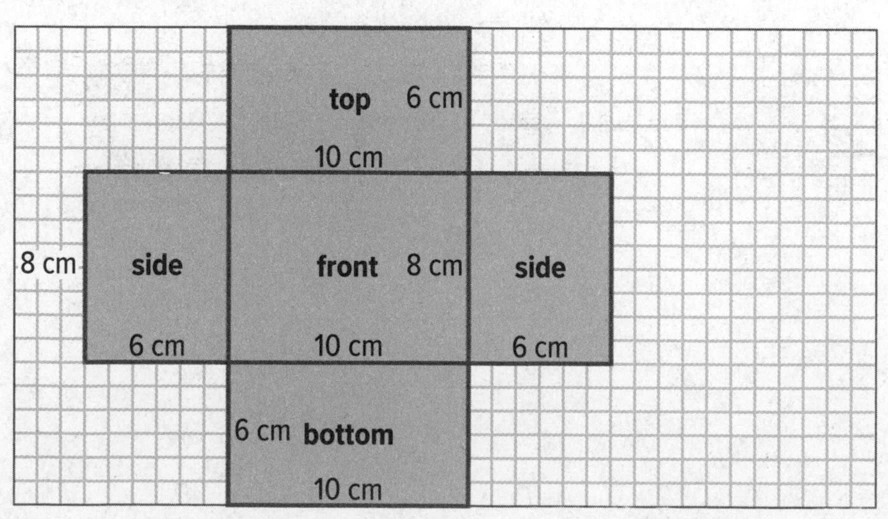

Step 3 Draw and label the back face.

The dimensions of the back are the same as the front, 10 centimeters by 8 centimeters.

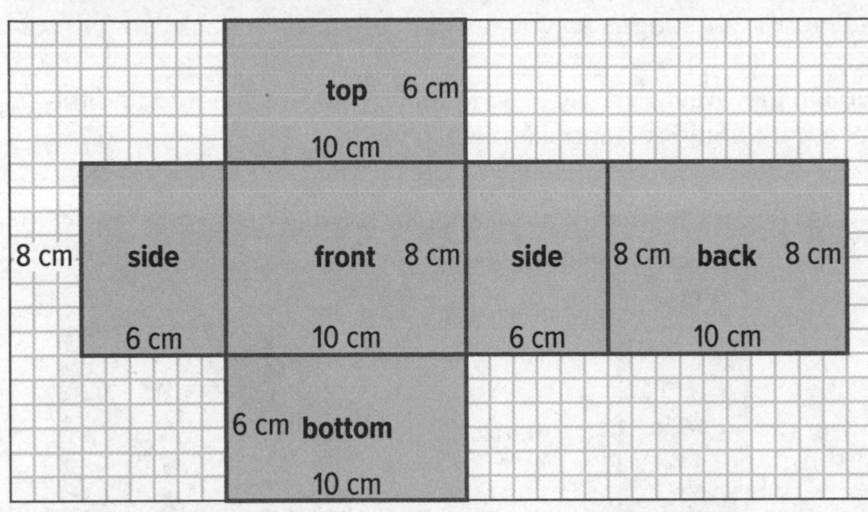

Check

Draw and label a net to represent the rectangular prism. Let each grid unit represent 1 inch.

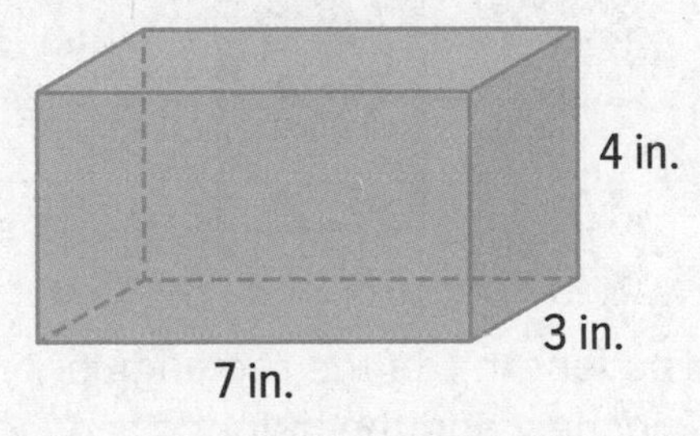

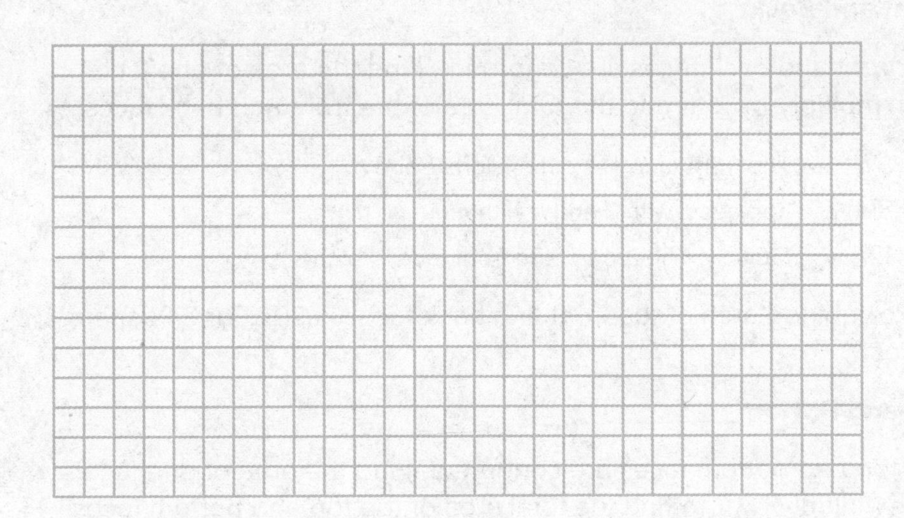

Talk About It!

Explain why there are only three measurements for a rectangular prism, with each face using two of the three measurements.

Go Online You can complete an Extra Example online.

Learn Surface Area of a Rectangular Prism

The **surface area** of a rectangular prism is the sum of the areas of the faces. Using a net can help you deconstruct the prism into two-dimensional shapes so you can find the area of each face.

Go Online Watch the video to learn how to use a net to find the surface area of the rectangular prism shown.

The video shows the net of a rectangular prism.

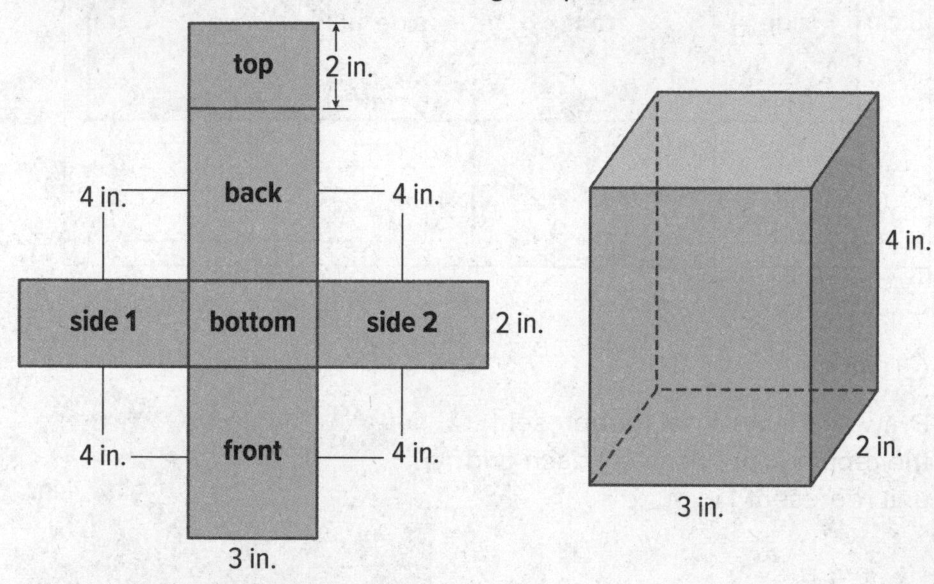

The length ℓ of the rectangular prism is 3 inches, the width w is 2 inches, and the height h is 4 inches.

Step 1 Find the area of each face.

Front and Back

The front and back faces are congruent. Find the area of one face. Then multiply by 2 to find the total area of the front and back faces.

$A = \ell h$	The front face has dimensions ℓ and h.
$= 3(4)$	Replace ℓ with 3 and h with 4.
$= 12$	Multiply. The area of the front face is 12 in^2.

The combined area of the front and back faces is 2(12), or 24 square inches.

Top and Bottom

The top and bottom faces are congruent. Find the area of one face. Then multiply by 2 to find the total area of the top and bottom faces.

$A = \ell w$	The top face has dimensions ℓ and w.
$= 3(2)$	Replace ℓ with 3 and w with 2.
$= 6$	Multiply. The area of the top face is 6 in^2.

(continued on next page)

Talk About It!

In the video, the student measured the top and bottom, front and back, and side 1 and side 2. What shortcut can you use when finding the surface area of a rectangular prism?

The combined area of the top and bottom faces is 2(6), or 12 square inches.

Sides

The two side faces are congruent. Find the area of one. Then multiply by 2 to find the total area of the side faces.

$A = wh$	Each side face has dimensions w and h.
$= 2(4)$	Replace w with 2 and h with 4.
$= 8$	Multiply. The area of each side face is 8 in^2.

The combined area of the two side faces is 2(8), or 16 square inches.

Step 2 Add the areas to find the total surface area.

$24 + 12 + 16 = 52$

So, the total surface area of the rectangular prism is ______ square inches.

Example 2 Surface Area of a Rectangular Prism

Jon is covering the faces of the gift box shown with wrapping paper.

8 cm
6 cm
10 cm

Use the net to determine the minimum amount of wrapping paper he will need to cover the box.

Step 1 Find the area of each face.

Front and Back

The front and back faces are congruent. Find the area of one face. Then multiply by 2 to find the total area of the front and back faces.

$A = \ell h$	The front face has dimensions ℓ and h.
$= 10(8)$	Replace ℓ with 10 and h with 8.
$= 80$	Multiply. The area of the front face is 80 cm^2.

The combined area of the front and back faces is 2(80), or 160 square centimeters.

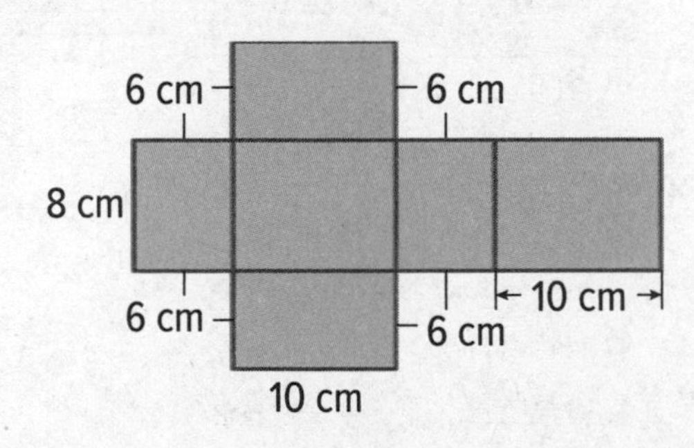

Think About It!

How many different-sized faces are there?

(continued on next page)

Top and Bottom

The top and bottom faces are congruent. Find the area of one face. Then multiply by 2 to find the total area of the top and bottom faces.

$A = \ell w$	The top face has dimensions ℓ and w.
$= 10(6)$	Replace ℓ with 10 and w with 6.
$= 60$	Multiply. The area of the top face is 60 cm^2.

The combined area of the top and bottom faces is 2(60), or 120 square centimeters.

Sides

The two side faces are congruent. Find the area of one face. Then multiply by 2 to find the total area of the side faces.

$A = wh$	Each side face has dimensions w and h.
$= 6(8)$	Replace w with 6 and h with 8.
$= 48$	Multiply. The area of each side face is 48 cm^2.

The combined area of the two side faces is 2(48), or 96 square centimeters.

Step 2 Add the areas to find the total surface area.

$160 + 120 + 96 = 376$

So, Jon will need a minimum of ________ square centimeters of wrapping paper.

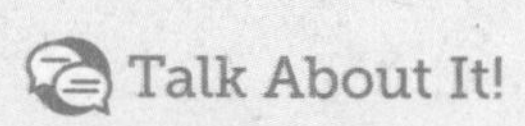

Talk About It!

Why is the unit of measure square centimeters rather than centimeters or cubic centimeters?

Check

A moving crate that is shaped like a rectangular prism with the dimensions shown needs to be painted. Use the net to determine the area that is to be painted.

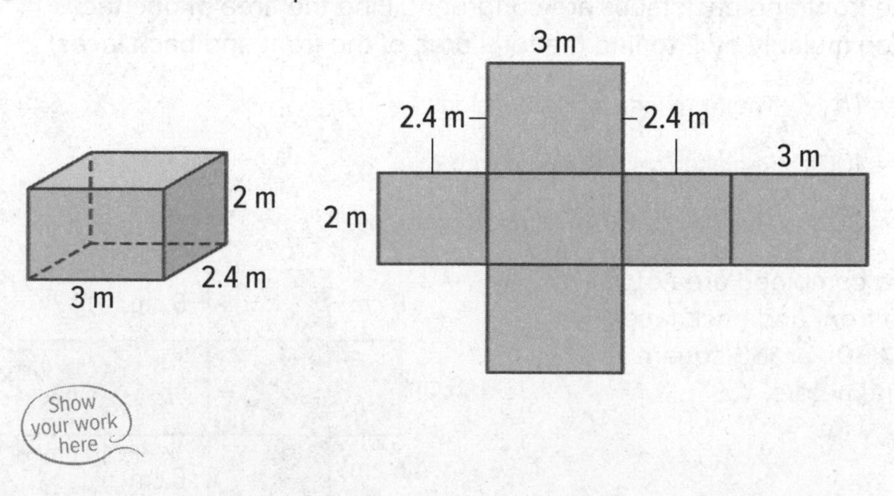

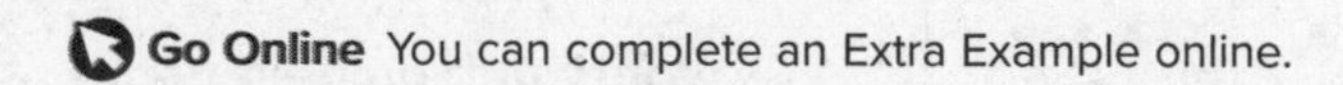

Go Online You can complete an Extra Example online.

Apply Home Improvement

Takeru is planning to paint the walls of his bedroom, which is in the shape of a rectangular prism. The bedroom has one window and two doors. The dimensions of the window and doors are shown in the table. If one gallon of paint covers about 150 square feet, how many gallons of paint are needed to cover the walls of a room that is 20 feet long, 15 feet wide, and 8 feet high?

Part	Height (ft)	Width (ft)
door	$6\frac{3}{4}$	$2\frac{3}{4}$
window	3	5

Go Online Watch the animation.

1 What is the task?

Make sure you understand exactly what question to answer or problem to solve. You may want to read the problem three times. Discuss these questions with a partner.

First Time Describe the context of the problem, in your own words.
Second Time What mathematics do you see in the problem?
Third Time What are you wondering about?

2 How can you approach the task? What strategies can you use?

3 What is your solution?

Use your strategy to solve the problem.

Talk About It!
Why were the floor and ceiling not included?

4 How can you show your solution is reasonable?

Write About It! Write an argument that can be used to defend your solution.

Check

Mrs. Hernandez is redesigning her craft room which is in the shape of a rectangular prism. She wants to add wainscoting, which is a wood wall covering, from the floor to halfway up the walls. There are two doors that are each 3 feet wide. How many square feet of wainscoting will she need to cover the space?

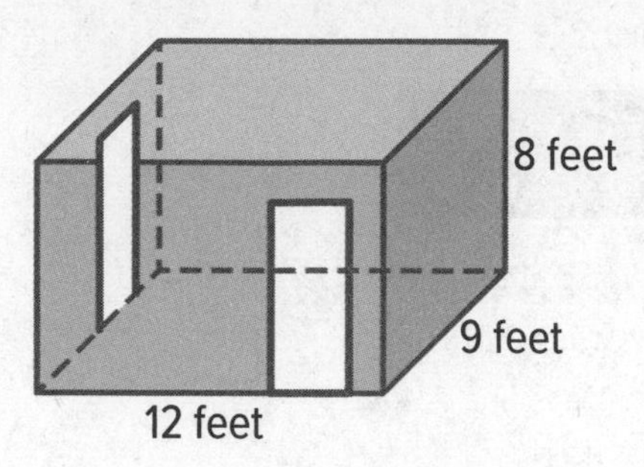

Go Online You can complete an Extra Example online.

Foldables It's time to update your Foldable, located in the Module Review, based on what you learned in this lesson. If you haven't already assembled your Foldable, you can find the instructions on page FL1.

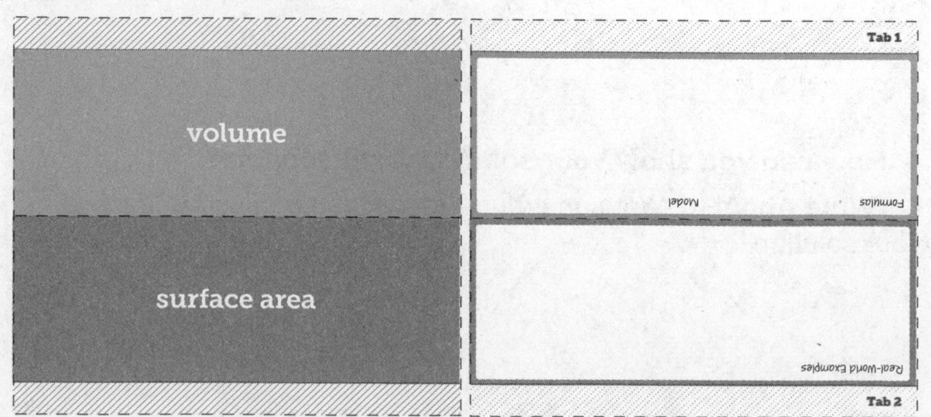

Name ______________________ Period __________ Date __________

Practice

Go Online You can complete your homework online.

1. Draw and label a net to represent the rectangular prism. Let each grid unit represent 1 inch. **(Example 1)**

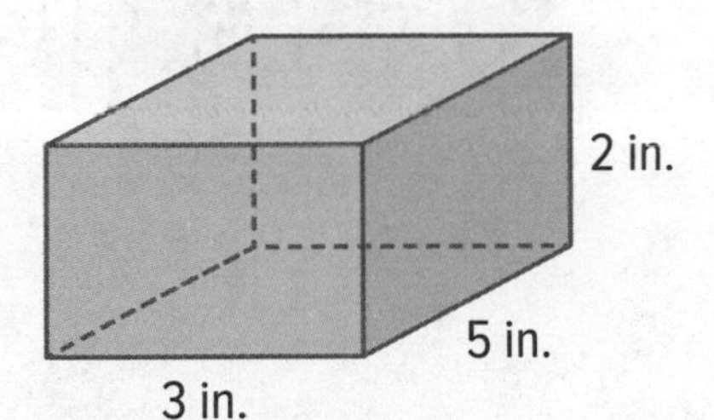

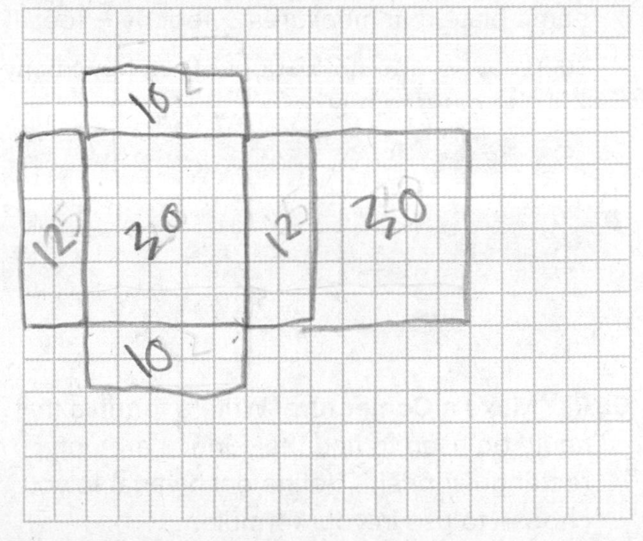

2. Trey is using cardboard to construct building blocks that are shaped like rectangular prisms. Use the net to determine the minimum amount of cardboard he will need to construct one block. **(Example 2)**

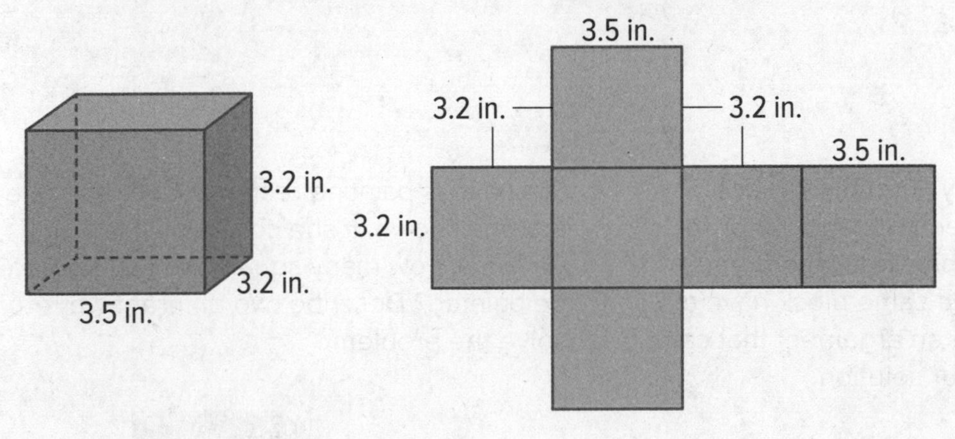

Test Practice

3. **Open Response** Cody is painting the box shown for part of his art project. If he paints all of the surfaces, how many square centimeters will he paint? Use the net to find the surface area of the rectangular prism.

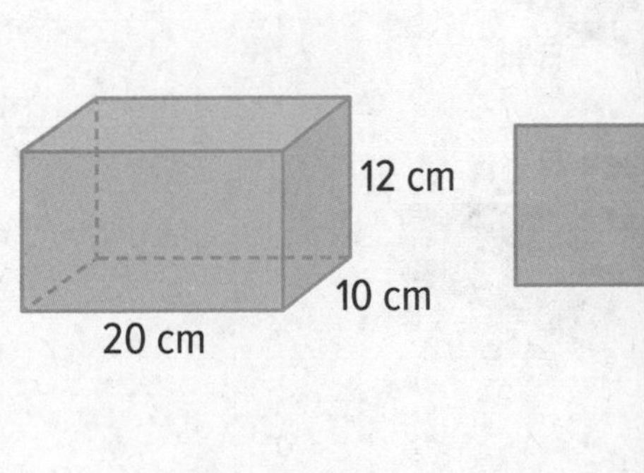

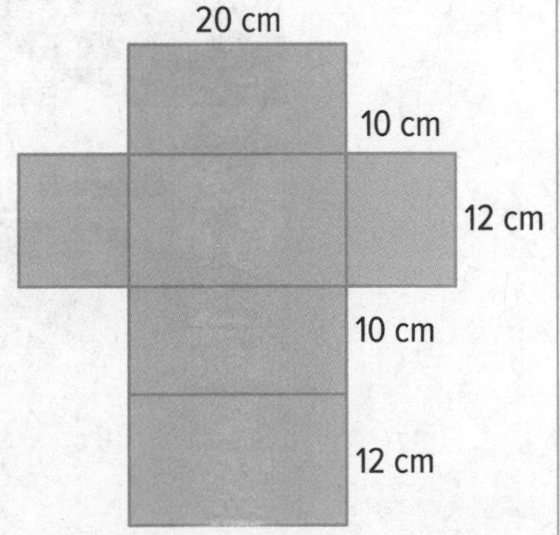

Apply

4. Jing is putting a special restorative stain on the entire surface of her rectangular prism shaped hope chest, except for her name plate that measures $\frac{1}{2}$ foot by $\frac{3}{4}$ foot. If one can of stain covers about 35 square feet, how many cans of stain will she need to buy?

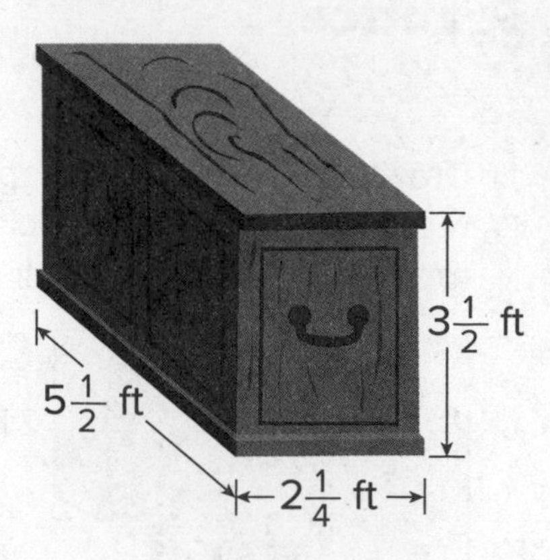

5. **Make a Conjecture** Write a formula that could be used to find the surface area of a rectangular prism. Define each variable you choose to use in your formula.

6. **Create** Draw and label a rectangular prism that has a surface area that is greater than its volume.

7. **Reason Abstractly** Find the surface area and volume of each rectangular prism shaped block. Which block has the greater surface area? Does the same block have a greater volume? Write an argument that can be used to defend your solution.

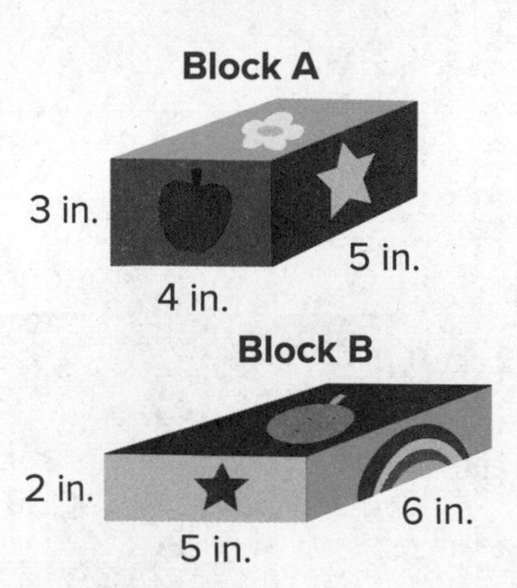

8. Meredith is painting rectangular prisms like the one shown. If she covers all the surfaces, how many square inches need to be painted? Describe two different ways to solve the problem.

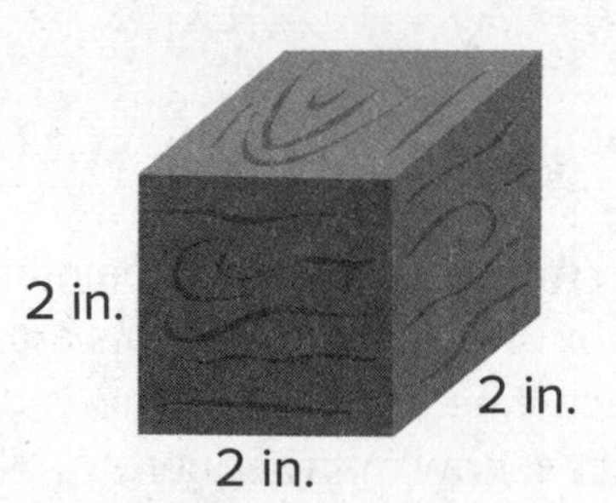

Surface Area of Triangular Prisms

I Can... create a net to represent a triangular prism and use the net to find the surface area of the prism.

What Vocabulary Will You Learn?
triangular prism

Explore Non-Rectangular Prism Nets

Online Activity You will use Web Sketchpad to explore the relationship between the shape of the base of a prism and the number of faces in the prism.

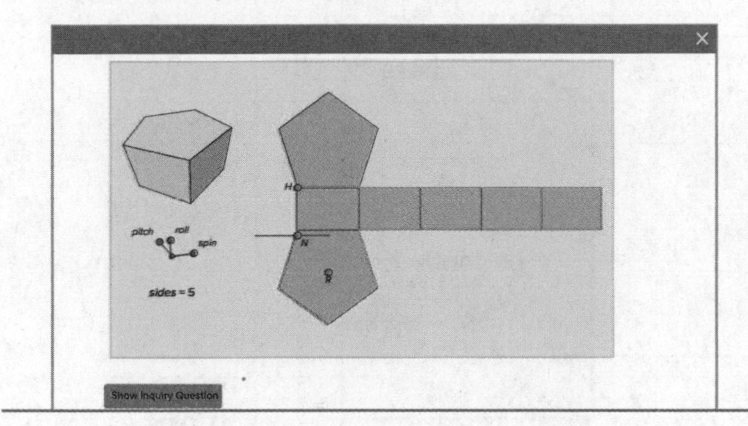

Learn Make a Net to Represent a Triangular Prism

A **triangular prism** is a prism that has triangular bases. The net of a right triangular prism is composed of two congruent triangles, called the bases, and three rectangles, which are the faces or sides.

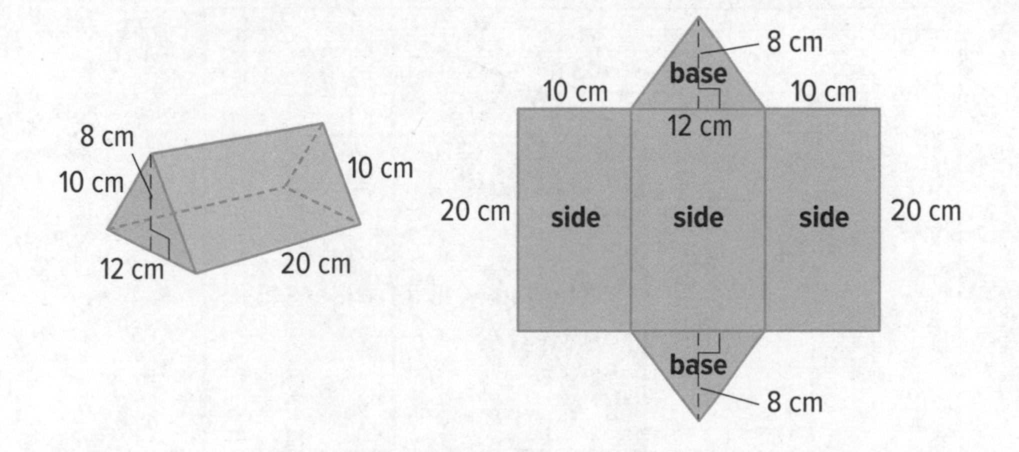

Compare and contrast the net of a rectangular prism and the net of a triangular prism.

Your Notes

Think About It!

How can the prism be unfolded to make a two-dimensional net?

Example 1 Make a Net to Represent a Triangular Prism

Draw and label a net to represent the triangular prism.

Step 1 Draw and label the bottom rectangular face and the two triangular bases.

Let each grid unit represent 1 foot. The bottom face is a rectangle with side lengths of 12 feet and 15 feet. The bases are triangles that have a base length of 12 feet and a height of 5 feet.

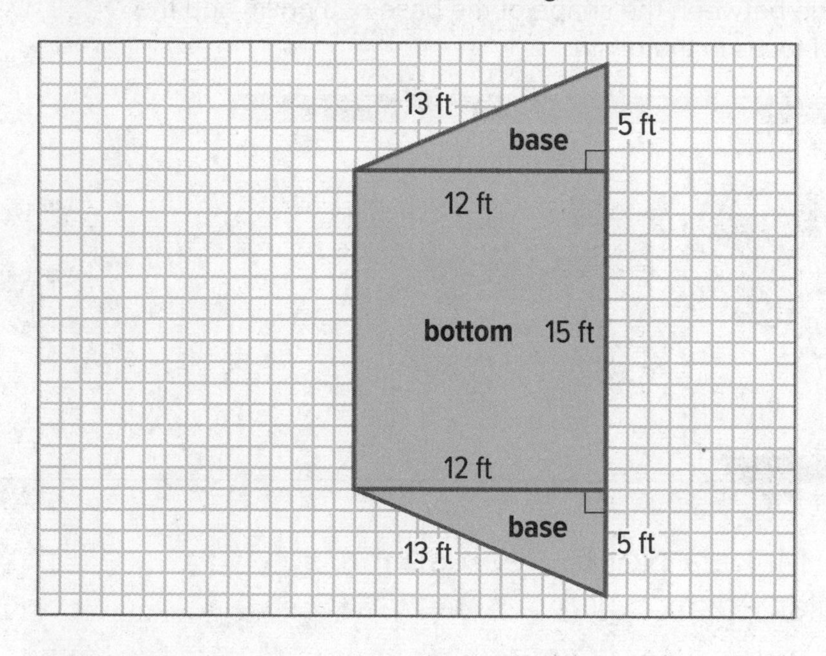

Step 2 Draw and label the side rectangular face.

The side face is a rectangle with side lengths of 15 feet and 5 feet.

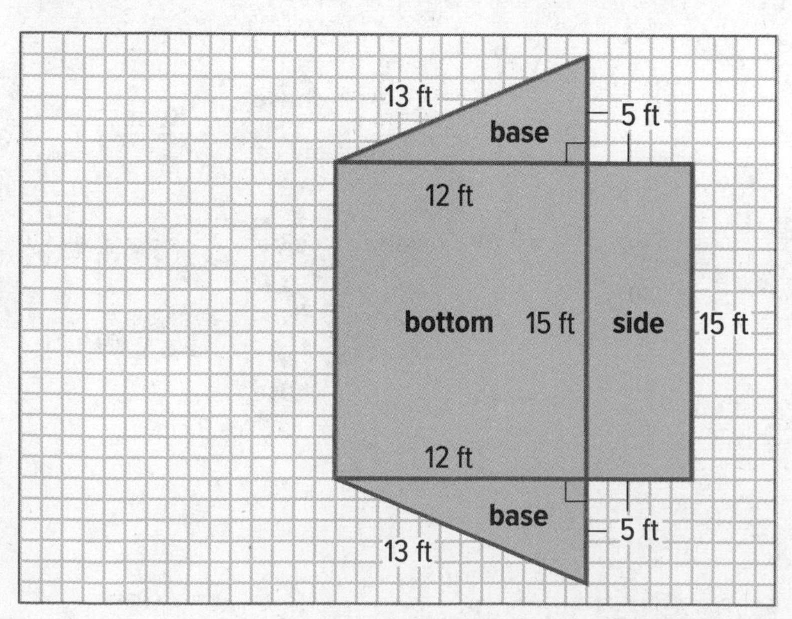

(continued on next page)

Step 3 Draw and label the remaining rectangular face.

The top face is a rectangle with side lengths of 15 feet and 13 feet.

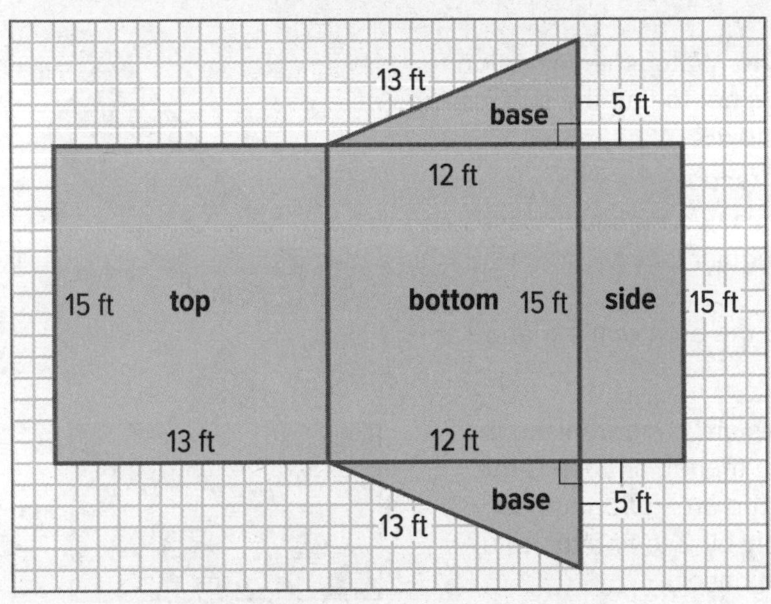

Check

Draw and label a net to represent the triangular prism. Let each grid unit represent 1 centimeter.

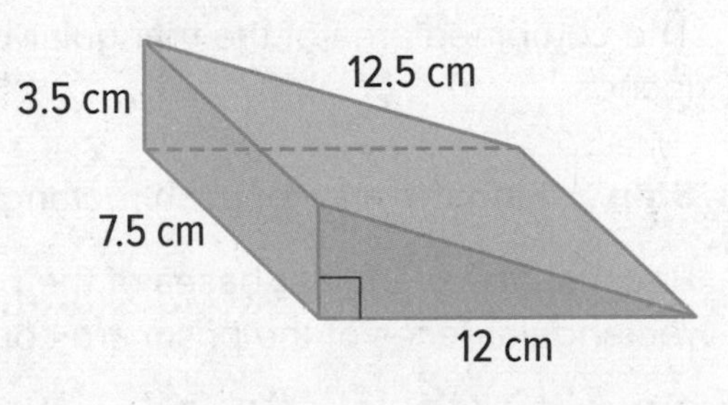

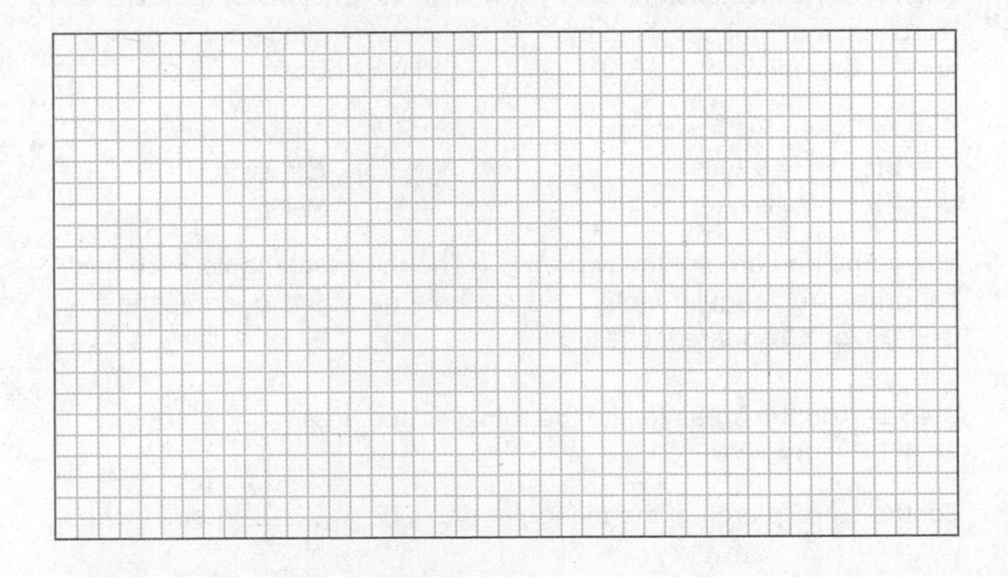

Go Online You can complete an Extra Example online.

Talk About It!

A rectangular prism has pairs of faces that have the same dimensions. This triangular prism has three rectangular faces that have different dimensions. Explain why there are no pairs of faces with the same dimensions in this prism.

Learn Surface Area of a Triangular Prism

You can use the net of a prism to find the surface area of the prism.

Go Online Watch the animation to learn how to use a net to find the surface area of the prism shown.

The prism has two triangular bases and three rectangular faces.

Step 1 Find the area of the triangular bases.

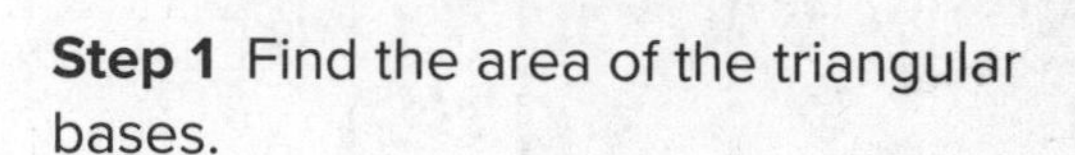

The triangles are congruent, so the area of each triangular base is the same. Find the area of one base. Then multiply by 2 to find the total area of both bases.

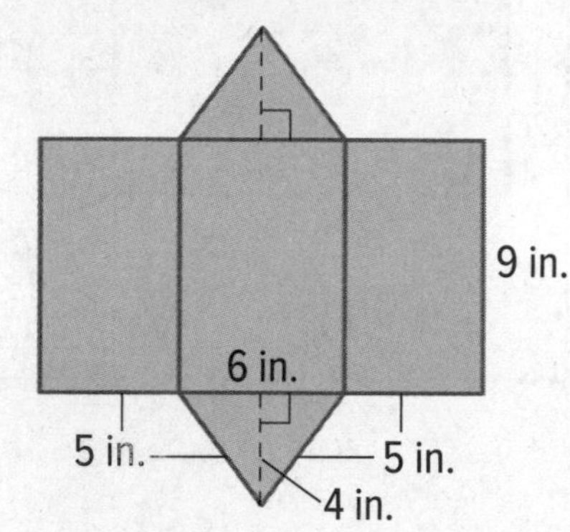

$A = \frac{1}{2}bh$	Area of a triangle
$A = \frac{1}{2}(6)(4)$	$b = 6$ and $h = 4$
$A = 12$	Multiply.

The combined area of the triangular bases is 2(12), or 24 square inches.

Talk About It!
Because the bases are isosceles triangles, two of the three rectangular faces are congruent. Is there a way that all three rectangular faces could be congruent? Explain.

Step 2 Find the area of each rectangular face.

Because the triangular bases of the prism are isosceles, two of the rectangular faces of the prism are congruent.

2 Congruent Rectangular Faces
Each face has dimensions of 9 inches and 5 inches. Find the area of one face.

$A = \ell w$	Area of a rectangle
$= 9(5)$	$\ell = 9$ and $w = 5$
$= 45$	Multiply.

The combined area of the two congruent rectangular faces is 2(45), or 90 square inches.

3rd Rectangular Face
The dimensions of the third rectangular face are 9 inches and 6 inches.

$A = \ell w$	Area of a rectangle
$= 9(6)$	$\ell = 9$ and $w = 6$
$= 54$	Multiply.

The third rectangular face has an area of 54 square inches.

Step 3 Add the areas to find the total surface area.
$24 + 90 + 54 = 168$

So, the surface area of the triangular prism is ______ square inches.

Example 2 Surface Area of a Triangular Prism

Use the net to find the surface area of the triangular prism.

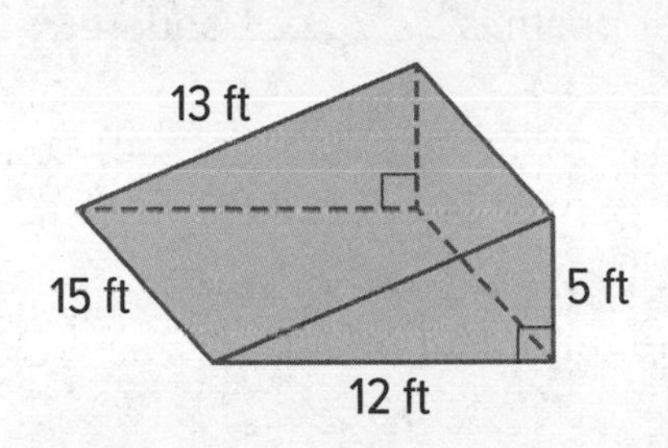

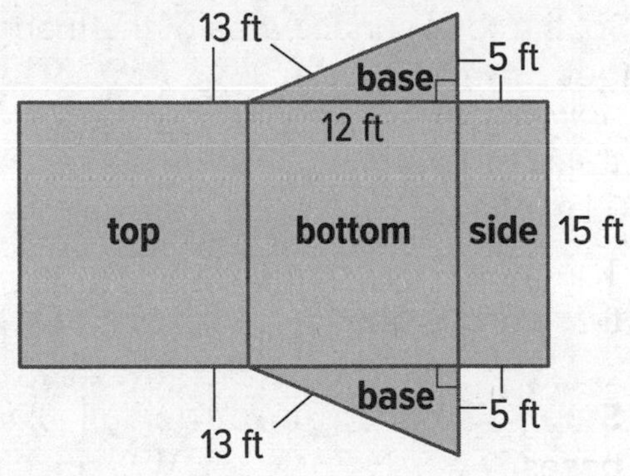

> **Think About It!**
> What shapes are the faces and bases? What formulas can you use to find the area of each face and base?

Step 1 Find the area of the triangular bases.

The triangles are congruent, so the area of each triangular base is the same. Find the area of one base. Then multiply by 2 to find the total area of both bases.

$A = \frac{1}{2}bh$ Area of a triangle

$A = \frac{1}{2}(12)(5)$ $b = 12$ and $h = 5$

$A = 30$ Multiply.

The combined area of the triangular bases is 2(30), or 60 square feet.

Step 2 Find the area of each rectangular face.

Because the triangular bases of the prism are scalene, all three rectangular faces have different dimensions.

Bottom

The length ℓ of the bottom face is 12 feet and the width w is 15 feet.

$A = \ell w$ Area of a rectangle

$= 12(15)$ $\ell = 12$ and $w = 15$

$= 180$ Multiply.

The area of the bottom face is 180 square feet.

Top

The length ℓ of the top face is 13 feet and the width w is 15 feet.

$A = \ell w$ Area of a rectangle

$= 13(15)$ $\ell = 13$ and $w = 15$

$= 195$ Multiply.

The area of the top face is 195 square feet.

Side

The length ℓ of the side face is 15 feet and the width w is 5 feet.

$A = \ell w$ Area of a rectangle

$= 15(5)$ $\ell = 15$ and $w = 5$

$= 75$ Multiply.

The area of the side face is 75 square feet. *(continued on next page)*

Talk About It!

Explain why the bases of the triangular prism have the same area.

Step 3 Add the areas to find the total surface area.

$60 + 180 + 195 + 75 = 510$

So, the total surface area of the triangular prism is ________ square feet.

Check

Use the net to find the surface area of the triangular prism.

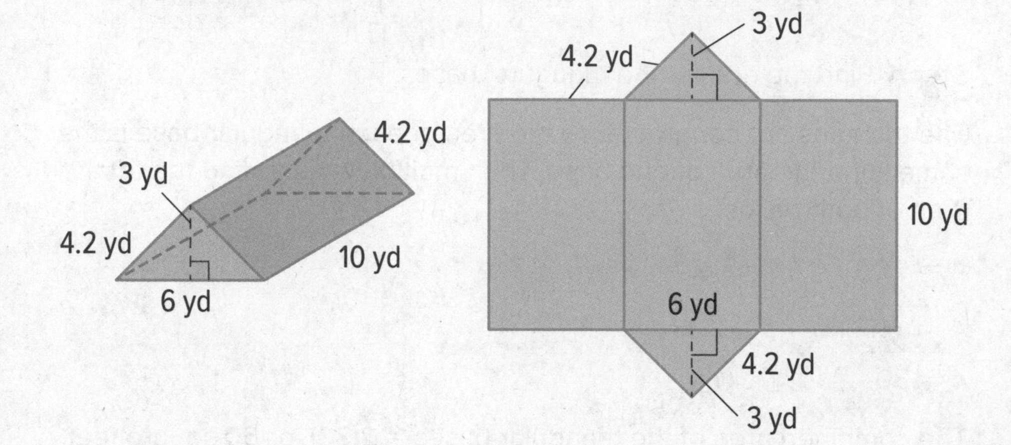

Go Online You can complete an Extra Example online.

Example 3 Find Surface Area of a Triangular Prism

Use the net to find the surface area of the triangular prism.

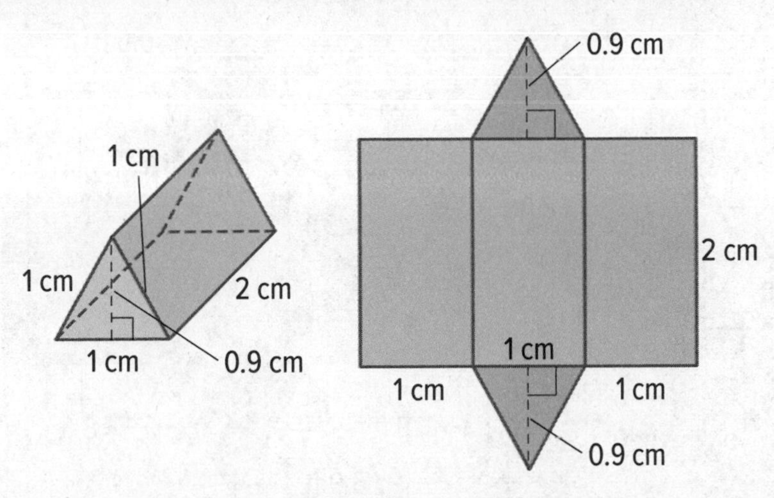

Step 1 Find the area of the triangular bases.

The triangles are congruent, so the area of each triangular base is the same. Find the area of one base. Then multiply by 2 to find the total area of both bases.

$A = \frac{1}{2}bh$ Area of a triangle

$A = \frac{1}{2}(1)(0.9)$ $b = 1$ and $h = 0.9$

$A = 0.45$ Multiply.

The combined area of the triangular bases is 2(0.45), or 0.9 square centimeter.

Step 2 Find the area of each rectangular face.

Because the triangular bases of the prism are equilateral, the rectangular faces of the prism are congruent.

The length ℓ of each rectangular face is 2 centimeters and the width w is 1 centimeter.

$A = \ell w$ Area of a rectangle
$= 2(1)$ $\ell = 2$ and $w = 1$
$= 2$ Multiply.

The combined area of the three rectangular faces is 3(2), or 6 square centimeters.

Step 3 Add the areas to find the total surface area.

$0.9 + 6 = 6.9$

So, the total surface area of the triangular prism is ________ square centimeters.

Think About It!

How many different-sized faces are there?

Talk About It!

Explain why a prism with an equilateral triangle base has only two sets of congruent faces.

Check

Use the net to find the surface area of the triangular prism.

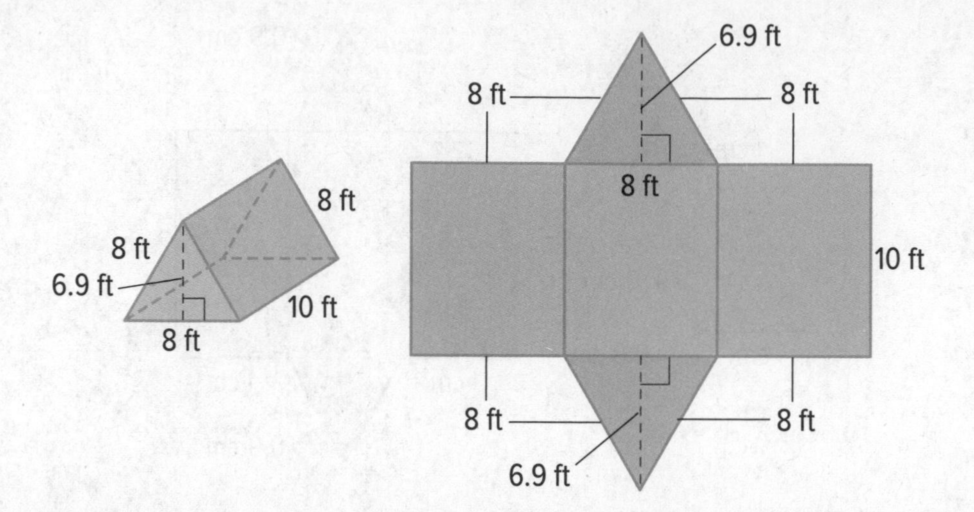

Go Online You can complete an Extra Example online.

Apply Food

The Flying Pizza food truck serves their individual slices of pizza in boxes that are shaped like triangular prisms. The box for a small piece of pizza costs $0.25 to make and the box for the large piece costs $0.32 to make. Which box has the greater cost per square inch?

Large

9 in.
6 in.
1 in.
9.5 in.

Small

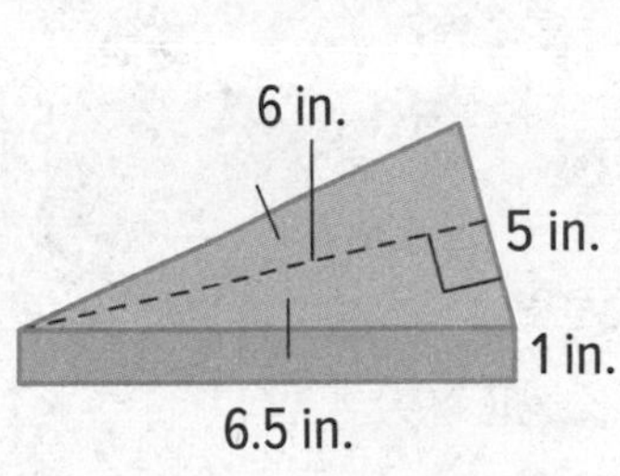

Go Online Watch the animation.

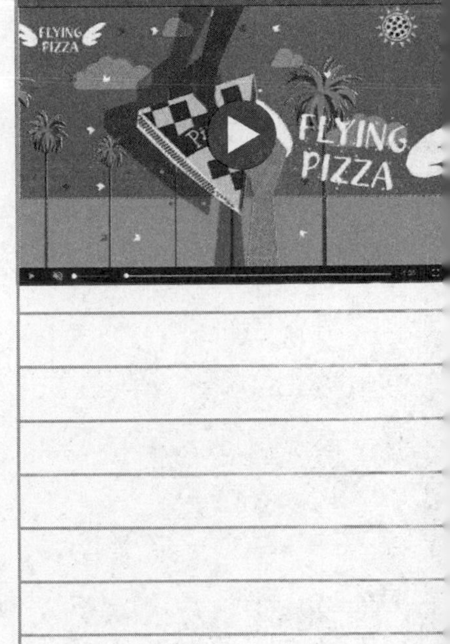

1 What is the task?

Make sure you understand exactly what question to answer or problem to solve. You may want to read the problem three times. Discuss these questions with a partner.

First Time Describe the context of the problem, in your own words.
Second Time What mathematics do you see in the problem?
Third Time What are you wondering about?

2 How can you approach the task? What strategies can you use?

3 What is your solution?

Use your strategy to solve the problem.

Talk About It!

Why is it important to find the surface area of each box first?

4 How can you show your solution is reasonable?

Write About It! Write an argument that can be used to defend your solution.

Check

The dimensions of two climbing walls that are in the middle of an obstacle course are shown. How much greater is the surface area of Wall B than Wall A?

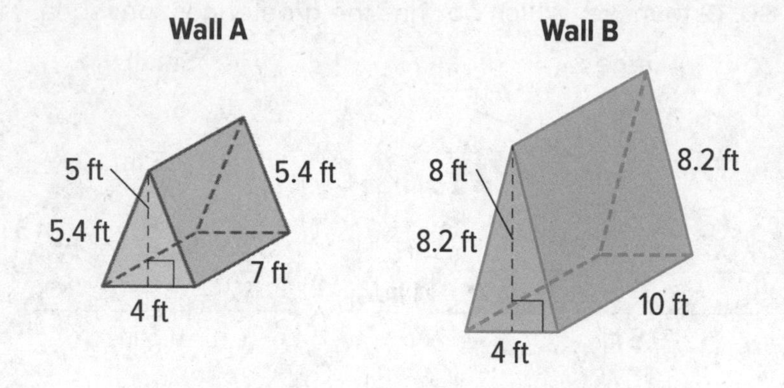

Go Online You can complete an Extra Example online.

Foldables It's time to update your Foldable, located in the Module Review, based on what you learned in this lesson. If you haven't already assembled your Foldable, you can find the instructions on page FL1.

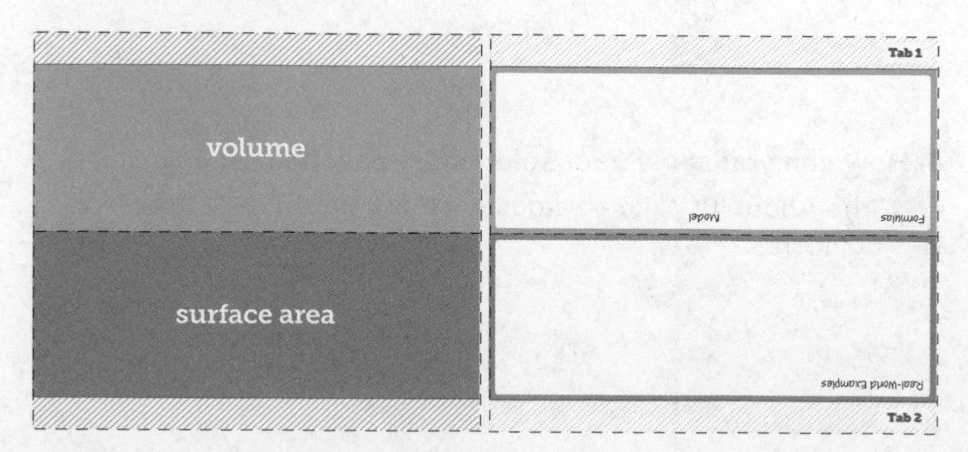

Name ______________________ Period ____________ Date ____________

Practice

Go Online You can complete your homework online.

1. Draw and label a net to represent the triangular prism. Let each grid unit represent 1 foot. **(Example 1)**

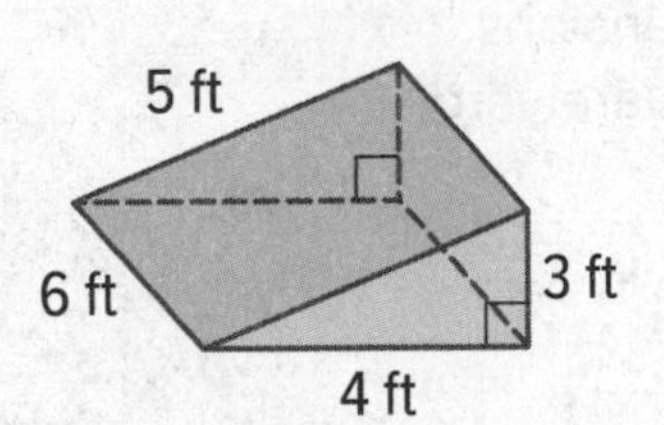

2. Use the net to find the surface area of the triangular prism. **(Example 2)**

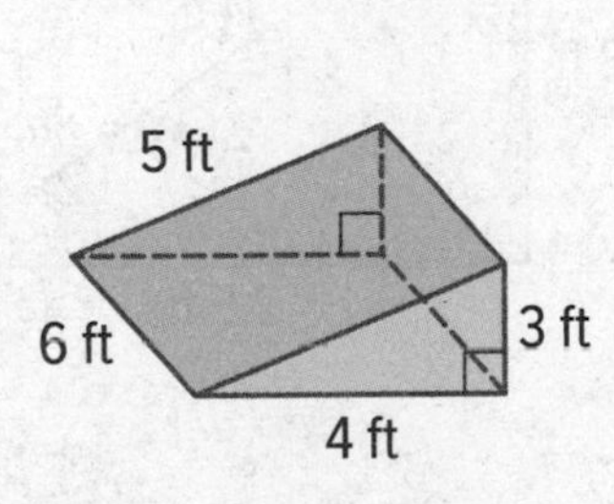

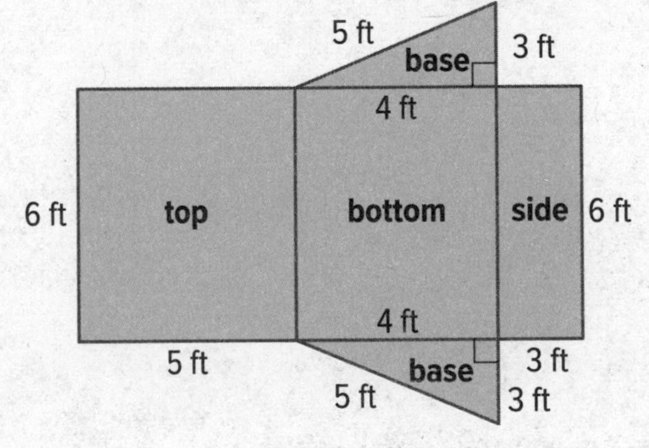

Test Practice

3. **Open Response** Use the net to find the surface area of the triangular prism in square meters. **(Example 3)**

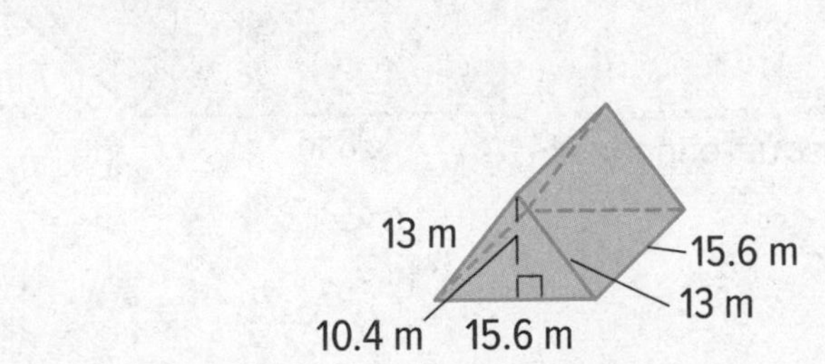

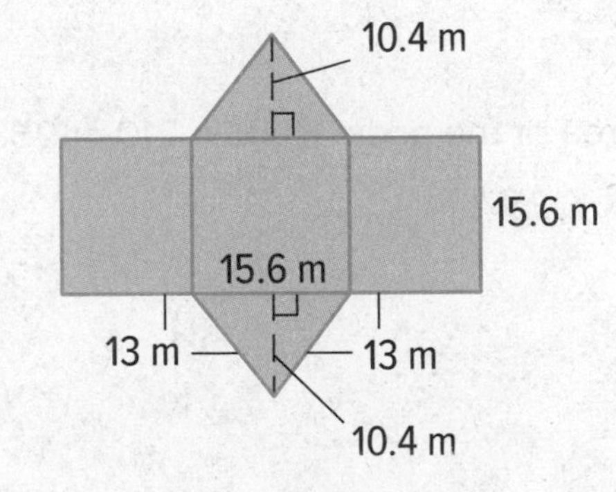

Apply

4. Mr. Saldivar is building a ramp in the shape of a triangular prism with the dimensions shown. Sheets of plywood are 8 feet long and 4 feet wide. What is the minimum number of sheets of plywood he needs to buy in order to have enough to build the ramp?

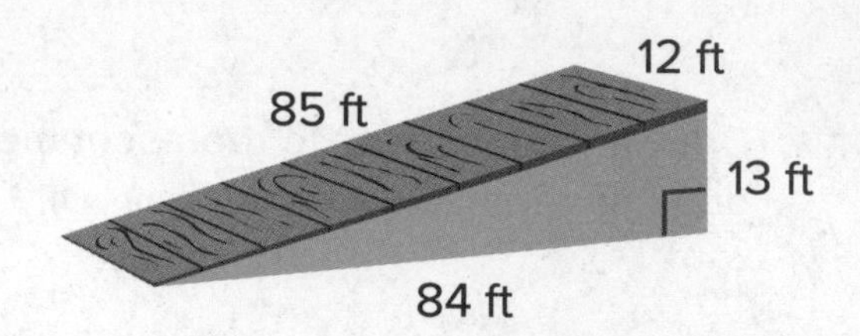

5. A tent is in the shape of the triangular prism with the dimensions shown. If the canvas to make the tent costs \$4.99 per square yard, how much will it cost for the fabric to make the tent?

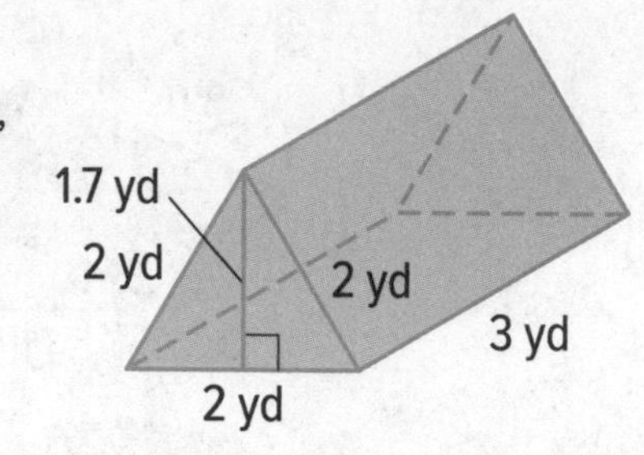

6. **Reason Abstractly** Why is the surface area of a triangular prism measured in square units rather than in cubic units? Explain.

7. Find the surface area of a triangular prism that has the base triangle shown and a prism height of 7 feet.

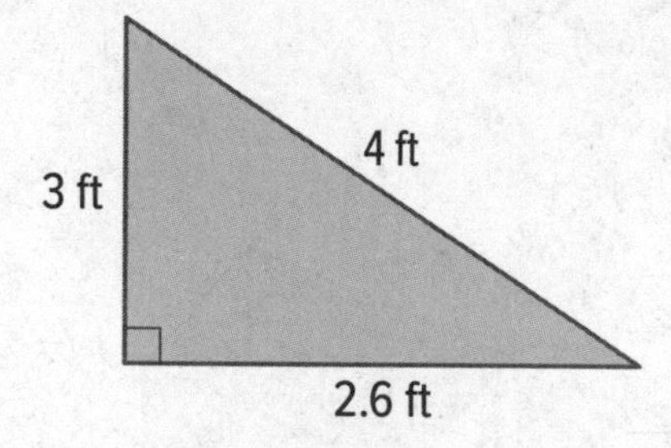

8. **Find the Error** A classmate found the surface area of the triangular prism shown. Find the error and correct it.

Area of Bases
$A = 2\left(\frac{1}{2}\right)(6)(8)$
$A = 48$

Area of Faces
$A = 3(7)(10)$
$A = 210$

The surface area of the prism is 48 + 210 or 258 square inches.

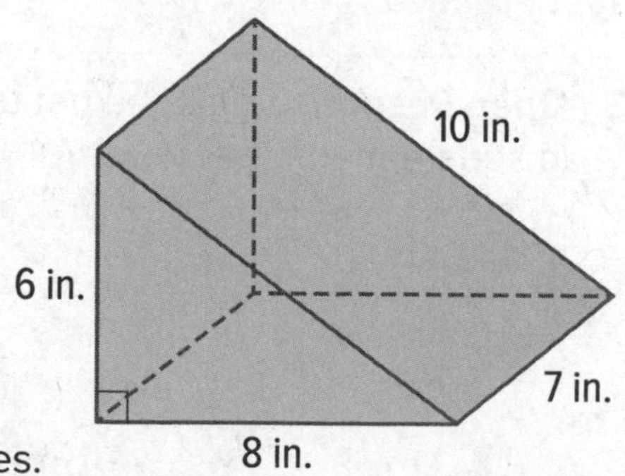

Surface Area of Pyramids

I Can... represent a triangular or square pyramid with a net made up of squares and triangles, and then use that net to find the surface area of the given figure.

What Vocabulary Will You Learn?
lateral faces
pyramid
slant height

Learn Make a Net to Represent a Pyramid

A **pyramid** is a three-dimensional figure that has one polygonal base and triangular sides that meet at a point. The sides are called **lateral faces.**

A *regular pyramid* has a base that is a regular polygon and lateral faces that are all congruent. The height of one of the lateral faces in a regular pyramid is the **slant height** of the pyramid. The slant height also divides the base of the triangular face in half, creating two congruent segments.

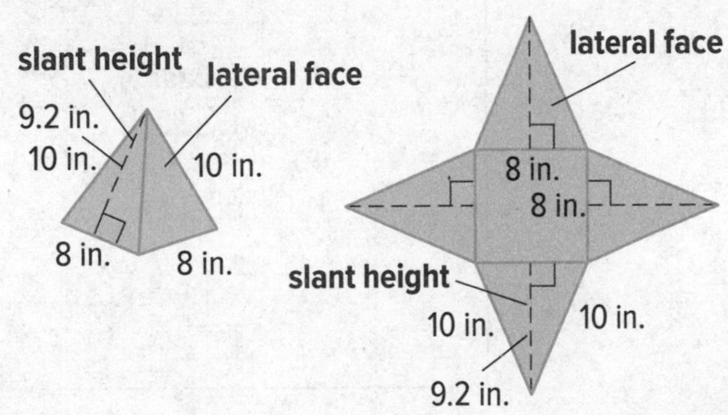

Talk About It!
Compare and contrast pyramids and prisms.

A *square pyramid* is a pyramid with a square base and four triangular faces.

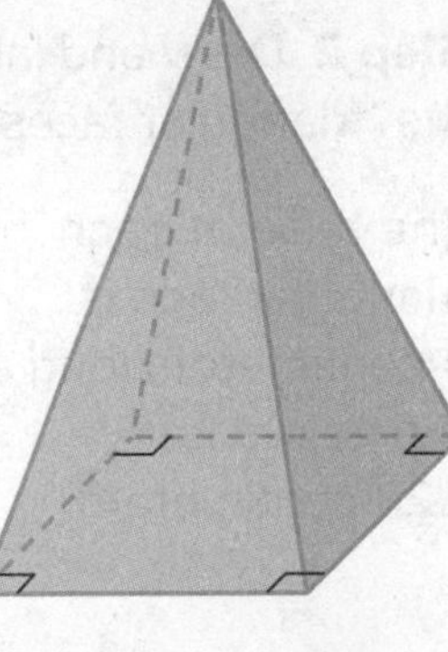
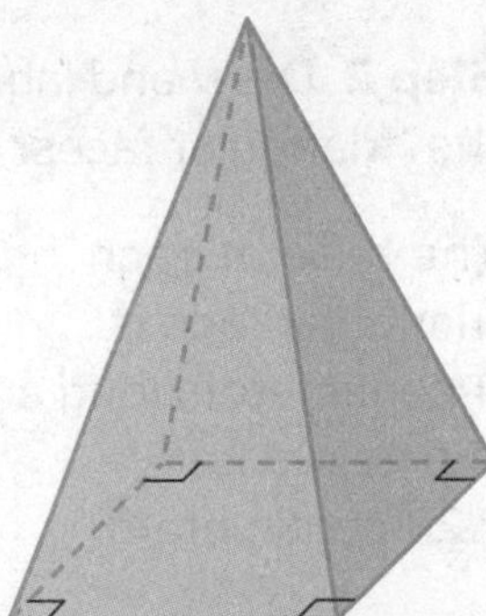

A *triangular pyramid* is a pyramid with a triangular base and three triangular faces. The base of a regular triangular pyramid is an equilateral triangle.

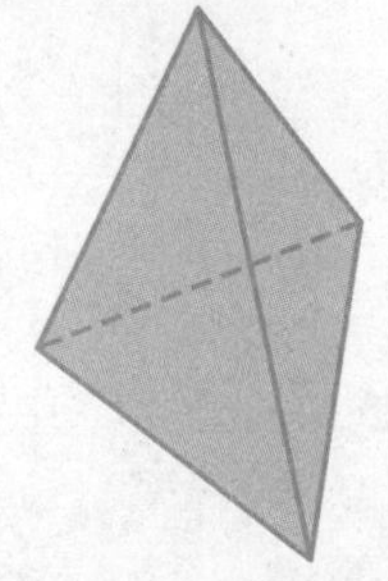

Your Notes

Think About It!

How can the pyramid be unfolded to make a two-dimensional net?

Talk About It!

Explain why the slant heights of the triangles are equal.

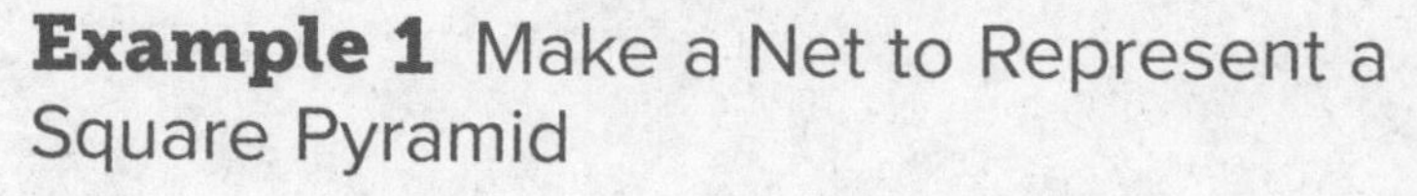

Example 1 Make a Net to Represent a Square Pyramid

Draw and label a net to represent the square pyramid.

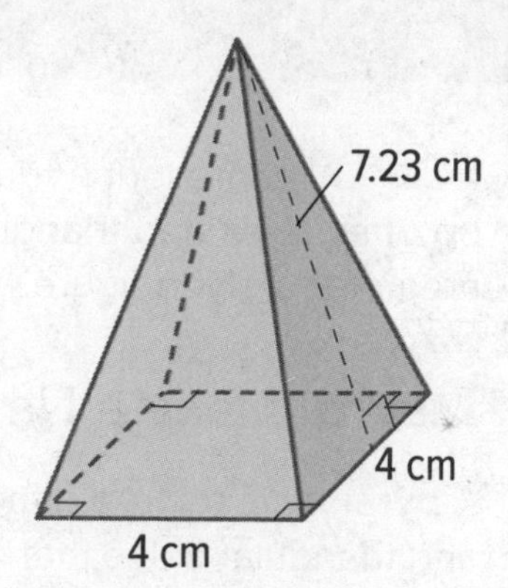

Step 1 Draw and label the square base.

The base is a square with 4-centimeter sides. Let each grid unit represent 1 centimeter.

base
4 cm

Step 2 Draw and label the triangular faces.

The base of each triangular face is 4 centimeters long and the height is 7.23 centimeters.

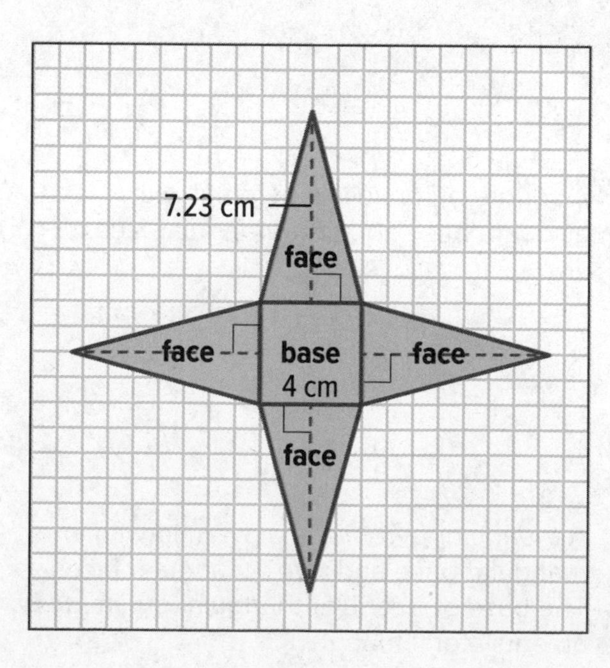

Check

Draw and label a net to represent the pyramid shown. Let each grid unit represent 1 foot.

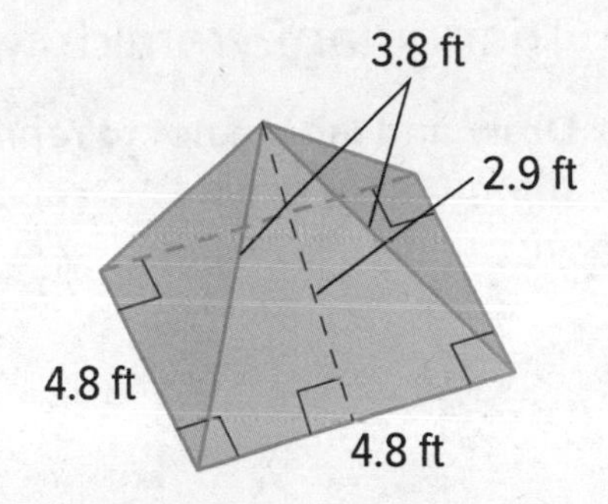

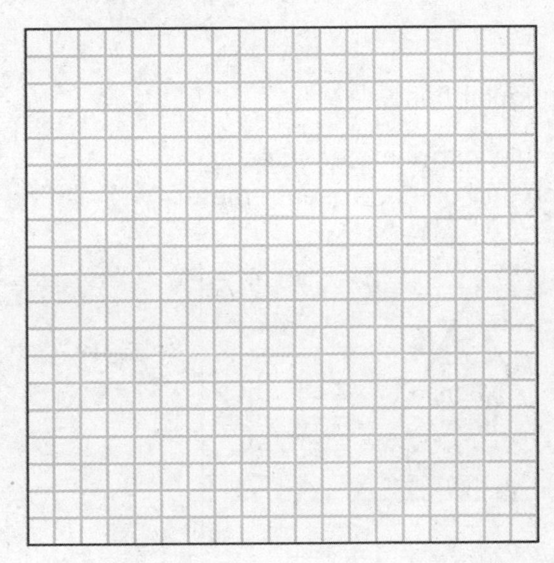

Go Online You can complete an Extra Example online.

Pause and Reflect

Draw a square pyramid in the space below, one that is different from the ones in Example 1 and Check. Trade your drawing with a partner. Draw and label a net that can be used to represent your partner's pyramid.

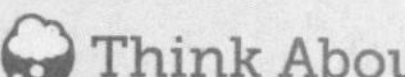

Think About It!

How can the pyramid be unfolded to make a two-dimensional net?

Example 2 Make a Net to Represent a Triangular Pyramid

Draw and label a net to represent the triangular pyramid.

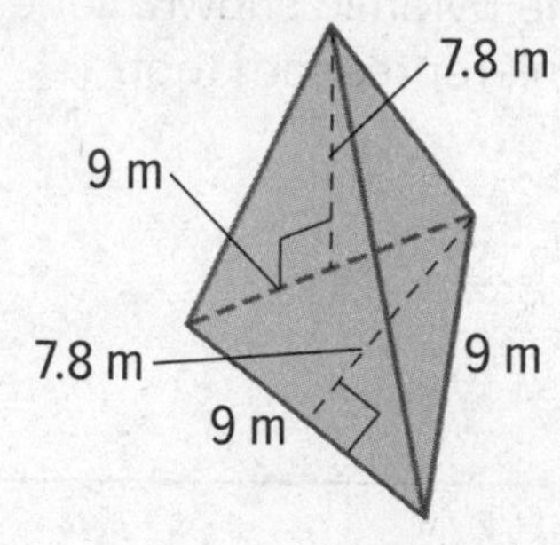

Step 1 Draw and label the triangular base.

The base is an equilateral triangle with 9-meter sides and a height of 7.8 meters. Let each grid unit represent 1 meter.

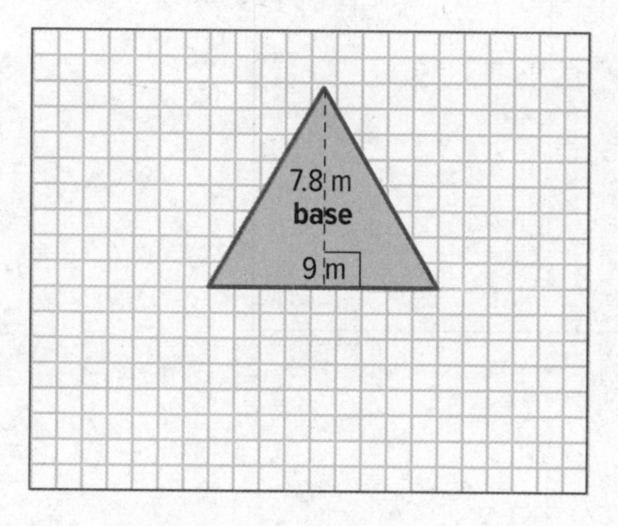

Talk About It!

What do you notice about each face and base of this pyramid? Are all triangular pyramids like this? If so, explain why. If not, give a counterexample.

Step 2 Draw and label the lateral faces.

The faces are also equilateral triangles with 9-meter sides and heights of 7.8 meters.

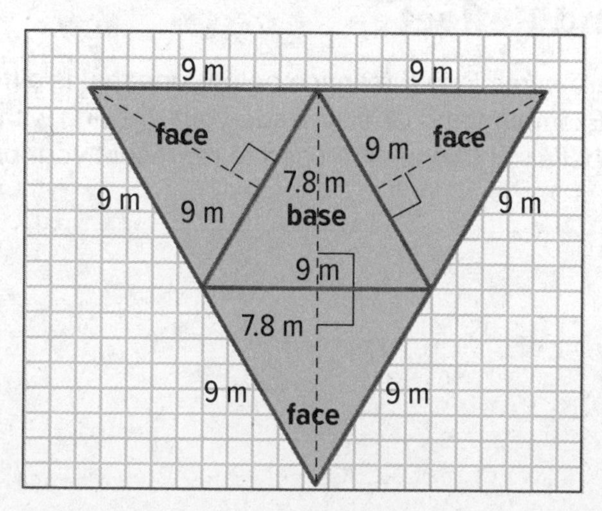

Check

Draw and label a net to represent the pyramid shown.

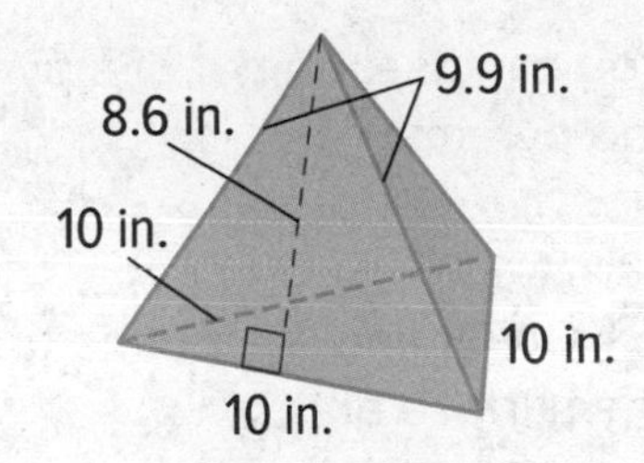

Go Online You can complete an Extra Example online.

Pause and Reflect

Draw a triangular pyramid in the space below, one that is different from the ones in Example 2 and Check. Trade your drawing with a partner. Draw and label a net that can be used to represent your partner's pyramid.

Learn Surface Area of a Pyramid

You can use the net of a pyramid to find the surface area of the pyramid.

Go Online Watch the animation to learn how to use a net to find the surface area.

You can use a net to find the surface area of the pyramid shown.

The pyramid has a square base and four triangular lateral faces.

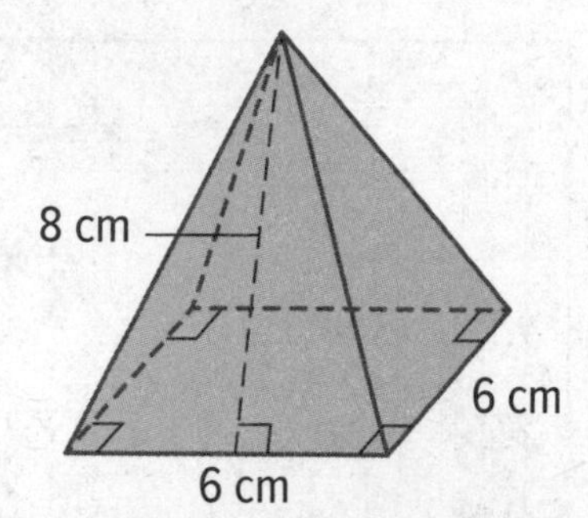

Step 1 Find the area of the square base.

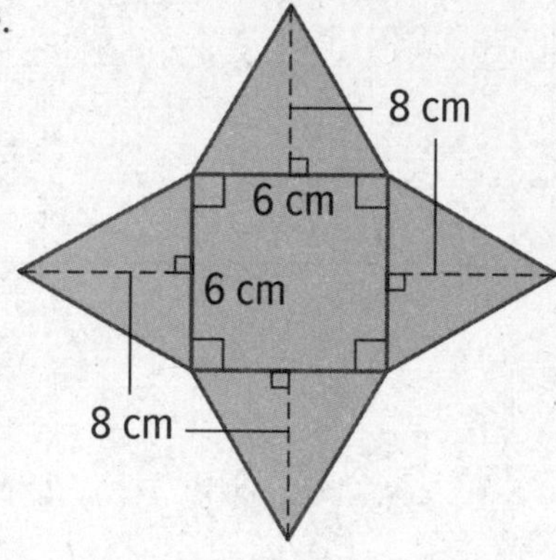

$A = s^2$	Area of a square
$A = (6)^2$	Replace s with 6.
$A = 36$	Simplify.

The area of the base is 36 square centimeters.

Talk About It!

If the base of a pyramid was a regular octagon, how many lateral faces would the pyramid have? Would they all be congruent? Explain.

Step 2 Find the area of each lateral face.

The base is a square, so the area of each triangular face is the same.

$A = \frac{1}{2}bh$	Area of a triangle
$A = \frac{1}{2}(6)(8)$	Replace b with 3 and h with 8.
$A = 24$	Multiply.

The combined area of the four lateral faces is 4(24), or 96 square centimeters.

Step 3 Add the areas to find the total surface area.

$36 + 96 = 132$

So, the total surface area of the pyramid is ________ square centimeters.

Example 3 Find Surface Area of a Square Pyramid

Use the net to find the surface area of the square pyramid.

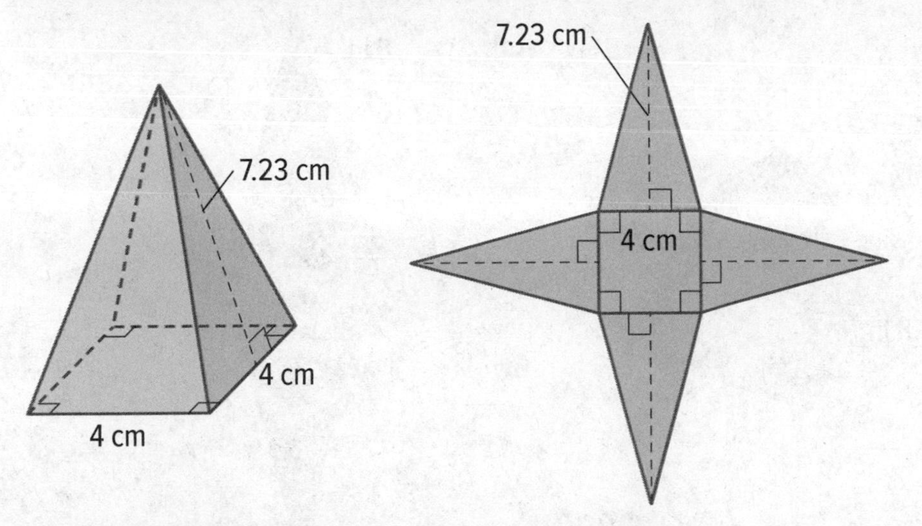

Step 1 Find the area of the square base.

The base of the pyramid is a square.

$A = s^2$	Area of a square
$A = (4)^2$	Replace s with 4.
$A = 16$	Simplify.

The area of the square base is 16 square centimeters.

Step 2 Find the area of each lateral face.

Because the base is a square, the lateral faces are congruent. The faces are congruent triangles with a base length of 4 centimeters and a height of 7.23 centimeters.

$A = \frac{1}{2}bh$	Area of a triangle
$A = \frac{1}{2}(4)(7.23)$	Replace b with 4 and h with 7.23.
$A = 14.46$	Multiply.

The combined area of the four lateral faces is 4(14.46), or 57.84 square centimeters.

Step 3 Add the areas to find the total surface area.

$16 + 57.84 = 73.84$

So, the total surface area of the square pyramid is ________ square centimeters.

Think About It!

What shapes are the different faces and base? What formulas can you use to find the area of each face and base?

Talk About It!

Can you think of a different type of pyramid where the faces are not congruent triangles? Explain its characteristics.

Check

Use the net to find the surface area of the square pyramid.

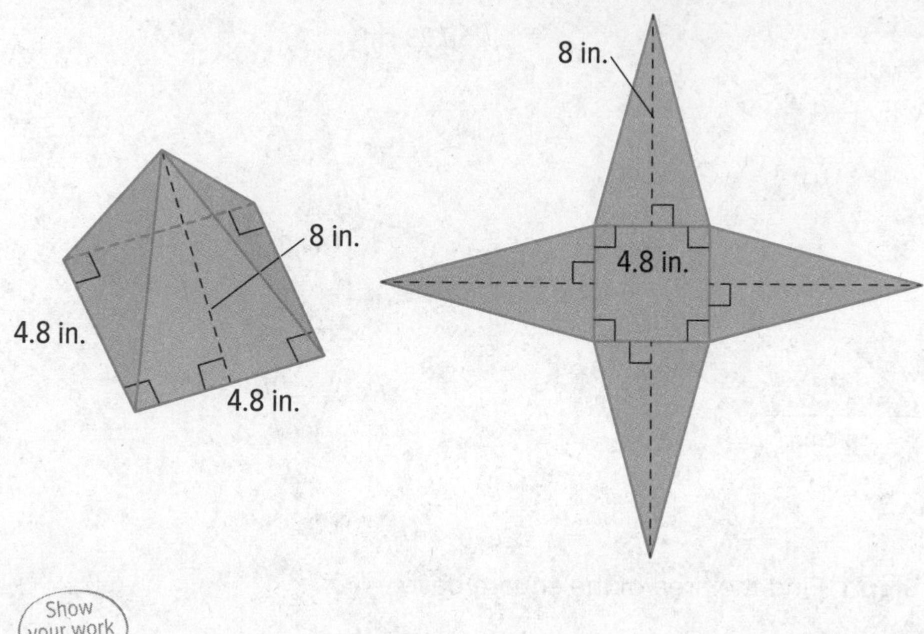

Go Online You can complete an Extra Example online.

Pause and Reflect

How might you explain how to find the surface area of a square pyramid to a classmate who is encountering difficulty? What vocabulary and steps might be important to include in your explanation?

Example 4 Find Surface Area of a Triangular Pyramid

Use the net to find the surface area of the triangular pyramid.

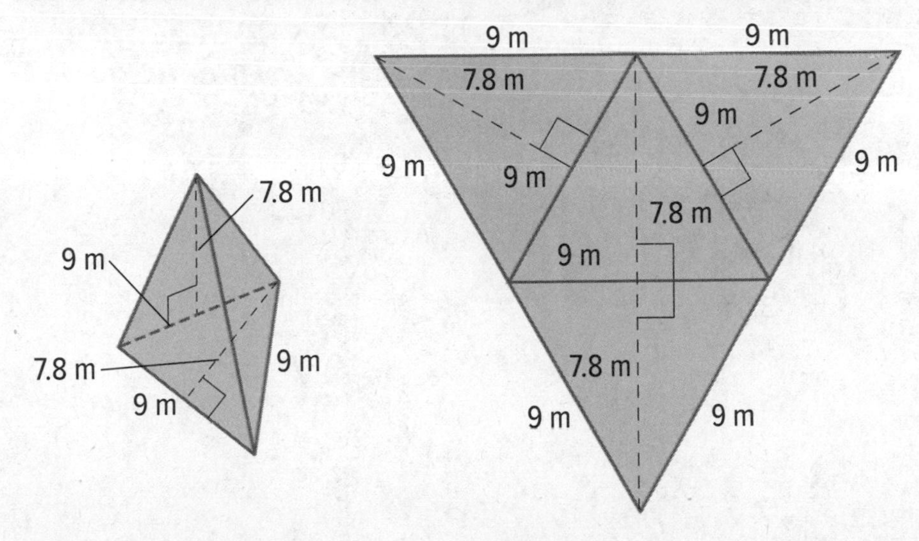

Step 1 Find the area of the triangular base.

The base is an equilateral triangle with 9-meter sides and a height of 7.8 meters.

$A = \frac{1}{2}bh$ Area of a triangle

$A = \frac{1}{2}(9)(7.8)$ Replace b with 9 and h with 7.8.

$A = 35.1$ Multiply.

The area of the triangular base is 35.1 square meters.

Step 2 Find the area of each lateral face.

Because the base is an equilateral triangle, the lateral faces are congruent. The faces are congruent triangles with 9-meter sides and heights of 7.8 meters.

$A = \frac{1}{2}bh$ Area of a triangle

$A = \frac{1}{2}(9)(7.8)$ Replace b with 9 and h with 7.8.

$A = 35.1$ Multiply.

The combined area of the three lateral faces is 3(35.1), or 105.3 square meters.

Step 3 Add the areas to find the total surface area.

$35.1 + 105.3 = 140.4$

So, the total surface area of the triangular pyramid is ________ square meters.

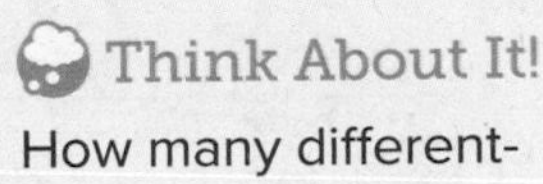

Think About It!

How many different-sized faces are there?

Talk About It!

In this pyramid, the base and the faces are all congruent triangles. Explain another way you could have solved the problem. Will this method work for all regular triangular pyramids? Explain your reasoning.

Check

Use the net to find the surface area of the triangular pyramid.

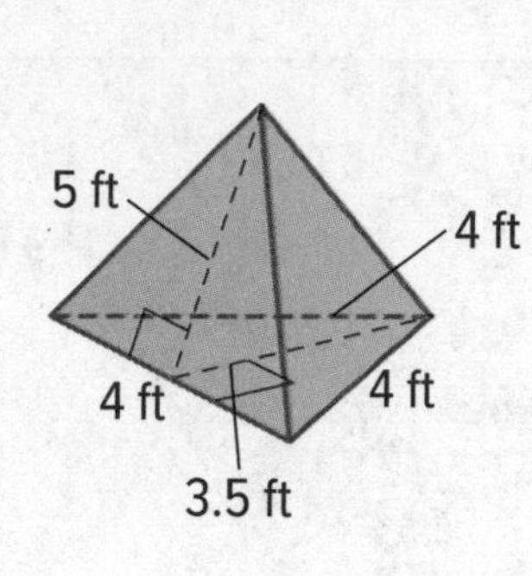

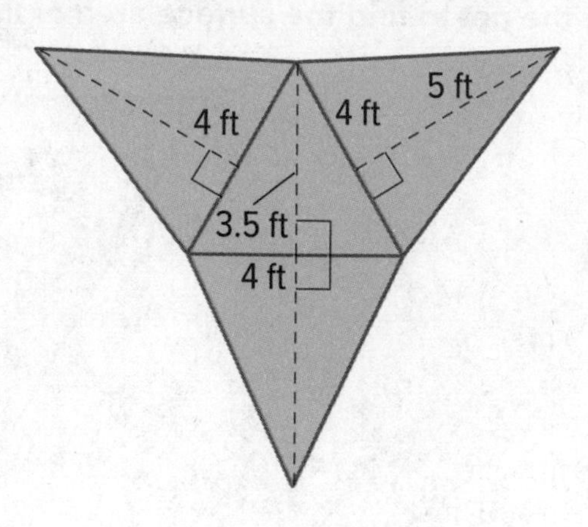

Go Online You can complete an Extra Example online.

Pause and Reflect

Did you make any errors when finding the surface area of triangular pyramids? What can you do to make sure you don't repeat that error in the future?

Apply Set Design

Morgan needs to construct three different square pyramids for the school play. The dimensions of the pyramids are shown in the table. The cost of materials to build the pyramids is $0.29 per square foot. How much will Morgan spend on materials for all three pyramids?

Pyramid	Base Edge (ft)	Height of Faces (ft)
A	2	5
B	5	12
C	3.5	9

1 What is the task?

Make sure you understand exactly what question to answer or problem to solve. You may want to read the problem three times. Discuss these questions with a partner.

First Time Describe the context of the problem, in your own words.
Second Time What mathematics do you see in the problem?
Third Time What are you wondering about?

2 How can you approach the task? What strategies can you use?

Record your observations here

3 What is your solution?

Use your strategy to solve the problem.

Show your work here

4 How can you show your solution is reasonable?

Write About It! Write an argument that can be used to defend your solution.

Talk About It!

Suppose Morgan needed to construct triangular pyramids instead of square pyramids. What information would we need to know to solve the problem?

Check

Lin is constructing three different square pyramids for a classroom display about Egypt. The dimensions of the pyramids are shown in the table. How much more surface area does the pyramid with the greatest surface area have than the pyramid with the least surface area?

Pyramid	Base Edge (in.)	Height of Faces (in.)
A	5	8.5
B	8	5
C	6	10

Go Online You can complete an Extra Example online.

Foldables It's time to update your Foldable, located in the Module Review, based on what you learned in this lesson. If you haven't already assembled your Foldable, you can find the instructions on page FL1.

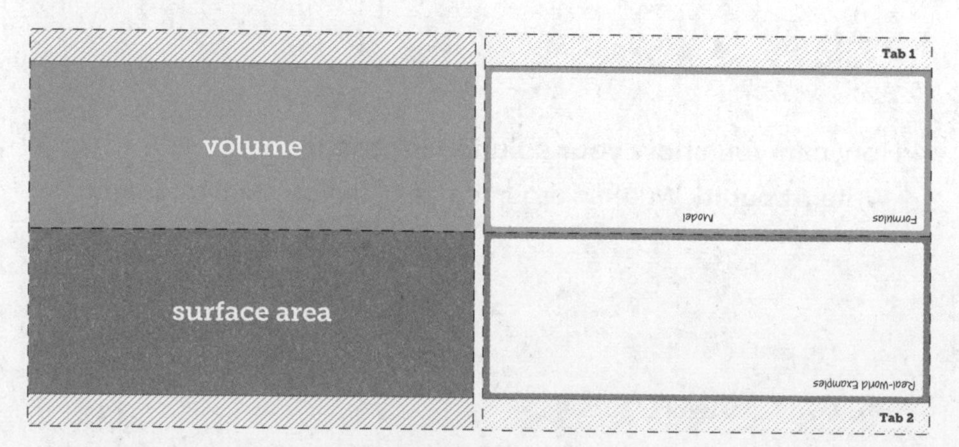

Name ______________________ Period __________ Date __________

Practice

Go Online You can complete your homework online.

1. Draw and label a net to represent the square pyramid. **(Example 1)**

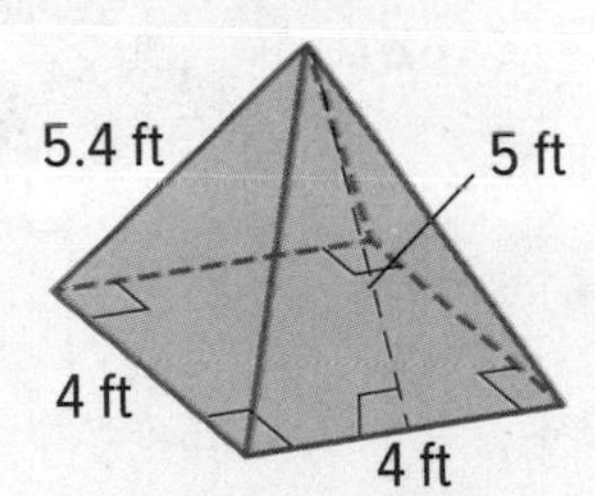

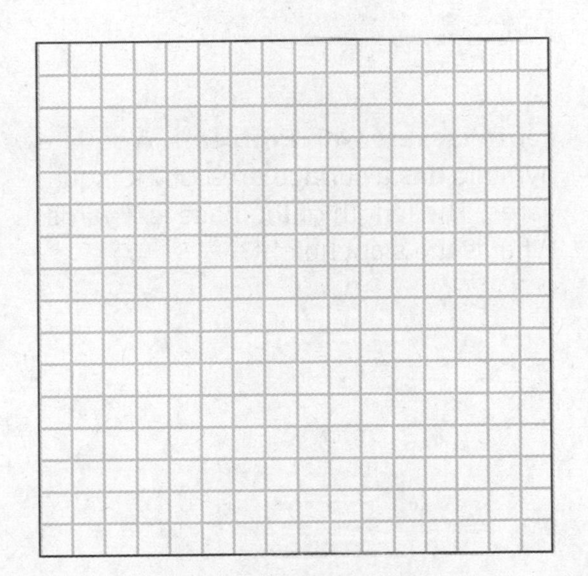

2. Draw and label a net to represent the triangular pyramid. **(Example 2)**

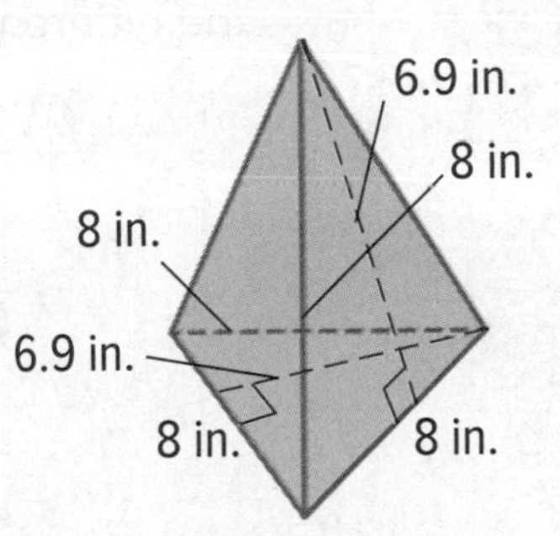

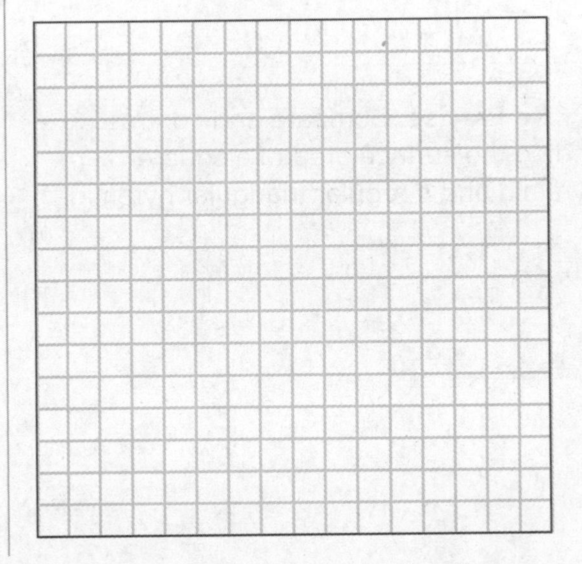

3. Use the net to find the surface area of the pyramid. **(Example 3)**

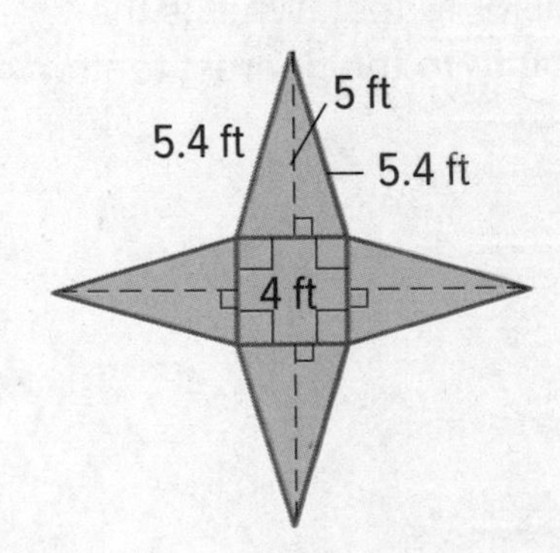

Test Practice

4. **Open Response** Use the net to find the surface area of the pyramid in square inches. **(Example 4)**

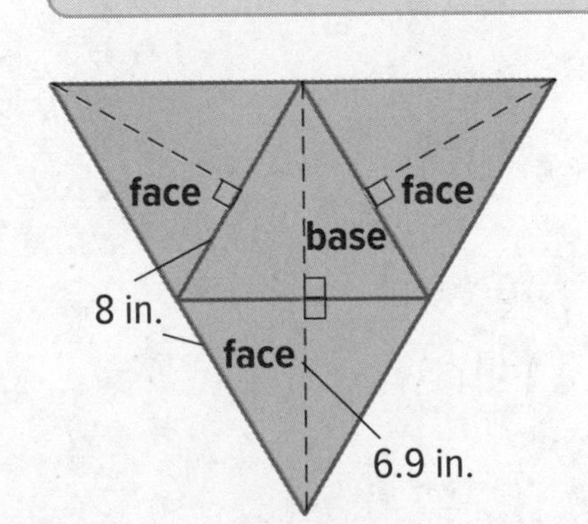

Apply

5. Mr. Potter makes two types of wooden pyramid puzzles. The base of Puzzle 1 is a square with side lengths of 5 inches and a slant height of 7 inches. Puzzle 2 is shown. If the cost of materials to build the puzzles is $0.16 per square inch, what is the difference in cost to make the puzzles?

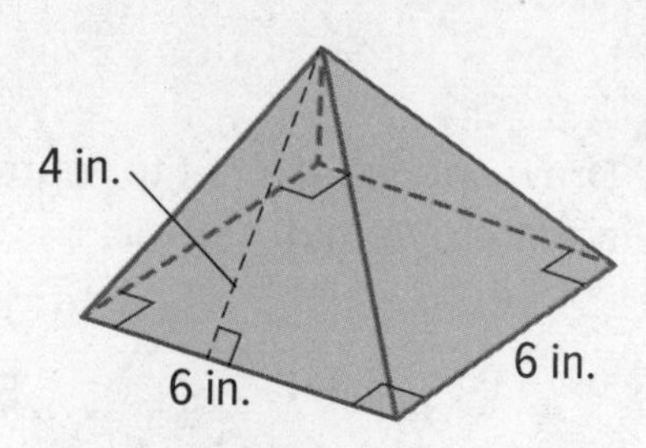

6. **MP Be Precise** Compare and contrast finding the surface area of a square pyramid and a regular triangular pyramid.

7. **MP Persevere with Problems** A square pyramid has a surface area of 210 square yards. The length of the base is 7 yards. What is the slant height?

8. **Create** Draw and label a square pyramid that has a surface area that is less than 100 square meters. Then find the surface area of the pyramid.

9. **MP Persevere with Problems** A triangular pyramid has a surface area of 174 square feet. It is made up of equilateral triangles with side lengths of 10 feet. What is the slant height? Round to the nearest tenth.

Review

Foldables Use your Foldable to help review the module.

Tab 1	
Real-World Examples	
Formulas	Model
Tab 2	

Rate Yourself!   

Complete the chart at the beginning of the module by placing a checkmark in each row that corresponds with how much you know about each topic after completing this module.

Write about one thing you learned.

____

____

____

____

Write about a question you still have.

____

____

____

____

Reflect on the Module

Use what you learned about volume and surface area to complete the graphic organizer.

Essential Question

How can you describe the size of a three-dimensional figure?

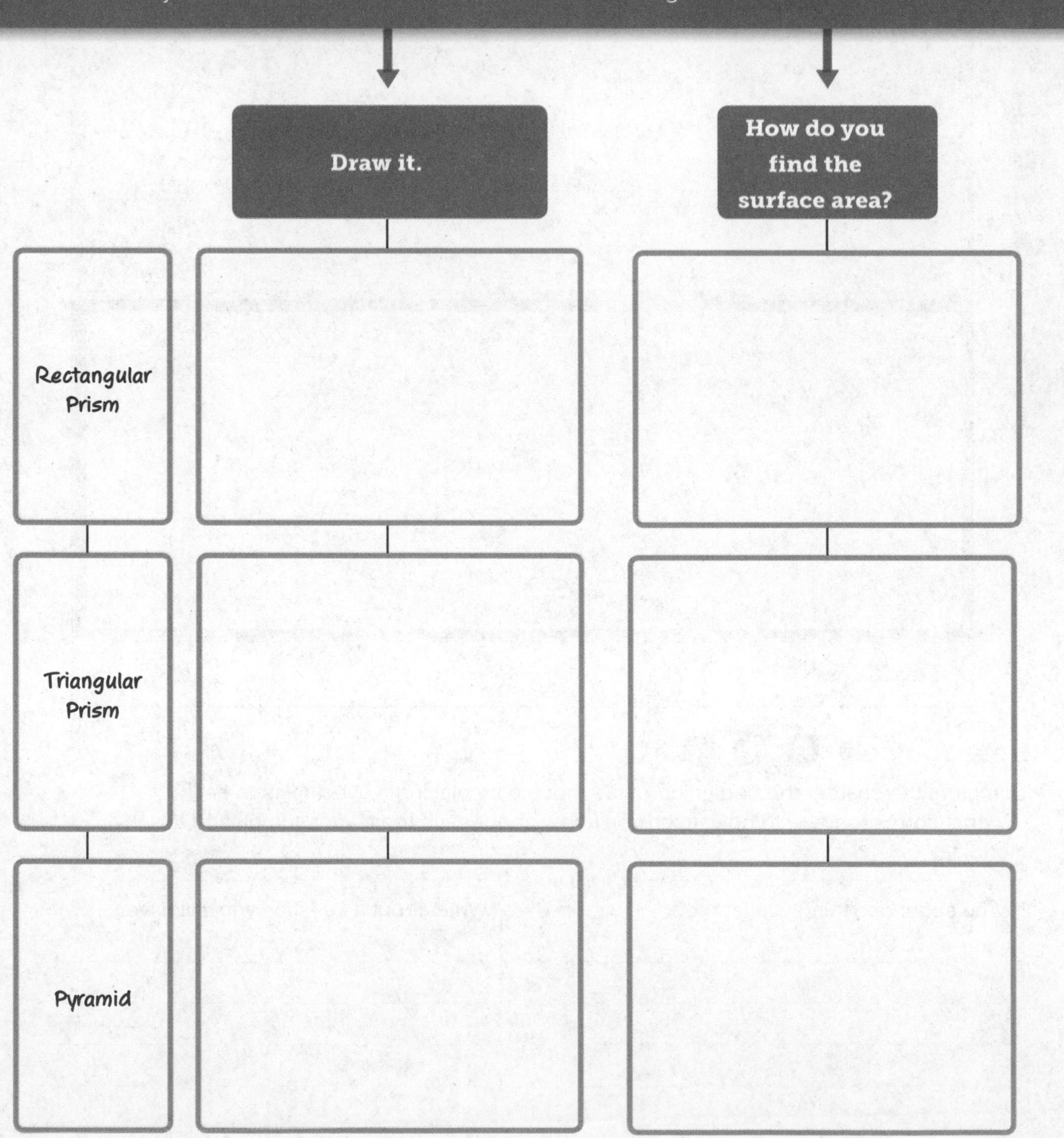

Name ______________________ Period ____________ Date ____________

Test Practice

1. Multiple Choice What is the volume of the prism? **(Lesson 1)**

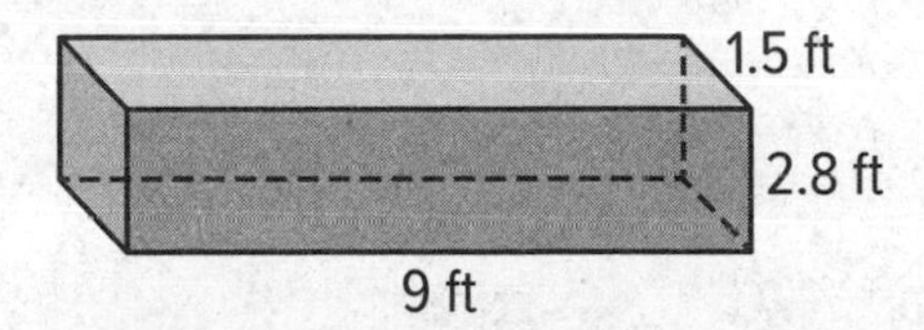

Ⓐ 13.3 ft^3

Ⓑ 13.5 ft^3

Ⓒ 25.2 ft^3

Ⓓ 37.8 ft^3

2. Open Response A grocery store offers two different-sized boxes of cereal. If the boxes are rectangular prisms, which box of cereal is the better buy? Justify your answer. **(Lesson 1)**

Box	Length (in.)	Width (in.)	Height (in.)	Price ($)
A	6	$1\frac{1}{2}$	11	2.99
B	$8\frac{1}{2}$	2	14	5.00

3. Open Response The volume of the prism shown is 520 cubic centimeters. Find the height of the prism. **(Lesson 1)**

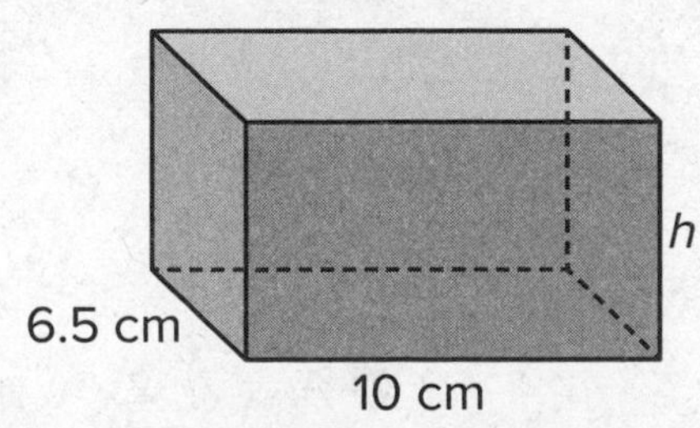

4. Open Response Use the net to find the surface area of the rectangular prism. **(Lesson 2)**

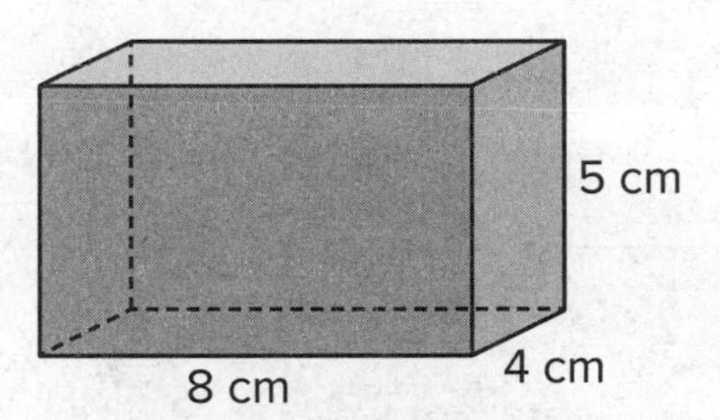

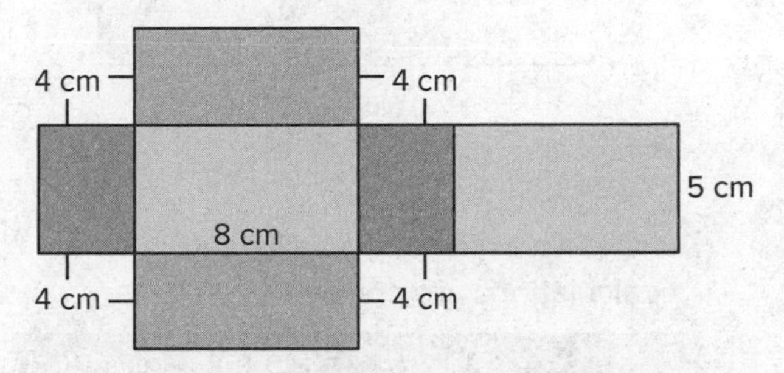

5. Multiselect Which of the following statements accurately describes the net of a rectangular prism with a length of 9 inches, a width of 4 inches, and a height of 11 inches? Select all that apply. **(Lesson 2)**

☐ The net will be made up of 4 parts, representing the top, bottom, and both sides of the rectangular prism.

☐ The net will be made up of 6 parts, representing the top, bottom, front, back, and both sides of the rectangular prism.

☐ Two parts of the net will have dimensions 4 inches by 11 inches.

☐ Two parts of the net will have dimensions 4 inches by 9 inches.

☐ Two parts of the net will have dimensions 11 inches by 13 inches.

6. Table Item Consider the prism and the net shown. **(Lesson 3)**

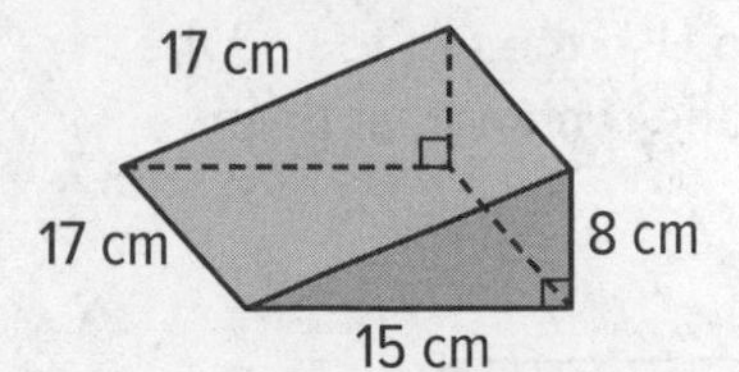

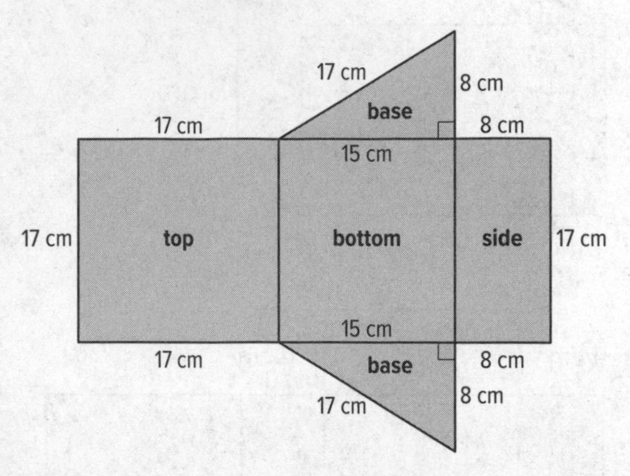

A. Indicate which of the following calculations are correct.

	Correct	Incorrect
Area of top face: 255 cm^2		
Area of bottom face: 120 cm^2		
Area of base: 70 cm^2		
Area of base: 60 cm^2		
Area of face: 136 cm^2		

B. What is the surface area of the prism?

Ⓐ 800 cm^2

Ⓑ 885 cm^2

Ⓒ 886 cm^2

Ⓓ 2,040 cm^2

7. Multiple Choice Select the net that represents the pyramid shown. **(Lesson 4)**

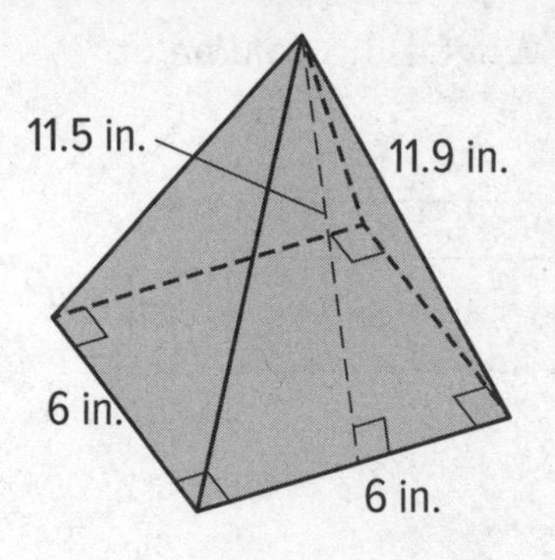

Ⓐ

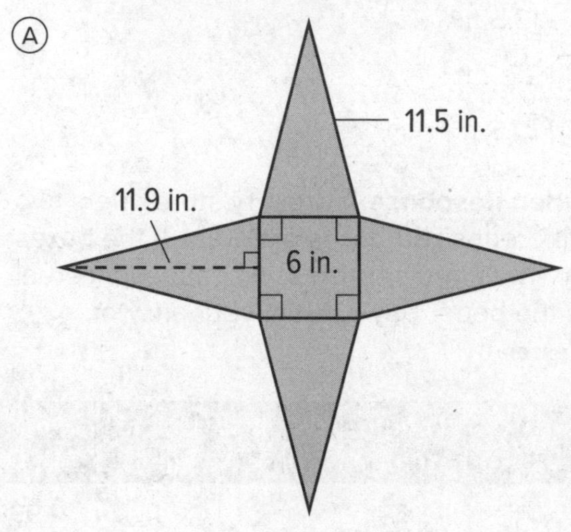

Ⓑ

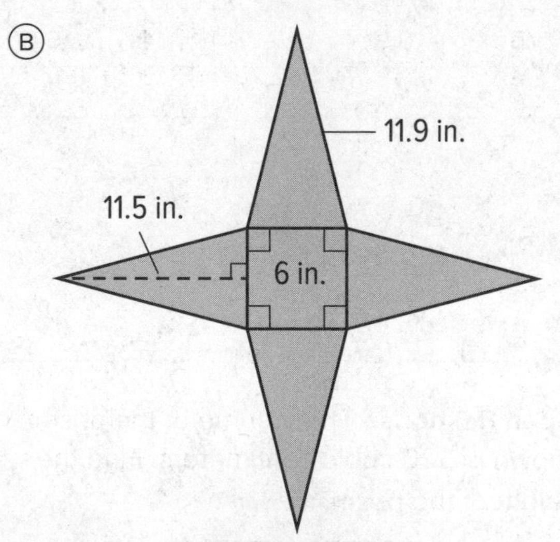

Ⓒ

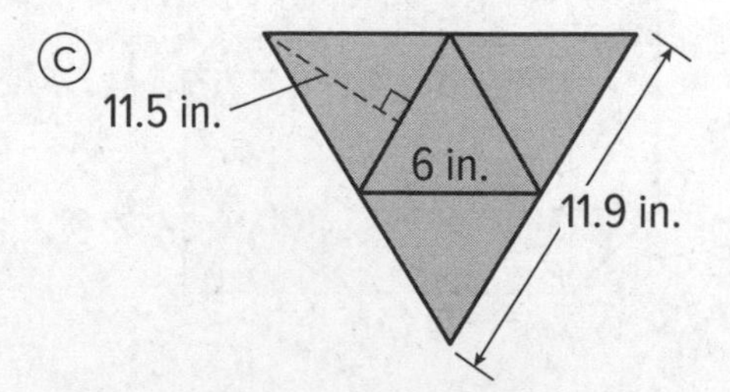

Ⓓ

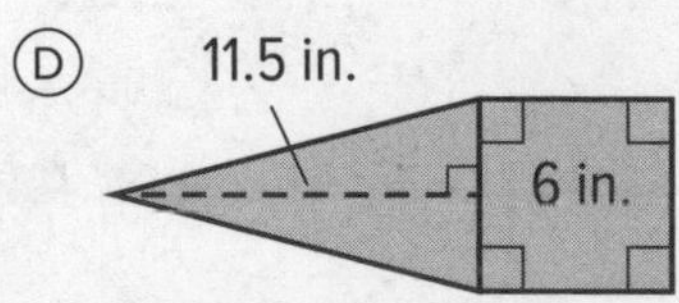

Module 10

Statistical Measures and Displays

Essential Question

Why is data collected and analyzed and how can it be displayed?

What Will You Learn?

Place a checkmark (✓) in each row that corresponds with how much you already know about each topic **before** starting this module.

KEY

□ — I don't know. ◇ — I've heard of it. ☆ — I know it!

	Before			After		
	□	◇	☆	□	◇	☆
identifying statistical questions						
displaying data in a table						
constructing dot plots						
constructing histograms						
finding the mean and median of a data set						
finding the range and interquartile range of a data set						
constructing box plots						
finding the mean absolute deviation of a data set						
identifying outliers of a data set and identifying their effect on the measures of center and variation						
interpreting the distribution of a data set						

Foldables Cut out the Foldable and tape it to the Module Review at the end of the module. You can use the Foldable throughout the module as you learn about statistical measures.

What Vocabulary Will You Learn?

Check the box next to each vocabulary term that you may already know.

- ☐ average
- ☐ box plot
- ☐ cluster
- ☐ distribution
- ☐ dot plot
- ☐ first quartile
- ☐ gap
- ☐ histogram
- ☐ interquartile range (IQR)
- ☐ mean
- ☐ mean absolute deviation
- ☐ measures of center
- ☐ measures of variation
- ☐ median
- ☐ outlier
- ☐ peak
- ☐ quartiles
- ☐ range
- ☐ second quartile
- ☐ statistical question
- ☐ statistics
- ☐ symmetric distribution
- ☐ third quartile

Are You Ready?

Study the Quick Review to see if you are ready to start this module. Then complete the Quick Check.

Quick Review

Example 1

Add rational numbers.

Find 11.83 + 8.76 + 13.28 + 16.38.

$$\begin{array}{r} 11.83 \\ 8.76 \\ 13.28 \\ +\ 16.38 \\ \hline 50.25 \end{array}$$

Add.

Example 2

Divide rational numbers.

Lydia typed 105.2 words in 4 minutes. How many words did Lydia average typing each minute?

$105.2 \div 4 = 26.3$ Divide the total number of words typed by the number of minutes.

Lydia averaged 26.3 words each minute.

Quick Check

1. Find 7.68 + 5.25 + 2.99 + 3.18.

2. A pilot flew 1,308.3 miles this week. The pilot flew the same number of miles each of 3 days this week. How many miles did the pilot fly each day?

How Did You Do?

Which exercises did you answer correctly in the Quick Check? Shade those exercise numbers at the right.

Statistical Questions

I Can... understand that a statistical question anticipates a variety of responses.

What Vocabulary Will You Learn?
statistical question
statistics

Learn Statistical Questions

Statistics involves collecting, organizing, and interpreting pieces of information, or data. One way to collect data is by asking statistical questions. A **statistical question** is a question that is answered by collecting data. Answers to a statistical question will vary based on the data collected.

The table gives some examples of statistical questions and examples that are not statistical questions.

Statistical Questions	Not Statistical Questions
How many text messages do middle school students typically send each day?	What is the height in feet of the tallest mountain in Colorado?
How many hours per night does a typical teenager spend watching television?	How many people attended last night's jazz concert?

Talk About It!
Why is *How many people attended last night's jazz concert?* not a statistical question? How can you rewrite the question so it is a statistical question?

In the table, the questions on the left are statistical questions because if you were to survey a group of students, you will likely get a variety of responses. The questions on the right are not statistical questions because each question has one specific response.

Constructing statistical questions is an important part of the process of using statistics to collect, organize, and interpret data. You will learn how to apply these steps in order to help answer a statistical question.

Step 1 Construct a statistical question.

Step 2 Use your question to collect data.

Step 3 Summarize the data using tables or graphical displays.

Step 4 Use the data to answer the statistical question.

Your Notes

Example 1 Identify Statistical Questions

Determine whether or not each question is a statistical question.

How many states are there in the United States?
This is not a statistical question, because it does not anticipate a variety of responses. There are 50 states in the United States.

How many states has the typical middle school student visited?
This is a statistical question, because it does anticipate a variety of responses. If you survey a group of students, you will likely get a variety of responses.

In what year did Alaska become a state?
This is not a statistical question, because it does not anticipate a variety of responses. Alaska became a state in 1959.

In how many states has the typical adult in your neighborhood lived?
This is a statistical question, because it does anticipate a variety of responses. If you survey a group of adults, you will likely get a variety of responses.

Check

Determine whether or not each question is a statistical question.

What is the height of the tallest roller coaster in the world?

How many roller coasters are typically found in an amusement park?

On average, how many roller coasters does the typical middle school student ride each summer?

In what year was the tallest roller coaster built?

Go Online You can complete an Extra Example online.

Explore Collect Data

Online Activity You will explore using a survey to collect data to explain how statistical questions anticipate a variety of answers.

Survey your classmates using a survey question based on your statistical question. Record your results in the table and find the total number of responses.

Survey Question	
Choose your question here.	
Range of Answer	Number of Responses
0	
1	
2	

Learn Display Data in a Table

A survey is one way to collect data to answer a statistical question. Once the data are collected, you can record the results in an organized way, such as a table, and then analyze the results.

Suppose a random group of adults were asked the question *How many hours do you exercise each week?* The results are shown in the table.

How many hours do you exercise each week?							
Number of Hours	0	1	2	3	4	5	6
Number of Responses	1	2	5	5	3	1	2

Based on the results in the table, one observation you can make is that more than half of the people who responded exercised fewer than four hours per week.

Talk About It!

What are some other observations you can make about the data in the table?

Example 2 Display Data in a Table

Suppose you want to answer the statistical question *How many hours per week does the typical sixth grade math student study?* You survey students in your math class using the question *How many hours do you typically spend studying each week?* The responses were 2, 4, 5, 4, 2, 1, 3, 1, 1, 4, 6, 3, 5, 2, 2, 1, 1, and 4 hours.

Organize the data in a table. Then analyze the results.

Part A Organize the data in a table.

Complete the table by recording the number of responses.

How many hours do you typically spend studying each week?	
Number of Hours	**Number of Responses**
1	
2	
3	
4	
5	
6	

(continued on next page)

Talk About It!
What are some other observations that can be made based on the data?

Part B Analyze the results.

Step 1 Find the total number of students surveyed.

Find the sum of the number of responses.

$5 + 4 + 2 + 4 + 2 + 1 = \square$ students

Step 2 Summarize the data.

Study the responses to determine if there is an overall trend.

One observation you can make is that half of the students in the survey studied fewer than 3 hours per week.

Check

Suppose you want to answer the statistical question *How many times does the typical middle school student exercise each month?* You survey your friends using the question *How many times each month do you typically exercise?*

The responses were 14, 12, 6, 2, 1, 0, 10, 6, 3, 4, and 5 times.

Organize the data in a table. Then analyze the results.

Part A

Organize the data by completing the table.

Number of Times Spent Exercising	Number of Responses
fewer than 4	
4–7	
8–11	
12 or more	

Part B

Select the statement that best represents the data.

Ⓐ Most students surveyed typically exercise at least 8 times each month.

Ⓑ Most students surveyed typically exercise more than 7 times each month.

Ⓒ Most students surveyed typically exercise 7 or fewer times each month.

Ⓓ Exactly half of the students surveyed typically exercise 4 or more times each month.

Go Online You can complete an Extra Example online.

Name ______________________ Period __________ Date __________

Practice

Go Online You can complete your homework online.

Determine whether or not each question is a statistical question. **(Example 1)**

1. *How many continents are there?*

2. *How many continents has the average student visited?*

3. *How many sporting events did the average student attend last year?*

4. *In what year was the first World Series?*

5. Suppose you want to determine the number of siblings each of your classmates have. You survey them using the question *How many siblings do you have?*. The responses were 1, 4, 2, 3, 0, 1, 0, 5, 1, 2, 2, 3, 0, 1, 2, 0, 1, 1, 6, and 2 siblings. Organize the data by completing the table and analyze the results. **(Example 2)**

Number of Siblings	Number of Responses
0–1	
2–3	
4–5	
6 or more	

6. You survey your classmates using the question *How many toppings do you like on an ice cream sundae?*. The responses were 2, 3, 7, 4, 5, 5, 4, 4, 1, 2, 4, 3, 4, 3, 6, 0, 4, 5, 6, and 5 toppings. Organize the data by completing the table and analyze the results. **(Example 2)**

Number of Toppings	Number of Responses
0–1	
2–3	
4–5	
6 or more	

7. You survey your classmates using the question *How many sports do you play?*. The responses were 2, 2, 1, 3, 1, 2, 4, 1, 2, 1, 3, 2, 2, and 2 sports. Organize the data by completing the table and analyze the results. **(Example 2)**

Number of Sports	Number of Responses
1	
2	
3	
4	

Test Practice

8. Multiselect Which of the following are statistical questions? Select all that apply.

- ☐ How many DVDs does a typical student own?
- ☐ How many oceans are there in the world?
- ☐ How many times did a typical student go to the zoo last year?
- ☐ How many classes does each student take?
- ☐ How many pets does a typical student own?
- ☐ How many continents are there?

9. Create Write a survey question that is a statistical question. Then write a survey question that is *not* a statistical question. Explain why each question is or is not a statistical question.

10. MP Find the Error Pete surveyed his friends as to the amount of their weekly allowance. The responses were $5, $0, $8, $10, $8, $10, $0, $0, and $1. Pete analyzed the results and stated that more than half of his friends earned $8 or more per week. Find his mistake and correct it.

11. MP Reason Abstractly Mara surveyed her friends as to the number of tablets their family owns. The responses were 1, 2, 2, 1, 0, 3, 1, 2, 4 and 2 tablets. Mara concludes that of her friends' families, most own 1 or 2 tablets. Is she correct? Explain.

12. Refer to Exercise 8. Choose one of the questions that is not a statistical question and rewrite it so that it is a statistical question.

Dot Plots and Histograms

I Can... use dot plots and histograms to display and analyze data.

Learn Construct Dot Plots

One way to represent a data set is to construct a **dot plot**. A dot plot is a visual display of a distribution of data values where each data value is shown as a dot above a number line.

The number of wins in a recent year by several football teams is 10, 9, 6, 6, 5, 5, 2, 12, 12, 8, 8, 7, 5, 5, and 4 wins. The dot plot shown organizes the data and shows possible patterns.

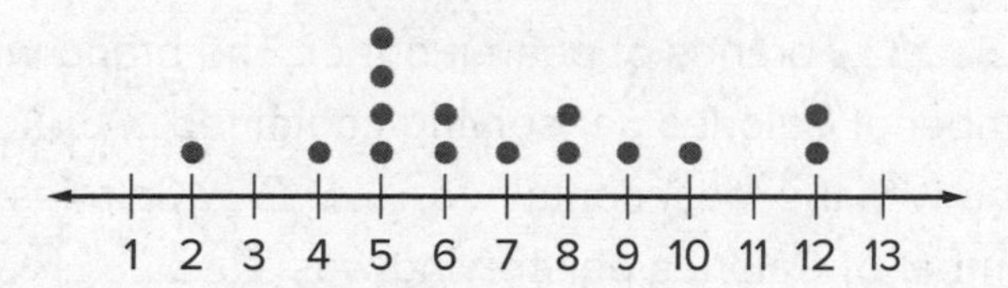

What Vocabulary Will You Learn?
dot plot
histogram

Talk About It!
How does using the visual representation allow you to make observations more easily?

Example 1 Construct Dot Plots

Jasmine surveyed the students in her class using the question *How many pets do you own?*. The results are shown in the table.

Number of Pets											
3	0	0	1	2	1	1	2	2	0	1	2
1	2	3	4	3	3	2	0	1	4	2	2

Construct a dot plot of the data. Then summarize the results.

Part A Construct a dot plot.

Draw and label a number line from 0 to 4, because the least data value is 0 and the greatest data value is 4. For each data value, place a dot above the corresponding number on the number line.

Number of Pets

0 1 2 3 4

Part B Summarize the results.

A total of 24 students responded to the survey. Students are more likely to own 1 or 2 pets than 3 or 4 pets. No student in the survey owned more than 4 pets.

Talk About It!
Which representation – the table or the dot plot – helps you visualize the results of the survey? Explain.

Your Notes

Check

Leah researched the number of Calories in a serving of peanut butter from various brands of peanut butter. The results are shown in the table. Construct a dot plot of the data. Then summarize the results.

Calories in a Serving of Peanut Butter			
190	160	210	210
200	185	190	190
185	200	190	210
190	185	200	200

Part A

Construct a dot plot.

Part B

There are ________ brands of peanut butter. The brand with the greatest number of Calories per serving contained ________ Calories and the brand with the least contained ________ Calories. The most common number of Calories per serving was ________.

Learn Construct Histograms

Data from a frequency table can be displayed as a **histogram**, a type of bar graph used to display numerical data that have been organized into equal intervals. This allows you to see the frequency distribution of the data, or the quantity of data that are in each interval.

When constructing a histogram, it is important that the intervals are equal and consecutive, so that you can make accurate observations about the data, based on the heights of the bars. The intervals should leave no gaps so that the entire range of data can be represented.

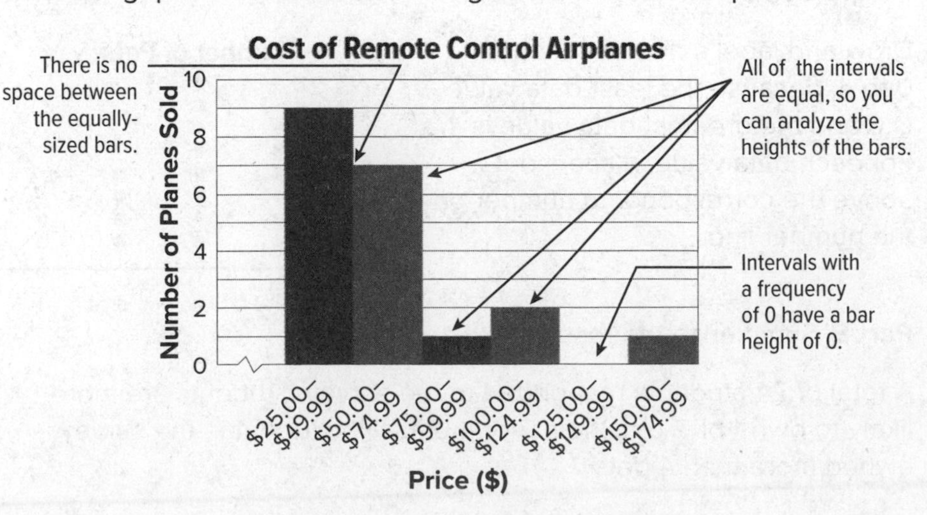

Talk About It!

For which representation – a dot plot or a histogram – can you see all of the individual data values? When might you choose to use a histogram as opposed to a dot plot?

Example 2 Construct Histograms

A park ranger at a state park was asked the question *How many daily visitors attended the park each day for 20 days?*. The table shows the results.

Daily Visitors				
108	209	171	152	236
165	244	263	212	161
327	185	192	226	137
193	235	207	382	241

Construct a histogram to represent the data.

Step 1 Make a frequency table.

Use a scale to include all of the values, 100 through 399, with equally-spaced intervals.

Complete the frequency table to organize the data.

Daily Visitors	
Visitors	Frequency
100–149	
150–199	
200–249	
250–299	
300–349	
350–399	

Step 2 Draw and label the axes.

When you construct the histogram, first draw the axes. Label the horizontal axis using the intervals from the frequency table, 100–149 through 350–399. Label the vertical axis with the frequencies, 1–10.

Step 3 Graph the intervals.

For each interval, draw a bar with a height that is indicated by the frequency table. Complete the histogram by drawing and shading the correct bar heights.

Think About It!
What intervals will you select for the histogram?

Talk About It!
What observations can you make about the data by studying the histogram?

Check

The students in Mrs. Angelo's class were asked the question *How many books did you read over summer vacation?*. The responses are shown in the table.

Number of Books Read					
3	6	4	2	8	0
6	3	9	3	1	4
5	12	10	4	11	0
7	3	7	5	12	6
7	13	14	5	1	2

Construct a histogram to represent the data.

Go Online You can complete an Extra Example online.

Foldables It's time to update your Foldable, located in the Module Review, based on what you learned in this lesson. If you haven't already assembled your Foldable, you can find the instructions on page FL1.

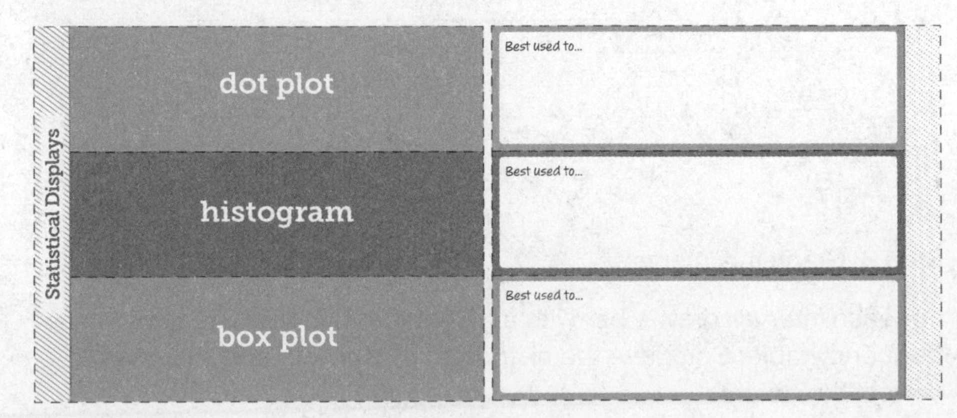

Name ________________ Period ________ Date ________

Practice

Go Online You can complete your homework online.

1. Chris surveyed the members of his tennis team by asking the question *In how many tennis tournaments have you played?*. The results are shown in the table. Construct a dot plot of the data and summarize the results. **(Example 1)**

Number of Tennis Tournaments					
0	2	1	4	0	1
1	0	3	2	6	0

2. The table shows the results of asking a group of teachers the question *How many students are in your homeroom?*. Construct a histogram to represent the data. **(Example 2)**

Homeroom Class Size						
17	26	20	23	19	23	22
22	24	19	20	21	20	23

3. The table shows the results of asking a group of students the question *How many hours per month do you volunteer?*. Construct a histogram to represent the data. **(Example 2)**

Hours Spent Volunteering						
48	30	21	10	1	40	19
10	5	40	39	20	9	40
31	45	29	40	18	49	31
24	32	15	0	15	27	12

Test Practice

4. Open Response Petra surveyed the members of her dance class by asking the question *How many hours outside of class do you usually practice dance each week?*. The results are shown in the table. Construct a dot plot of the data.

Number of Hours				
1	3	4	5	2
2	2	4	3	1
3	3	2	4	2

Apply

5. Lou wanted to determine how much his friends pay for video games. He surveyed them using the question *How much did you pay for the last video game you bought?* The responses were \$29, \$45, \$50, \$55, \$34, \$28, \$35, \$35, \$45, \$30, \$34, and \$55. How many more games cost between \$30 and \$39 than between \$40 and \$49?

6. Provide a data set that can be represented by the histogram shown.

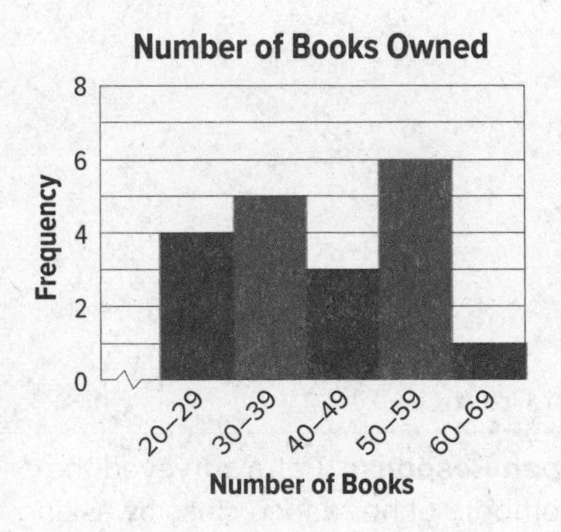

7. MP **Make a Conjecture** Refer to the histogram. In one or two sentences, write a conclusion you can make about the data.

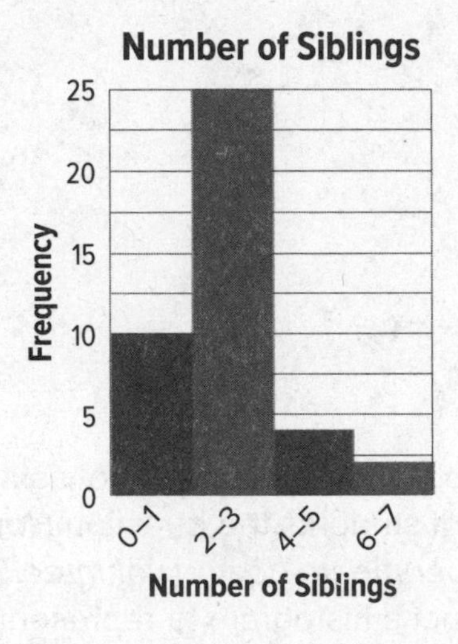

8. MP **Reason Abstractly** Laura recorded the daily temperatures, in degrees Fahrenheit, during January in Minnesota. What changes might she have to make in a number line for a dot plot that starts at zero and goes to 20, so that it could be used to make a dot plot of the temperatures? Explain.

9. MP **Justify Conclusions** Determine if the statement is *true* or *false*. Justify your conclusion.

Histograms display individual data values.

Measures of Center

I Can... use the measures of center to summarize a numerical data set with a single number, and find a missing data value given the mean.

What Vocabulary Will You Learn?
average
mean
measures of center
median

Explore Mean

Online Activity You will use interactive workmats to explore how to find the mean of a data set.

Talk About It!

Compare your strategy with a partner. If your strategies were the same, is there another way to find the value?

Each box represents a student. Drag the icons to model the apps a typical student opens before school.

Friend 1 Friend 2 Friend 3 Friend 4 Friend 5

Learn Measures of Center

A data set can contain many values, but sometimes it is beneficial to find a single value that can represent, or summarize, the entire data set. **Measures of center** are numbers used to describe the center of a numerical data set. The measures of center you will learn about in this lesson are the mean and median.

One measure of center used to describe a numerical data set is the **mean**. The mean, or **average**, of a data set is the sum of the data divided by the number of data values.

Suppose you have 4 test scores, 86%, 90%, 72%, and 88%. You can find the mean by adding the test scores and then dividing by the total number of scores, 4.

$$\frac{86 + 90 + 72 + 88}{4} = \square$$

Add the test scores. Then divide by the total number of scores.

The mean score is 84%.

(continued on next page)

Your Notes

Talk About It!

What are some ways you have seen mean used in real life?

Think About It!

Without calculating the mean, what temperature do you think best describes the center of the data? Explain your reasoning.

Talk About It!

How would the mean change if the data value 0°F was included?

The mean is the *balance point* of the data. The dot plot displays the test scores. The total distance between the values below the mean and the mean must be equal to the total distance between the values above the mean and the mean.

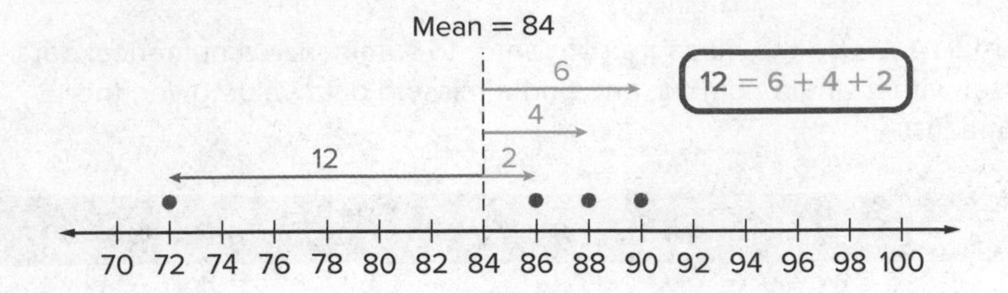

Example 1 Find the Mean

The table shows the recorded high temperatures in degrees Fahrenheit for six days in Little Rock, Arkansas.

Find the mean temperature to summarize the data.

High Temperatures in Little Rock, Arkansas (°F)					
Mon.	Tues.	Weds.	Thurs.	Fri.	Sat.
45	52	45	50	49	47

$$\text{mean} = \frac{\text{sum of the data values}}{\text{number of data values}}$$ Definition of mean

$$= \frac{45 + 52 + 45 + 50 + 49 + 47}{6}$$ There are 6 data values.

$$= \frac{288}{6}$$ Add.

$$= 48$$ Divide.

The mean temperature for the selected days is ______°F. The dot plot confirms that the mean temperature of 48°F is the balance point of the data. The total distance between the values above the mean and the mean, 7, is equal to the total distance between the values below the mean and the mean.

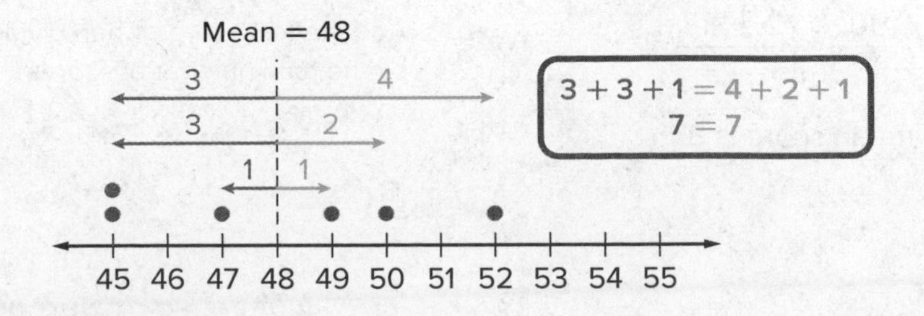

Check

The table shows the number of headphones sold at an electronics store during a sale. Find the mean number of headphones sold to summarize the data.

Headphones Sold					
Mon.	Tues.	Weds.	Thurs.	Fri.	Sat.
9	18	7	7	10	15

Go Online You can complete an Extra Example online.

Learn Find a Missing Data Value Using the Mean

You can use dot plots and bar diagrams to find a missing data value given the mean and the other data values. Consider the following problem.

Caitlin's first four quiz scores are shown in the table. What score does Caitlin need to earn on her fifth quiz to have a mean quiz score of 90?

Caitlin's Quiz Scores				
88	95	93	80	?

Method 1 Use the mean as a balance point.

Plot the four known quiz scores and label the mean.

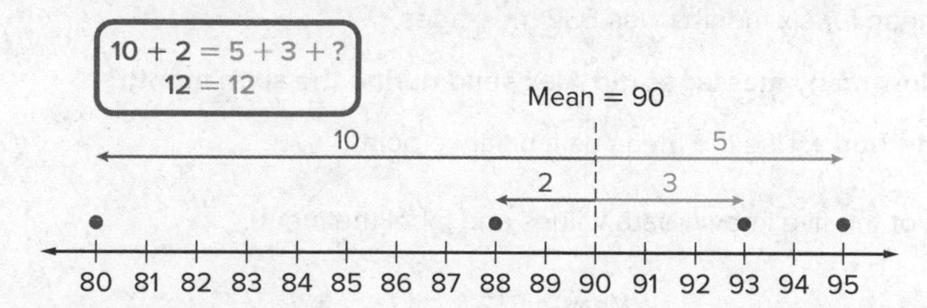

distance below the mean = **10 + 2**, or **12**

distance above the mean = **5 + 3**, or **8**

The distances are not the same because the fifth quiz score is not plotted. There is a greater distance below the mean. This means the missing value must be above the mean. In order for the total distance above the mean to equal 12, the missing value must be 4 units above the mean, because 8 + 4 = 12. The missing value is 90 + 4, or 94.

(continued on next page)

Method 2 Use an equation.

Draw a bar diagram to represent the situation. To find the total amount needed to achieve a mean score of 90, multiply the mean, 90, by the number of data values, 5.

$90(5) = 450$

The sum of the known data values is $88 + 95 + 93 + 80$, or 356. Let q represent Caitlin's score on her fifth quiz.

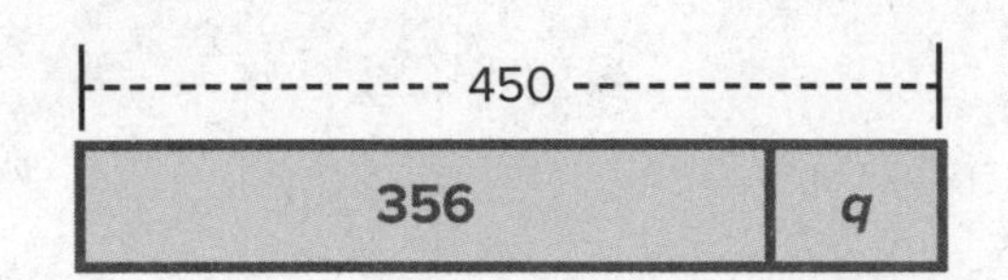

The equation that can be used to find the missing data value is $356 + q = 450$. Solve the equation.

$356 + q = 450$	Write the equation.
$-356 \quad -356$	Subtraction Property of Equality
$q = 94$	Simplify.

The missing value is 94.

So, using either method, Caitlin needs a score of ______ on her fifth quiz to have a mean quiz score of 90.

Example 2 Find a Missing Data Value Using the Mean

The number of messages Alex sent on her phone each month for the past five months were 494, 502, 486, 690, and 478. Suppose the mean for six months was 532 messages.

How many messages did Alex send during the sixth month?

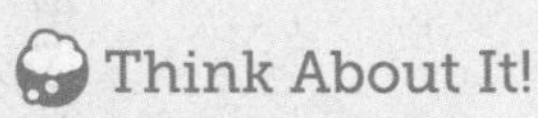

Think About It!

Do you think the missing value is less than, greater than, or equal to the mean? Explain.

Method 1 Use the mean as a balance point.

Plot the five known data values and label the mean.

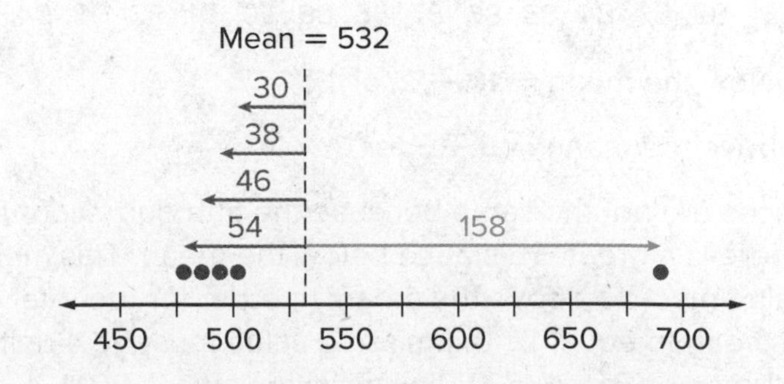

(continued on next page)

distance below the mean = **54** + **46** + **38** + **30**, or ☐

distance above the mean = **158**

The distances are not the same because the sixth amount is not plotted. There is a greater distance below the mean. This means the missing value must be above the mean. In order for the total distance above the mean to be equal to 158, the missing value must be 10 units above the mean, because 158 + 10 = 168.

The missing value is 532 + 10, or ☐.

Method 2 Use an equation.

Draw a bar diagram to represent the situation. To find the total amount needed for the mean number of messages to be 532, multiply the mean by the number of data values.

532(6) = 3,192

The sum of the known data values is 494 + 502 + 486 + 690 + 478 or 2,650. Let m represent the number of messages Alex sent during the sixth month.

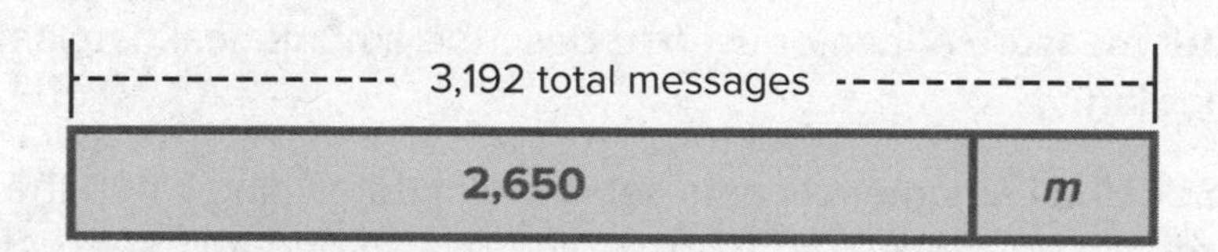

The equation that can be used to find the missing data value is $2{,}650 + m = 3{,}192$. Solve the equation.

$2{,}650 + m = 3{,}192$	Write the equation.
$-2{,}650 \qquad -2{,}650$	Subtraction Property of Equality
$m =$ ☐	Simplify.

The missing value is ☐.

So, using either method, Alex sent 542 messages during the sixth month.

Talk About It!

Compare and contrast Method 1 and Method 2. When might it be more advantageous to use Method 2?

Check

The table shows the greatest depths of four of Earth's five oceans. If the average greatest depth is 8.094 kilometers, what is the greatest depth of the Southern Ocean? Round to the nearest hundredth.

Ocean	Greatest Depth (km)
Pacific	10.92
Atlantic	9.22
Indian	7.46
Arctic	5.63
Southern	d

 **Go Online** You can complete an Extra Example online.

Learn Find the Median

Another measure of center used to describe a numerical data set is the **median**.

The median of a numerical data set is the middle value when the data are ordered from least to greatest. If there is an odd number of data values, the median is the middle data value. If there is an even number of data values, the median is the mean of the two values in the middle.

Just as the mean is a single value used to summarize a data set, the median also summarizes a data set with a single value.

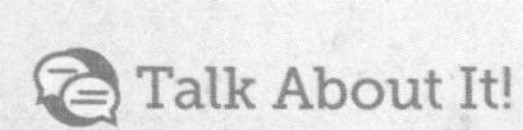

Talk About It!

Why must the data be ordered from least to greatest before finding the median?

Consider the following set of numerical data, which represents the ages of participants in a board game club.

8, 8, 8, 8, 9, 10, 10, 11, 12, 12, 16, 16, 19

There are 13 data values. Since the number of data values is odd, the median is the middle data value. Make sure the data values are ordered from least to greatest before finding the median.

The median is 10.

8, 8, 8, 8, 9, 10, 10, 11, 12, 12, 16, 16, 19

There are 6 data values below the median. There are 6 data values above the median.

Example 3 Find the Median Given an Odd Number of Data Values

Between 2009 and 2015, the number of Atlantic hurricanes each year were 3, 12, 7, 10, 2, 6, and 4.

Find the median of the data.

There are 7 data values. Since the number of data values is odd, the median is the middle data value.

Step 1 Order the values from least to greatest.

least **greatest**

Step 2 Find the median.

How many data values are below the median? ______

How many data values are above the median? ______

What is the median? ______

The center of the data can be represented by the single value, ______. So, the median number of hurricanes from 2009 to 2015 is 6 hurricanes.

Check

Dina's scores on recent science tests were 86, 98, 85, 90, 85, 91, 89, 88, and 89 points. Find the median of her test scores.

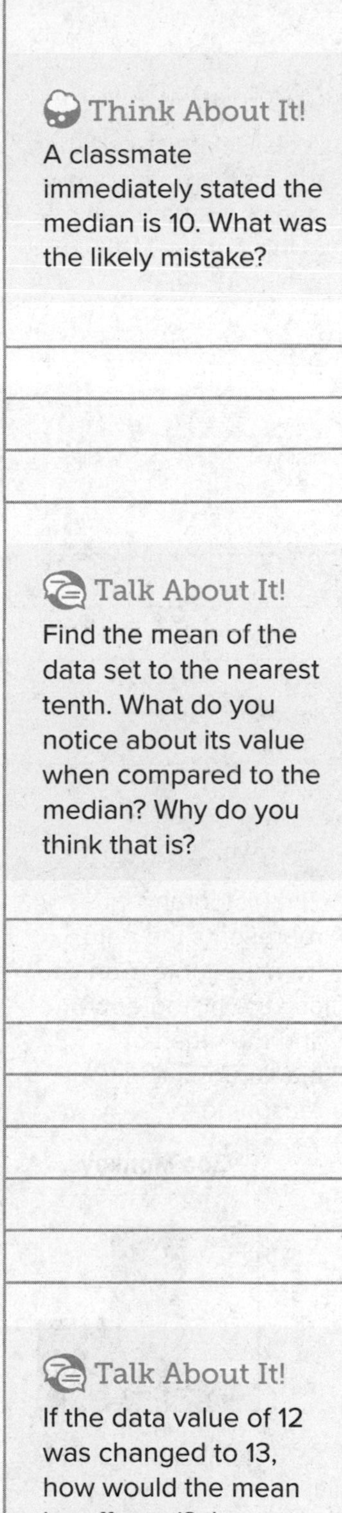

Think About It!
A classmate immediately stated the median is 10. What was the likely mistake?

Talk About It!
Find the mean of the data set to the nearest tenth. What do you notice about its value when compared to the median? Why do you think that is?

Talk About It!
If the data value of 12 was changed to 13, how would the mean be affected? the median?

Go Online You can complete an Extra Example online.

Think About It!

A classmate stated that the median will not be one of the data values. Is this correct? Why or why not?

Talk About It!

The histogram represents the data from the table. Can you use the histogram to find the mean or median? Explain your reasoning.

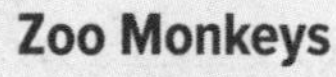

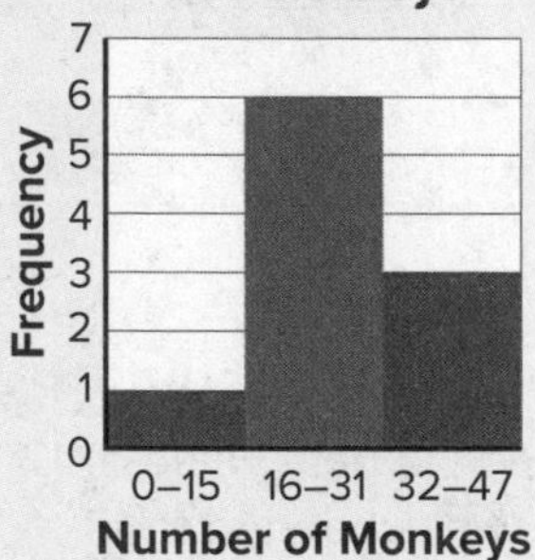

Example 4 Find the Median Given an Even Number of Data Values

The table shows the number of monkeys at ten different zoos.

Find the median of the data.

Number of Monkeys				
27	36	18	25	12
18	42	34	16	30

There are 10 data values. Because the number of data values is even, the median is the mean (average) of the two middle data values.

Step 1 Order the values from least to greatest.

In order from least to greatest, the values are 12, 16, 18, 18, 25, 27, 30, 34, 36, and 42.

Step 2 Find the median.

Because there is an even number of data values, find the two values closest to the middle.

The two values closest to the middle are ________ and ________.

Find the mean of the two middle data values.

$\text{mean} = \frac{25 + 27}{2}$ Find the mean of 25 and 27.

$= \frac{52}{2}$ Add.

$= 26$ Divide.

So, the median of the data is ________ monkeys. The data can be summarized by describing the center of the data as 26 monkeys.

Check

The table shows the prices of different packages of juice boxes at a local store. Find the median of the data.

Cost of Juice Boxes ($)			
1.65	1.97	2.45	2.87
2.35	3.75	2.49	2.87

Show your work here

Go Online You can complete an Extra Example online.

Apply Track

The table shows Kendra's 100-meter dash times. Kendra wants to record the measure of center that describes her times as the fastest. Which measure should she use, the mean or median? Why?

Kendra's 100-meter Dash Times (seconds)			
15.1	17.2	14.6	16.2
17.9	16.5	17.8	17.1
14.7	17.1	19.5	13.8

1 What is the task?

Make sure you understand exactly what question to answer or problem to solve. You may want to read the problem three times. Discuss these questions with a partner.

First Time Describe the context of the problem, in your own words.
Second Time What mathematics do you see in the problem?
Third Time What are you wondering about?

2 How can you approach the task? What strategies can you use?

3 What is your solution?

Use your strategy to solve the problem.

4 How can you show your solution is reasonable?

Write About It! Write an argument that can be used to defend your solution.

Talk About It!

In the next two races, Kendra had times of 14 seconds and 19 seconds. Does adding these times to the data set affect the measure that she should choose?

Check

Rosario recorded the number of hours she spent doing homework for five nights. She wants to use the greater measure of center to describe her time spent doing homework. Which measure should she use, the mean or median? Why?

Day	Time (h)
1	1.25
2	2.25
3	1.5
4	2
5	0.75

 **Go Online** You can complete an Extra Example online.

Pause and Reflect

Create a graphic organizer that compares and contrasts the two measures of center you studied in this lesson.

Name ______________________ Period __________ Date __________

Practice

Go Online You can complete your homework online.

1. The number of cans collected over the weekend by each sixth grade homeroom was 57, 59, 60, 58, 58, and 56 cans. Find the mean number of cans collected. **(Example 1)**

2. Grace and her friends are comparing the number of pets they own. They have 1, 2, 0, 5, 1, 1 and 4 pets. Find the mean number of pets owned. **(Example 1)**

3. The amount Lucy earned babysitting each month for the past five months was $225, $280, $240, $180, and $200. Suppose the mean for six months was $220. How much did Lucy earn babysitting during the sixth month? **(Example 2)**

4. The average high temperature last week was 65 degrees Fahrenheit. The high temperatures for Sunday through Friday were 68, 70, 73, 45, 68, and 71 degrees Fahrenheit. What was the high temperature on Saturday? **(Example 2)**

5. The table shows the results of a survey about the number of E-mails sent in one day. Find the median number of E-mails sent per day. **(Example 3)**

Number of E-mails Sent Per Day						
20	24	22	27	21	27	20
27	22	23	20	22	24	26
23	26	27	22	27	20	25

6. The table shows the number of students in each group on a school field trip. Find the median size of a group. **(Example 3)**

Number of Students in Each Group				
5	7	8	7	6
4	4	5	6	9
7	5	7	8	6
9	7	5	4	5

7. The table shows the number of points scored by a basketball team in each game last season. Find the median number of points scored. **(Example 4)**

Number of Points					
64	41	52	63	44	54
42	67	44	68	43	61

Test Practice

8. **Open Response** The number of points Seth has earned playing his favorite game is shown. Find the median of the data.

40, 28, 24, 37, 43, 26, 30, 36

Apply

9. The table shows the number of minutes Kenny spent practicing the piano. Kenny wants to record the greater measure of center that describes his time spent practicing. Which measure should he use, the mean or median? Why?

Number of Minutes			
38	30	26	25
20	24	25	60

10. The table shows the number of push-ups Jade completed each day this week. Jade wants to record the greater measure of center that describes her ability to do push-ups. Which measure should she use, the mean or median? Why?

Number of Push-ups			
65	70	67	38
55	68	64	

11. Create Generate a real-world data set that has a mean of 8.

12. MP **Use a Counterexample** Determine if the following statement is *true* or *false*. If *false*, provide a counterexample.

The mean and median of a data set cannot be the same value.

13. MP **Find the Error** A student said the mean of the data set shown is 17. Find the student's error and correct it.

number of texts sent in an hour: 15, 11, 25, 19, 11, 27

14. MP **Reason Abstractly** Ty worked 5 nights this week at an ice cream shop. He earned \$23, \$29, \$25, and \$16 in tips. The average amount he earned in tips for the 5 nights was \$22. Is the amount he earned in tips on night 5 more or less than the average amount? Explain.

Interquartile Range and Box Plots

I Can... understand how a measure of variation describes the variability of a data set with a single value, display a numerical data set in a box plot, and summarize the data.

What Vocabulary Will You Learn?
box plot
first quartile
interquartile range
measures of variation
quartiles
range
second quartile
third quartile

Learn Measures of Variation

Measures of variation are values that describe the variability, or spread, of a data set. They describe how the values of a data set vary with a single number.

One measure of variation is the **range**, which is the difference between the greatest and least data values in a data set. Consider the data set shown.

0, 0, 1, 1, 2, 2, 2, 3, 4, 5, 6, 6, 7, 7, 7, 8

The data values range from 0 to 8. The range is 8–0, or 8.

Another measure of variation is the *interquartile range*. Before you can find this measure, you first need to understand and find quartiles. **Quartiles** divide the data into four equal parts. The **first quartile**, Q_1, is the median of the data values less than the median. The **third quartile**, Q_3, is the median of the data values greater than the median. The median is also known as the **second quartile**, Q_2.

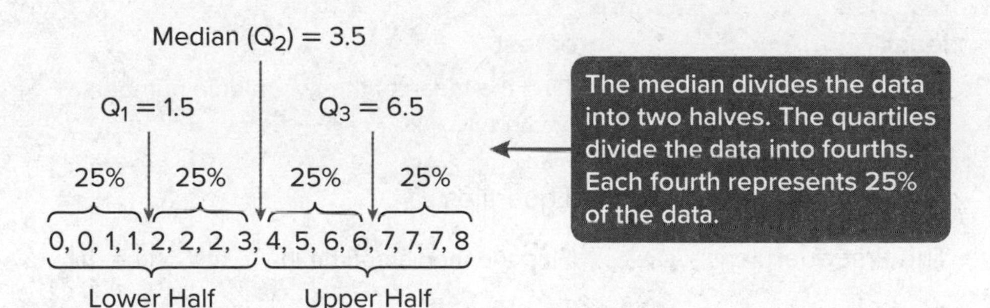

Talk About It!
How does knowing that the data is divided into four equal parts help you remember the vocabulary term quartile?

The **interquartile range (IQR)** is the distance between the first and third quartiles of the data set. To find the IQR, subtract the first quartile from the third quartile.

The interquartile range represents the middle half, or middle 50%, of the data. The lower the IQR is for a data set, the closer the middle half of the data is to the median.

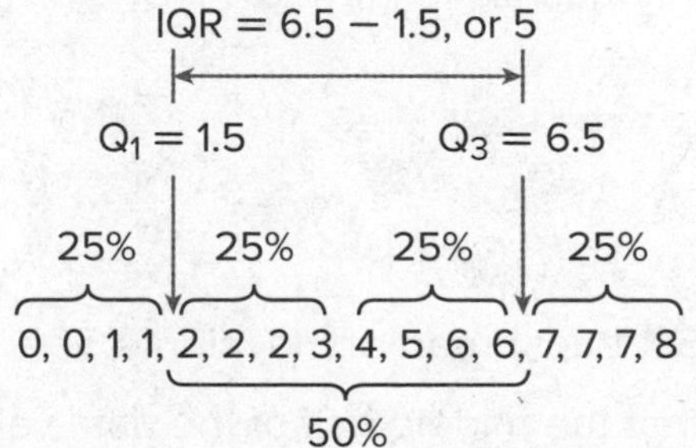

Talk About It!
If the median describes the center of a data set, what does the interquartile range describe?

In the given data set, the IQR is 6.5 − 1.5, or 5.

Your Notes

Think About It!

Do the data values need to be in numerical order? Why?

Talk About It!

Which value, the interquartile range, the first quartile, or the third quartile tells you more about the spread of the data values? Explain your reasoning.

Example 1 Find the Range and Interquartile Range

The table shows the approximate maximum speeds, in miles per hour, of different animals.

Animal	Speed (mph)
Housecat	30
Cheetah	70
Elephant	25
Lion	50
Mouse	8
Spider	1

Use the range and interquartile range to describe how the data vary.

Part A Describe the variation of the data using the range.

The greatest speed in the data set is 70 miles per hour. The least speed in the data set is 1 mile per hour.

The range is 70 − 1, or 69 miles per hour.

The speeds of animals vary by 69 miles per hour.

Part B Describe the variation of the data using the interquartile range.

Step 1 Find the median.

Write the speeds in order from least to greatest.

least **greatest**

The median is ______. Find the mean of the two middle numbers, 25 and 30.

Step 2 Find the first and third quartiles.

The first quartile is ______. Find the median of the lower half of the data.

The third quartile is ______. Find the median of the upper half of the data.

Step 3 Find the interquartile range.

Interquartile range = $Q_3 - Q_1$

= ☐ − ☐ $Q_3 = 50$; $Q_1 = 8$

= ☐ Subtract.

So, the spread of the middle 50% of the data is ______. This means that the middle half of the data values vary by ______ miles per hour.

Check

The average wind speeds for several cities in Pennsylvania are given in the table. Use the range and interquartile range to describe how the data vary.

Wind Speed	
City	Speed (mph)
Allentown	8.9
Erie	11.0
Harrisburg	7.5
Middletown	7.7
Philadelphia	9.5
Pittsburgh	9.0
Williamsport	7.6

Part A

Describe the variation of the data using the range.

Part B

Describe the variation of the data using the interquartile range.

Go Online You can complete an Extra Example online.

Learn Construct Box Plots

A **box plot**, or box-and-whisker plot, uses a number line to show the distribution of a data set by plotting the median, quartiles, and extreme values. The extreme values, or extremes, are the greatest and least values in the data set. The extremes, quartiles, and median are referred to as the *five-number summary*.

A box is drawn around the two quartile values. The whiskers extend from each quartile to the extreme data values, unless the extremes are very far apart from the rest of the data set. The median is marked with a vertical line, and separates the box into two boxes.

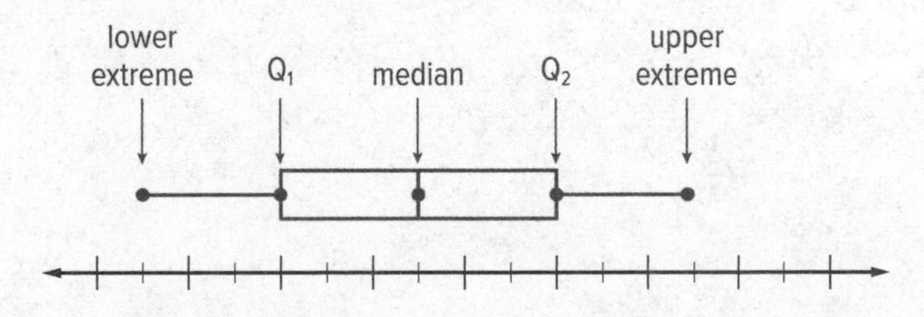

Box plots separate data into four sections. These sections are visual representations of quartiles. Even though the parts may differ in length, each contain 25% of the data. The two boxes represent the middle 50% of the data. A longer box or whisker indicates the data are more spread out in that section. A longer box or whisker does not mean there are more data values in that section. Each section contains the same number of values, 25% of the data.

Math History Minute

Florence Nightingale (1820–1910) used statistics to help improve the survival rates of hospital patients. She discovered that by improving sanitation, survival rates improved. She designed charts to display the data, as statistics had rarely been presented with charts before. She is known for inventing the *coxcomb graph,* which is a variation of the circle graph.

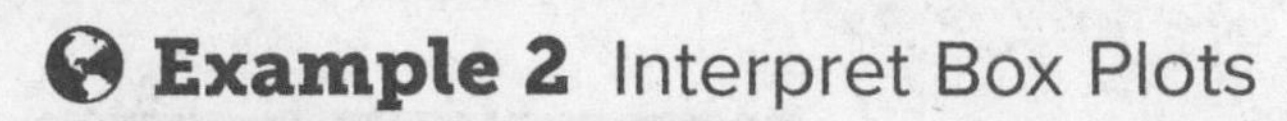

Example 2 Interpret Box Plots

The box plot shows the annual snowfall totals, in inches, for a certain city over a period of 20 years.

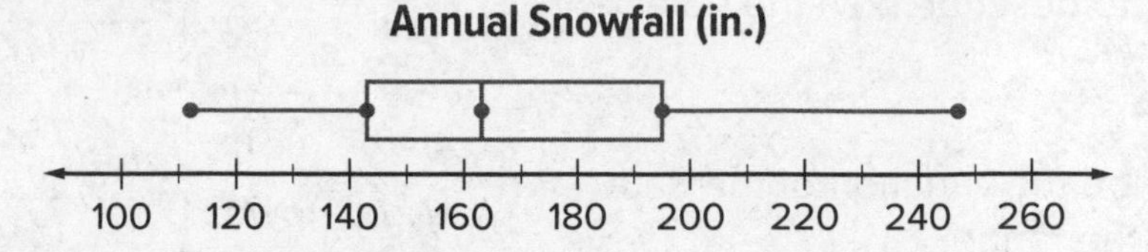

Describe the distribution of the data. What does it tell you about the snowfall in this city?

The annual snowfall ranges from about 110 inches to about 250 inches. The middle half of the data range from about 140 inches to about 195 inches. Because the boxes are shorter than the whiskers, there is less variation among the middle half of the data. Having less variation means there is a greater consistency among the middle 50% of the data than in either whisker.

Check

The average gas mileage, in miles per gallon, for various sedans is shown in the box plot. Describe the distribution of the data. What does it tell you about the average gas mileage for those sedans?

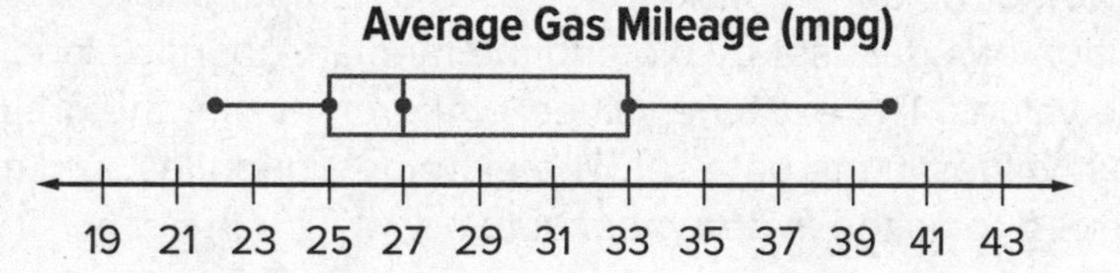

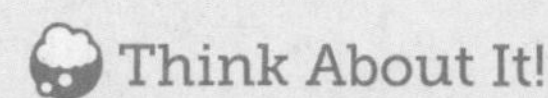

Think About It!

What does the length of the box and whiskers tell you about the spread of the data in the box plot?

Talk About It!

What does the interquartile range describe in the context of the problem?

Go Online You can complete an Extra Example online.

Example 3 Construct and Interpret Box Plots

The table shows the recorded speeds of cars traveling on a country road.

Car Speeds (mph)									
25	35	27	22	34	40	20	19	23	25

Construct a box plot to represent the data. Then describe the distribution of the data.

Part A Construct a box plot.

Step 1 Order the values from least to greatest.

In order from least to greatest, the speeds are 19, 20, 22, 23, 25, 25, 27, 30, 34, 35, and 40 miles per hour.

Step 2 Graph the values above a number line.

Find the median, the extremes, and the first and third quartiles. Graph the values above a number line.

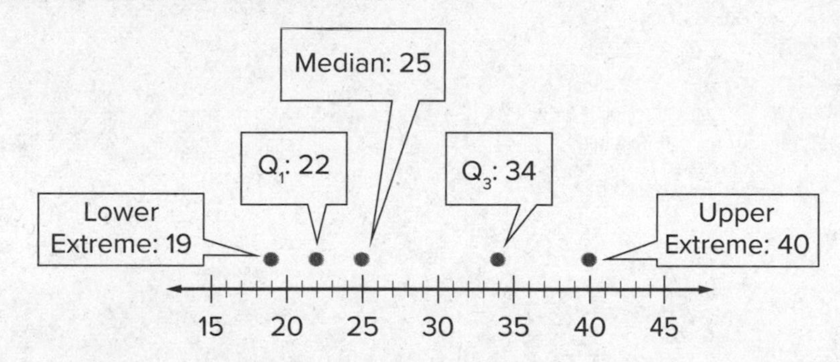

Step 3 Draw the box plot.

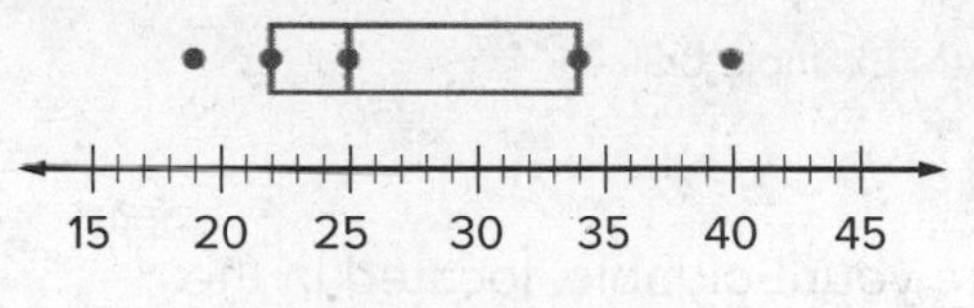

Draw a box around the first quartile and the third quartile. Draw a line through the median.

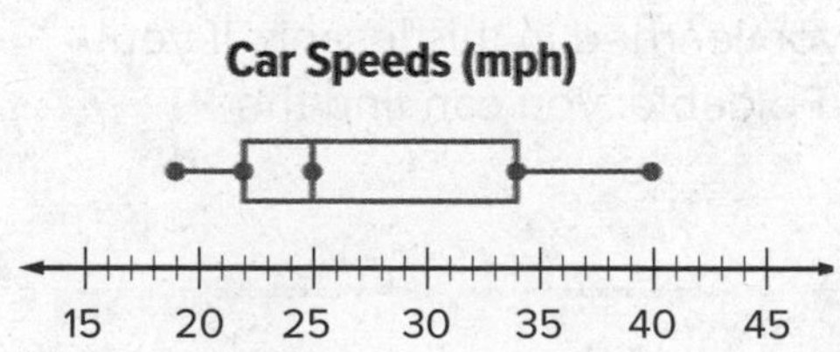

Draw a line from the first quartile to the least value. Draw a line from the third quartile to the greatest value. Add a title.

Part B Describe the distribution of the data.

The recorded speeds range from 19 miles per hour to 40 miles per hour. The middle half of the data range from 22 miles per hour to 34 miles per hour. Because the boxes are longer than the whiskers, there is more variation among the middle half of the data. Having more variation means there is a lesser consistency among the middle 50% of the data than in either whisker.

Think About It!

What are the different measures of variation you need to find in order to construct a box plot?

Talk About It!

How does constructing a box plot to represent the data help you to understand the distribution of the data?

Check

Earthquakes occur at different depths below Earth's surface. Stronger earthquakes happen at depths that are closer to the surface. The table shows the depths of recent earthquakes, in kilometers.

Depth of Recent Earthquakes (km)						
5	15	1	11	2	7	3
9	5	4	9	10	5	7

Part A Construct a box plot to represent the data.

Part B Describe the distribution of the data.

Go Online You can complete an Extra Example online.

Foldables It's time to update your Foldable, located in the Module Review, based on what you learned in this lesson. If you haven't already assembled your Foldable, you can find the instructions on page FL1.

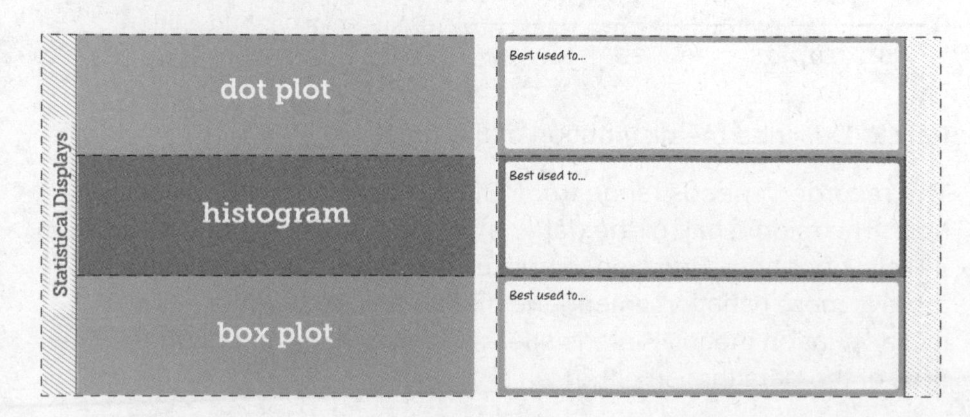

Name ______________________ Period __________ Date __________

Practice

Go Online You can complete your homework online.

1. Cameron surveyed her friends about the number of apps they use. The responses were 15, 16, 18, 9, 18, 4, 19, 20, 17, and 36 apps. Use the range and interquartile range to describe how the data vary. **(Example 1)**

2. The table shows the number of hours different animals spend sleeping per day. Use the range and interquartile range to describe how the data vary. **(Example 1)**

Time Animals Spend Sleeping (h)					
12	20	16	11	4	2

3. The box plot shows the ages of vice presidents when they took office. Describe the distribution of the data. What does it tell you about the ages of vice presidents? **(Example 2)**

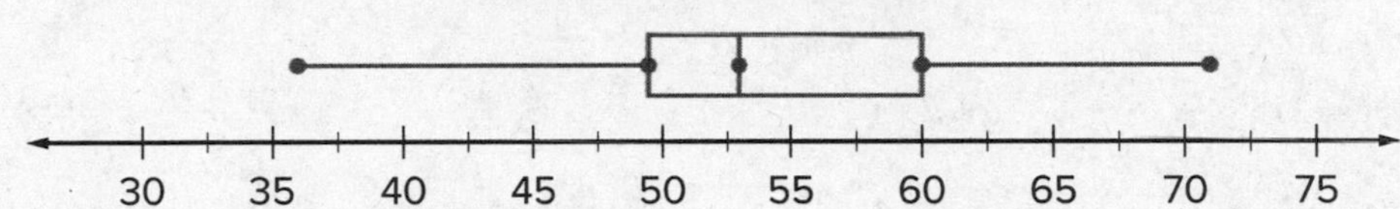

4. The ages of children taking a hip-hop dance class are 10, 9, 9, 7, 12, 14, 14, 9, and 16 years old. Construct a box plot of the data. Then describe the distribution of the data. **(Example 3)**

Test Practice

5. Open Response The cost of tents on sale at a sporting goods store are \$66, \$72, \$78, \$69, \$64, \$70, \$67, \$72, and \$66. Use the range and interquartile range to describe how the data vary.

Apply

6. The table shows the number of points scored by the seventh and eighth grade girls basketball teams in each of their games this season. Construct a box plot to represent the data for each team. Then use the box plots to compare the data.

Points Scored per Game							
Seventh Grade Team				**Eighth Grade Team**			
39	36	40	27	34	36	47	40
35	29	36	29	39	38	45	43
31	38	30	34	42	41	45	42

7. MP **Justify Conclusions** Determine if the following statement is *true* or *false*. If *false*, justify your reasoning.

You can determine the mean of a data set from a box plot.

8. **Create** Provide a set of real-world data and then construct a box plot of the data.

9. MP **Make an Argument** A student said that, in a box plot, if the box to the right of the median is longer than the box to the left of the median, there are more data values represented by the longer box. Is the student's reasoning correct? Construct an argument to defend your solution.

10. MP **Reason Inductively** What can you conclude about a data set shown in a box plot where the whiskers and boxes are all the same length?

Mean Absolute Deviation

I Can... understand how the mean absolute deviation describes the variation in a data set and interpret its value within the context of a given real-world scenario.

What Vocabulary Will You Learn?
mean absolute deviation

Learn Mean Absolute Deviation

You have learned how the range and interquartile range describe the spread of a data set. Another measure of variation is the **mean absolute deviation** (MAD). The MAD of a data set is the average distance between each data value and the mean. The lower the MAD is for a data set, the closer the data values are to the mean.

Go Online Watch the animation to learn about mean absolute deviation. The animation shows a data set of the points scored in each game. Follow the steps to find the mean absolute deviation.

Points Scored per Game					
46	58	50	53	48	57

Step 1 Find the mean.

$$\frac{46 + 58 + 50 + 53 + 48 + 57}{6} = 52$$

Calculate the sum of the values in the data set and then divide by the total number of values.

Step 2 Find the distance between each data value and the mean.

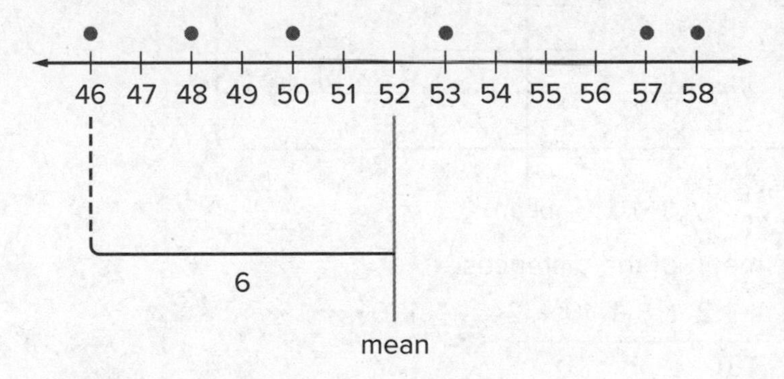

Plot the data on a number line. Find the distance between the mean and the first value. Distance is always positive.

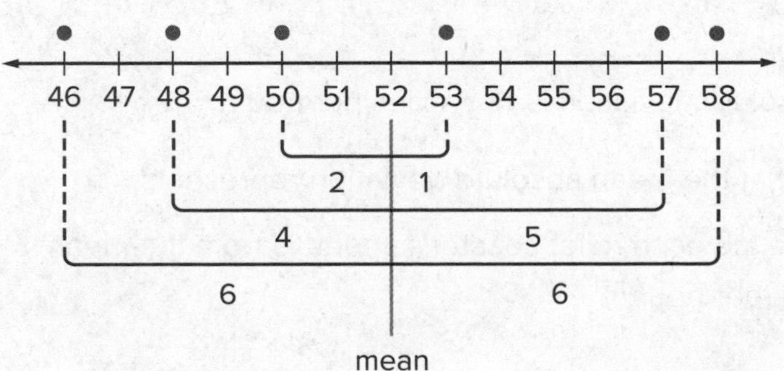

Continue to find the distance between the mean and each of the other data values.

(continued on next page)

Talk About It!
The term *absolute* in *mean absolute deviation* refers to the absolute value of a number. How do you think absolute value relates to mean absolute deviation?

Your Notes

Talk About It!

The mean absolute deviation is a measure of variability that compares each data value's distance to the mean. In the animation, the MAD of the team scores is 4. Do you think the MAD indicates the data set has a great deal of variability? Explain.

Talk About It!

Does the MAD indicate a large or small variation in the data? Explain your reasoning.

Step 3 Find the mean of the distances.

$$\frac{6+4+2+1+5+6}{6} = 4$$

Calculate the sum of the distances and divide by the number of distances, 6.

So, the mean absolute deviation of the data set is 4. In other words, the average distance each score is from the mean score of 52 is 4 points.

Example 1 Find Mean Absolute Deviation

The table shows the maximum speeds of eight roller coasters.

Maximum Speeds (mph)			
58	88	40	60
72	66	80	48

Find the mean absolute deviation of the data set. Explain what the mean absolute deviation represents.

Part A Find the mean absolute deviation.

Step 1 Find the mean.

$$\frac{58+88+40+60+72+66+80+48}{8} = \square \text{ mph}$$

Step 2 Find the distance between each data value and the mean.

Use a number line. Remember that distance is always positive.

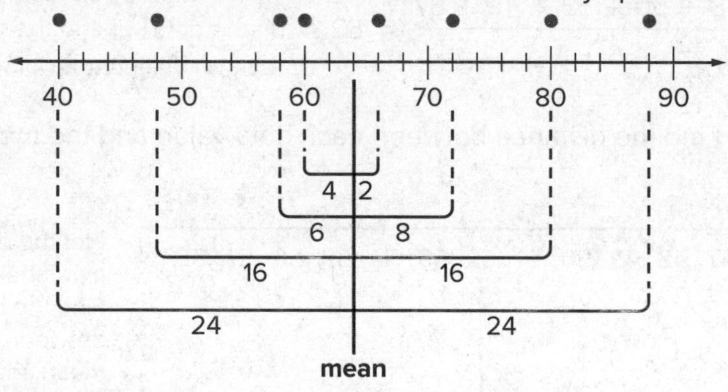

Step 3 Find the mean of the distances.

$$\frac{24+16+6+4+2+8+16+24}{8} = \frac{100}{8}$$

$$= \square$$

So, the mean absolute deviation is 12.5 miles per hour.

Part B Explain what the mean absolute deviation represents.

The average distance each roller coaster's speed is from the mean is ________ miles per hour.

Check

The table shows the number of daily visitors to a website on the Internet. Find the mean absolute deviation of the data set. Explain what the mean absolute deviation represents.

Number of Daily Visitors				
112	145	108	160	122

Part A Find the mean absolute deviation. Round to the nearest hundredth.

Part B Explain what the mean absolute deviation represents.

Go Online You can complete an Extra Example online.

Example 2 Compare Mean Absolute Deviations

Two driving schools use the same practice driver's test. Out of 100, School A had scores of 70, 79, 80, 82, and 95. School B had scores of 77, 83, 83, 81, and 82.

Find the mean absolute deviations. Then compare the variations.

Part A Find the means and the mean absolute deviations.

School A

Mean: ________

MAD: ________

School B

Mean: ________

MAD: ________

Show your work here

Part B Compare the variations.

The mean absolute deviation for School ________ is greater than that for School ________. This means the scores for School ________ are closer together and clustered around the mean. The scores for School ________ are more spread out and not as clustered around the mean.

Think About It!
What does a small value for the MAD tell you about the data set? What does a large value for the MAD tell you about the data set?

Talk About It!
Based on what you know about the mean absolute deviations, explain which school had more consistent test score results.

Check

The table shows the height of waterslides at two different water parks.

Height of Waterslides (ft)									
Splash Lagoon					Wild Water Bay				
75	95	80	110	88	120	108	94	135	128

Part A Find the mean absolute deviations.

Part B Compare the variations.

Go Online You can complete an Extra Example online.

Foldables It's time to update your Foldable, located in the Module Review, based on what you learned in this lesson. If you haven't already assembled your Foldable, you can find the instructions on page FL1.

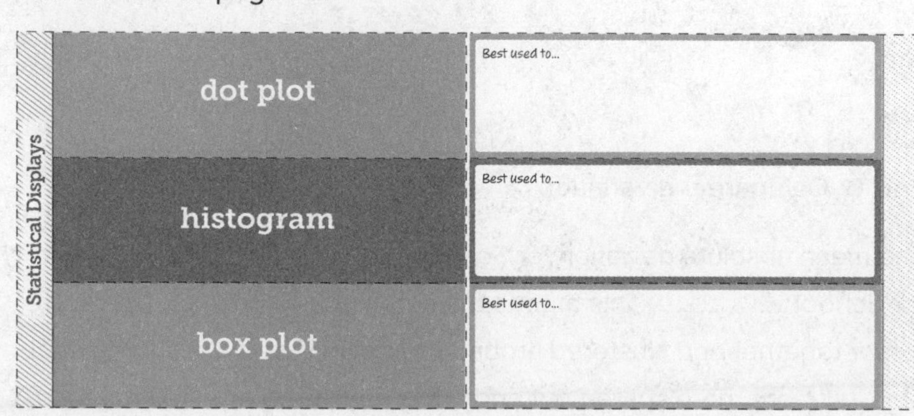

Name ______________________ Period ____________ Date ____________

Practice

Go Online You can complete your homework online.

1. The table shows the number of sunny days in various U.S. cities in the last month. Find the mean absolute deviation. Explain what the mean absolute deviation represents. **(Example 1)**

Number of Sunny Days in Various Cities Last Month			
15	27	10	19
24	21	28	16

2. The table shows the number of flowers sold by each sixth grade homeroom. Find the mean absolute deviation. Explain what the mean absolute deviation represents. **(Example 1)**

Number of Flowers Sold				
75	89	80	145	85
60	92	104	90	100

3. The table shows the number of wins of two school baseball teams over the last five years. Find the mean absolute deviation for each team. Then compare the variations. **(Example 2)**

	Number of Wins Per Season				
Bears	7	10	13	12	9
Saints	12	15	10	14	13

4. The table shows the number of canned goods each homeroom collected over seven days. Find the mean absolute deviation. Then compare the variations. Round to the nearest hundredth, if necessary. **(Example 2)**

	Number of Canned Goods Collected						
Room 101	57	52	40	42	37	54	47
Room 102	51	17	42	40	46	74	31

Test Practice

5. Open Response The table shows the number of Calories per serving of different snacks. What is the mean absolute deviation of the data set? Round to the nearest hundredth, if necessary.

Number of Calories					
61	42	52	27	35	23

Apply

6. The table shows the number of laps Candice and her two friends ran each day for five days. Which friend ran the most consistent number of laps each day? Use the mean absolute deviation to construct an argument to justify your response.

Girl	Day 1	Day 2	Day 3	Day 4	Day 5
Candice	5	6	8	5	7
Malaya	4	5	3	3	5
Zoe	7	8	6	8	8

7. MP **Persevere with Problems** The table shows the highway fuel economy of various popular vehicles. Find the mean absolute deviation. How many data values are closer than one mean absolute deviation away from the mean?

Fuel Economy (miles per gallon)				
34	48	25	35	33
37	32	34	23	30

8. MP **Justify Conclusions** The table shows the high temperatures for the last 6 days. If today's high temperature was 61°F, how is the mean absolute deviation affected? Justify your response.

High Temperature (°F)					
75	58	72	68	69	66

9. MP **Make an Argument** Use the meanings of the terms *mean*, *absolute*, and *deviation* to make an argument for why the mean absolute deviation of a data set is named using these terms.

10. MP **Reason Inductively** If the distance between the mean and a data value on a number line is 0, what do you know about the data value? Explain.

Outliers

I Can... understand how an outlier may affect a measure of center, and determine which measure of center is most appropriate to use when describing a data set that does or does not contain an outlier.

What Vocabulary Will You Learn?
outlier

Learn Outliers

An **outlier** is a data value that is very far away from the other data values. It can be much greater in value or much less than the other values. Consider the data set shown.

225, 245, 295, 305, 360, 387, 388, 420, 470, 480, 625, 780

How do you know if either of the extreme values, 225 or 780, are considered outliers?

An outlier is defined as a value that lies more than 1.5 times the interquartile range either above Q_3 or below Q_1.

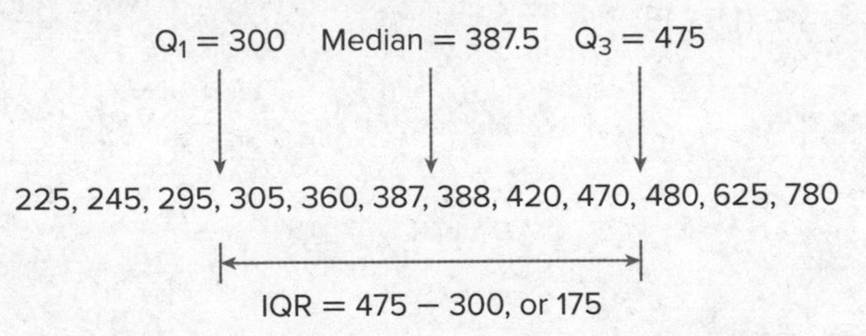

Determine the upper and lower limits for the outliers.

Upper Limit		Lower Limit
$Q_3 + (1.5 \cdot IQR)$		$Q_1 - (1.5 \cdot IQR)$
$= 475 + (1.5 \cdot 175)$	Substitute.	$= 300 - (1.5 \cdot 175)$
$= 475 + 262.5$	Multiply.	$= 300 - 262.5$
$= 737.5$	Simplify.	$= 37.5$

Any data values that are greater than 737.5 or less than 37.5 are outliers. So, the value 780 is an outlier. Because the data set does not contain any values that are less than 37.5, the only outlier is 780.

The box plot represents the data set. Outliers are indicated with an asterisk (*).

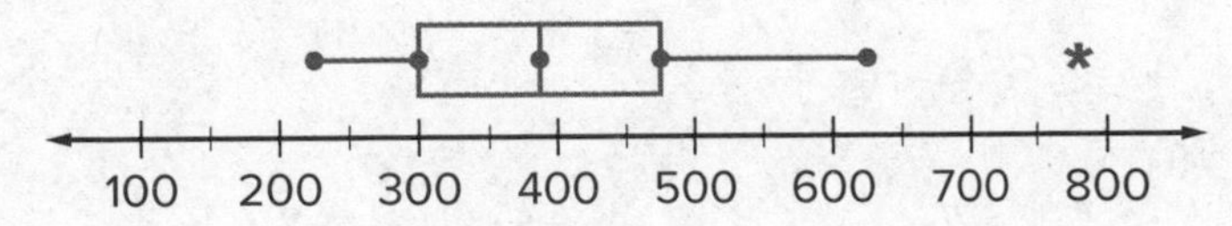

Talk About It!
If the outlier was removed from the data set, will the median still be 387.5? Why or why not?

Your Notes

Think About It!

What measures of variation do you need to find in order to identify any outliers?

Talk About It!

What does it mean that 23 is an outlier?

Example 1 Identify Outliers

The ages, in years, of the candidates in an election are 55, 49, 48, 57, 23, 63, and 72.

Identify any outliers in the data set.

Step 1 Find the quartiles and interquartile range.

List the data values from least to greatest.

least **greatest**

Find the quartiles and interquartile range.

Median = ____ Q_1 = ____ Q_3 = ____ IQR = ____

Step 2 Determine the upper and lower limits for the outliers.

Upper Limit

$Q_3 + (1.5 \cdot \text{IQR})$

$= \square + (1.5 \cdot 15)$ Substitute.

$= \square + \square$ Multiply.

$= \square$ Simplify.

Lower Limit

$Q_1 - (1.5 \cdot \text{IQR})$

$= \square - (1.5 \cdot 15)$

$= \square - \square$

$= \square$

Step 3 Identify any outliers.

Any data values that are greater than 85.5 or less than 25.5 are outliers. So, the value 23 is an outlier. Because the data set does not contain any values that are greater than 85.5, the only outlier is 23.

Check

The lengths, in feet, of various bridges are 354, 88, 251, 275, 727, and 1,121. Identify any outliers in the data.

Go Online You can complete an Extra Example online.

Explore Mean, Median, and Outliers

Online Activity You will use Web Sketchpad to explore how outliers affect the mean and median.

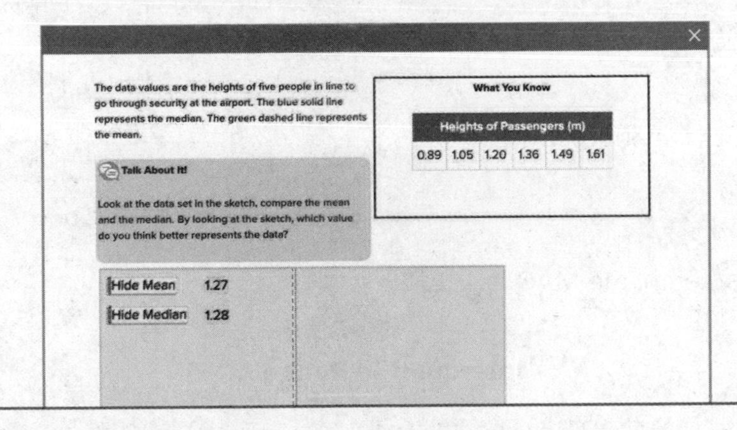

Learn Describe the Effect of Outliers

If a data set contains an outlier, the outlier may affect the measures of center and/or variation.

Suppose you track the daily high temperatures for one week and the results are recorded in the table shown.

High Temperatures (°F)	
Sunday	72
Monday	68
Tuesday	71
Wednesday	74
Thursday	75
Friday	72

Suppose that the high temperature on Saturday is 42°F. This temperature is much lower than the other temperatures in the data set. It is also an outlier, because 42 is less than the lower limit for outliers.

$Q_1 - (1.5 \cdot \text{IQR})$

$= 68 - (1.5 \cdot 6)$ Substitute.

$= 68 - 9$ Multiply.

$= 59$ Simplify.

Because $42 < 59$, 42 is an outlier.

Median = 72

$Q_1 = 68$ $Q_3 = 74$

42, 68, 71, 72, 72, 74, 75

IQR = 74 − 68, or 6

(continued on next page)

Talk About It!

Suppose Saturday's temperature had been 59°F, which does not qualify as an outlier, but is cooler than the rest. How does this affect the mean? the median?

To see how an outlier affects the measures of center and variation, calculate the measures both with and without the outlier.

Calculate the measures with the outlier.

Mean

$$\frac{42 + 68 + 71 + 72 + 72 + 74 + 75}{7} \approx 67.7$$

Mean Absolute Deviation (MAD)

Show your work here

Median = 72

$Q_1 = 68$ $Q_3 = 74$

42, 68, 71, 72, 72, 74, 75

IQR = 74 − 68, or 6

To the nearest tenth, the MAD is ______.

Median

The median is ______.

Interquartile Range (IQR)

The IQR is ______.

Calculate the measures without the outlier.

Mean

$$\frac{68 + 71 + 72 + 72 + 74 + 75}{6} = 72$$

Mean Absolute Deviation (MAD)

Median = 72

$Q_1 = 71$ $Q_3 = 74$

68, 71, 72, 72, 74, 75

IQR = 74 − 71, or 3

To the nearest tenth, the MAD is ______.

Median

The median is ______.

Interquartile Range (IQR)

The IQR is ______.

The median was not affected by the inclusion of the outlier. Without the outlier, the mean, MAD, and IQR all increased in value. With the outlier, the mean is not the best representation of center, because most of the values are higher than 67.7.

Use either the mean or median when the data does not contain any outliers. Use only the median when the data contains an outlier. While the median might change a little when an outlier is included or removed, it does not change as much as the mean.

Use the corresponding measure of variation to describe the spread of the data.

- If you choose the mean to describe the center, choose the MAD to describe the variation.
- If you choose the median to describe the center, choose the IQR to describe the variation.

Example 2 Describe the Effect of Outliers

The table shows the average lifespans of selected animals.

Calculate the mean and median with and without the outlier, 200. Then choose the measure that best describes the center.

Average Lifespan	
Animal	Lifespan (years)
Elephant	35
Dolphin	30
Chimpanzee	50
Tortoise	200
Gorilla	30
Gray Whale	70
Horse	20

Step 1 Calculate the mean and median with the outlier. Round to the nearest tenth, if necessary.

Mean

$$\frac{35 + 30 + 50 + 200 + 30 + 70 + 20}{7} \approx 62.1$$

The mean lifespan is about ______ years.

Median

The median lifespan is ______ years.

Step 2 Calculate the mean and median without the outlier. Round to the nearest tenth, if necessary.

Mean

$$\frac{35 + 30 + 50 + 30 + 70 + 20}{6} \approx 39.2$$

The mean lifespan is about ______ years.

Median

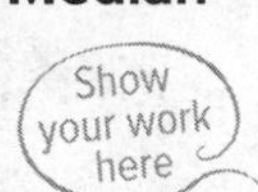

The median lifespan is ______ years.

Step 3 Choose the measure that best describes the center.

The __________ was most affected by the inclusion of the outlier.

The __________ changed very little.

So, the __________ best describes the center of the data.

Think About It!

Will the outlier affect the mean or the median more? Explain your reasoning.

Talk About It!

Explain why it makes sense that the lifespan of the animals listed in the table are centered around 32.5 or 35 years, rather than around 39 or 62 years.

Check

The table shows the cooking temperatures for different recipes. Calculate the mean and median with and without the outlier. Round to the nearest tenth, if necessary. Then choose the measure that best describes the center.

Cooking Temperature (°F)			
175	325	325	350
350	350	400	450

Go Online You can complete an Extra Example online.

Pause and Reflect

Create a graphic organizer that will help you study the concepts you learned today in class.

Name ______________________ Period __________ Date __________

Practice

Go Online You can complete your homework online.

1. Last week, Joakim spent 40, 25, 60, 30, 35, and 40 minutes practicing the piano. Identify any outliers in the data. **(Example 1)**

2. Last month, a basketball team scored 83, 84, 85, 87, 89, 88, 67, 79, and 81 points in their games. Identify any outliers in the data. **(Example 1)**

3. Abrianna sold 20, 23, 18, 4, 17, 21, 15, and 56 boxes of cookies after different football games. Identify any outliers in the data. **(Example 1)**

4. Last week a certain pet store had 52, 72, 96, 21, 58, 40, and 75 paying customers. Identify any outliers in the data. **(Example 1)**

5. The prices of trees that Sahana bought are $46, $39, $40, $45, $44, $68, and $51. Calculate the mean and median with and without the outlier. Round to the nearest tenth, if necessary. Choose the measure that best describes the center. **(Example 2)**

6. The prices of backpacks are $37, $43, $41, $36, $44, and $70. Calculate the mean and median with and without the outlier. Round to the nearest tenth, if necessary. Choose the measure that best describes the center. **(Example 2)**

7. The table shows the number of points scored by a football team. Calculate the mean and median with and without the outlier. Round to the nearest tenth, if necessary. Choose the measure that best describes the center. Explain. **(Example 2)**

Points Scored by a Football Team			
14	20	3	9
18	35	21	24
7	12	31	68

Test Practice

8. Open Response The table shows the number of points scored by the players in a trivia game. Which measure of center best represents the data? Explain your reasoning.

Points Scored in a Trivia Game			
12	9	5	11
6	0	14	7

9. Create Generate a set of real-world data that contains two outliers.

10. MP Justify Conclusions The ages, in years, of participants in a relay race are 12, 15, 14, 13, 15, 12, 22, 16, and 11. Identify any outliers in the data set. Justify your response.

11. MP Construct an Argument Explain how an outlier may or may not affect the mean and median.

12. MP Justify Conclusions Does an outlier affect the range of a data set? Explain.

Interpret Graphical Displays

I Can... determine the symmetry of data represented in different displays, determine the most appropriate measure of center and variation based on the symmetry, and use the measures to describe the data.

What Vocabulary Will You Learn?
cluster
distribution
gap
peak
symmetric distribution

Learn Interpret Dot Plots

The **distribution** of a data set shows the arrangement of data values. It can be described by its center, spread (variation), or overall shape. Determining the symmetry of a distribution is one way to describe its shape. If the left side of a distribution looks like the right side, then the distribution is a **symmetric distribution**. If there is an outlier, the distribution is usually not symmetric.

Symmetric	Not Symmetric
The left side looks like the right side.	The right side does not look like the left side.

Talk About It!
Why will the mean and median for a symmetric graph always be the same value?

The shape of a data distribution tells you which measure of center and measure of spread are most appropriate to use.

Is the data distribution symmetric?	
Yes	Use the mean to describe the center. Use the mean absolute deviation to describe the spread.
No	Use the median to describe the center. Use the interquartile range to describe the spread.

Your Notes

Think About It!
What do you notice about the shape of the distribution?

Talk About It!
What do you notice about the measure of center's location on the dot plot?

Example 1 Interpret Dot Plots

The results of a class survey about the number of states visited by students are shown in the dot plot.

Number of States Visited

0 1 2 3 4 5 6 7 8 9 10

Choose the appropriate measure of center and variation. Then use the measures to describe the distribution.

Part A Choose the appropriate measures.

The data are not evenly distributed between the left side and the right side.

There appears to be an outlier.

So, the distribution is not symmetric.

Which measure of center should you use? ________________

Which measure of variation should you use? ________________

Part B Describe the distribution.

A total of ________ students responded to the survey.

Find the measure of center you chose in Part A.

Show your work here

The measure of center indicates that the number of states visited by the students can be summarized by the single value of ________ states.

Find the measure of variation you chose in Part A.

The measure of variation indicates that the spread of the data around the center is ________ states. Other than the outlier, there is not a lot of variation among the data.

Check

The results of a class survey about the number of hours spent on the Internet each week by students are shown in the dot plot.

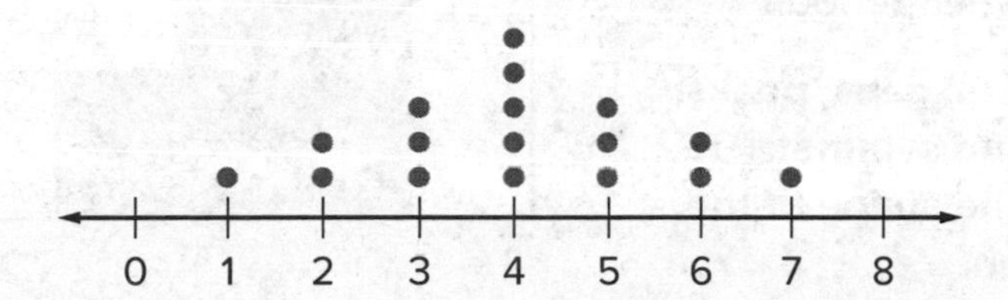

Part A Choose the appropriate measure of center and variability.

Part B Use the chosen measures to describe the distribution.

Go Online You can complete an Extra Example online.

Learn Interpret Histograms

You can also describe the distribution of histograms, including symmetry, clusters, gaps, and peaks. A **cluster** occurs when data values are grouped together. A **gap** occurs where there are no data values. A **peak** is the most frequently occurring value or interval of values in a data set.

Data were collected on the heights of some buildings in Seattle, Washington and are displayed in the histogram. The graph shows an example of a peak, a gap, and a cluster. This distribution is not symmetric and does not contain any outliers.

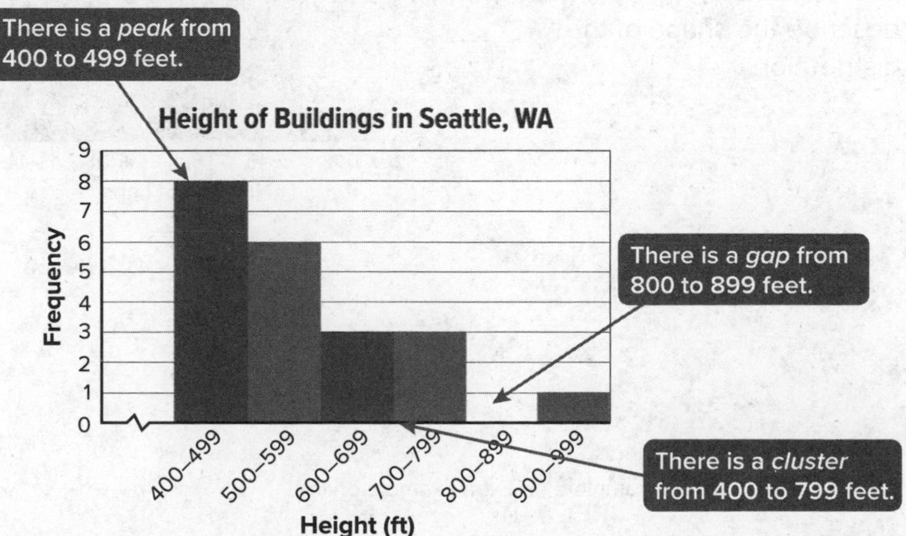

Talk About It!

Use the histogram to describe the heights of the buildings in Seattle.

Example 2 Interpret Histograms

The histogram shows the duration, in minutes and seconds, of solar eclipses over a 10-year period.

Use clusters, gaps, peaks, outliers, and symmetry to describe the shape of the distribution.

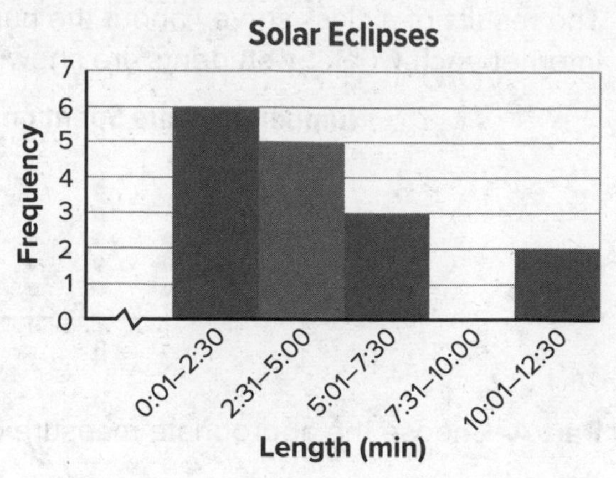

Step 1 Identify any symmetry, clusters, and outliers. The distribution is not symmetric. There is a cluster from 0:01–7:30. There are no outliers.

Step 2 Identify any peaks.
There is a peak from 0:01–2:30.

Step 3 Identify any gaps.
There is a gap from 7:31–10:00.

Step 4 Describe the distribution.

Summarize the information you found.

The distribution is not symmetric and does not contain any outliers. The data cluster around 1 second to 7 minutes and 30 seconds and have a peak at 1 second to 2 minutes and 30 seconds. There is a gap at 7 minutes and 31 seconds to 10 minutes.

Think About It!

How many solar eclipses are represented in the data set?

Talk About It!

What can you infer about solar eclipses using the cluster of data values?

Check

The histogram shows the number of laps each student walked while exercising. Use clusters, gaps, peaks, outliers, and symmetry to describe the shape of the distribution.

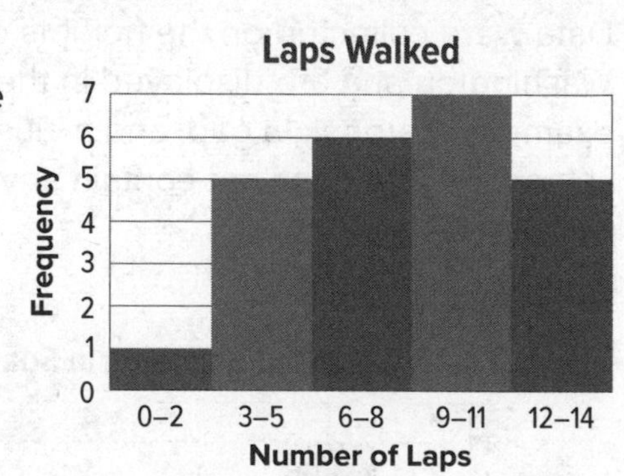

Go Online You can complete an Extra Example online.

Explore Interpret Box Plots

Online Activity You will use Web Sketchpad to explore how changes in a data set affect a box plot.

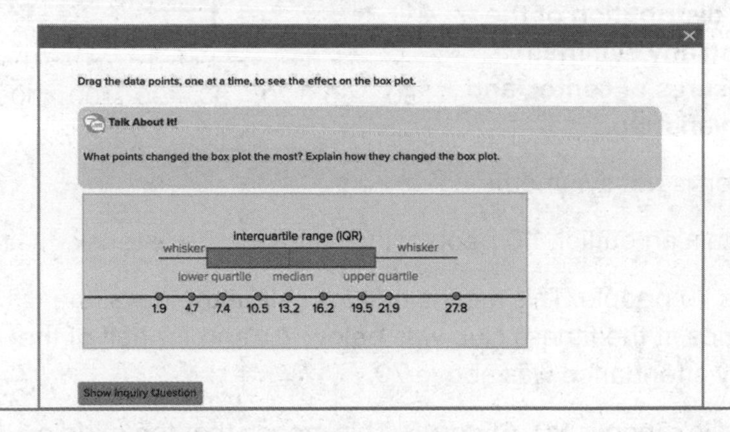

Learn Interpret Box Plots

Although a box plot does not show individual data values, you can still describe the distribution of data.

Box plots are constructed using the median and interquartile range, so use those measures to describe the center and variation of the data. Because a box plot does not show individual data values, the mean cannot be found, unless the data are perfectly symmetric. In this case, the mean and the median have the same value.

Box plots do indicate symmetry.

If the whiskers are all the same length, and the median line divides the box into two equal-sized boxes, then the distribution is symmetric.

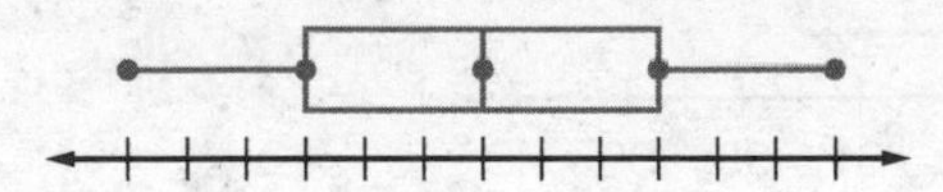

If the boxes and whiskers are of varying lengths, then the distribution is not symmetric.

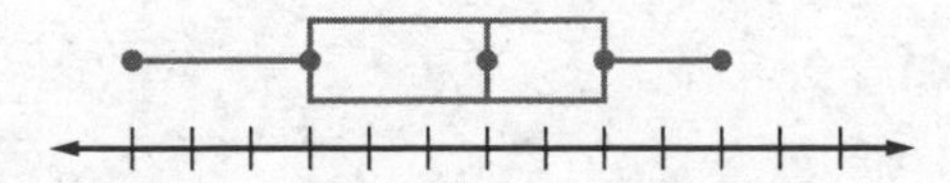

Outliers are represented by an asterisk (*) on a box plot. Whiskers will not extend to outliers, but instead to the previous or next data value.

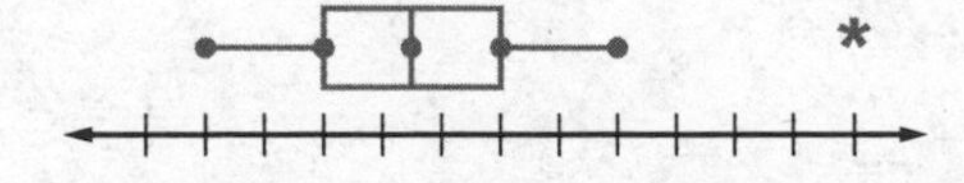

Talk About It!

What percent of the data is represented by each box and whisker? What do shorter boxes or whiskers indicate about the data? longer boxes or whiskers?

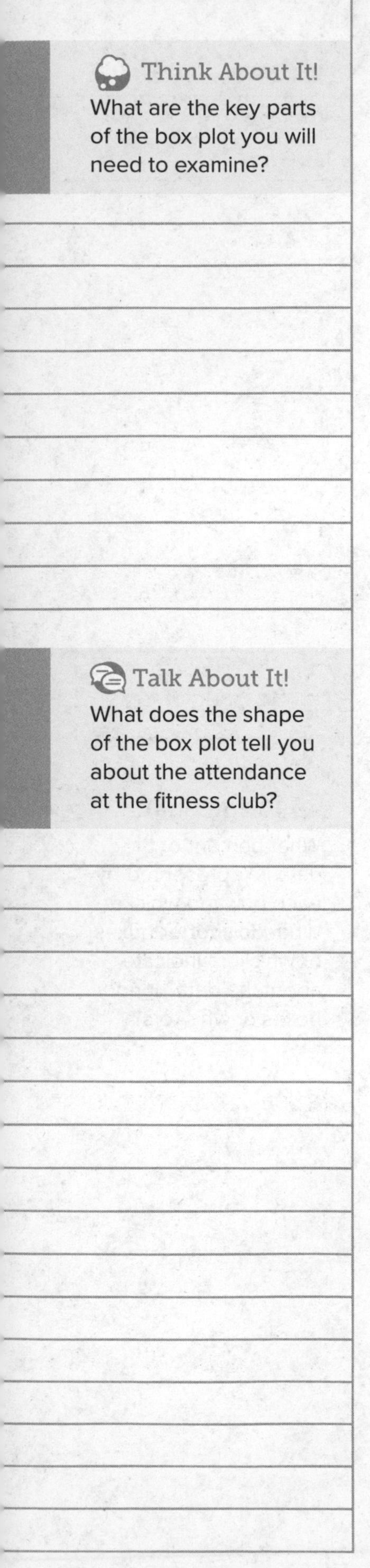

Example 3 Interpret Box Plots

The box plot shows the daily attendance at a fitness club.

Describe the distribution of the data, including any symmetry, outliers, measures of center, and measures of variation.

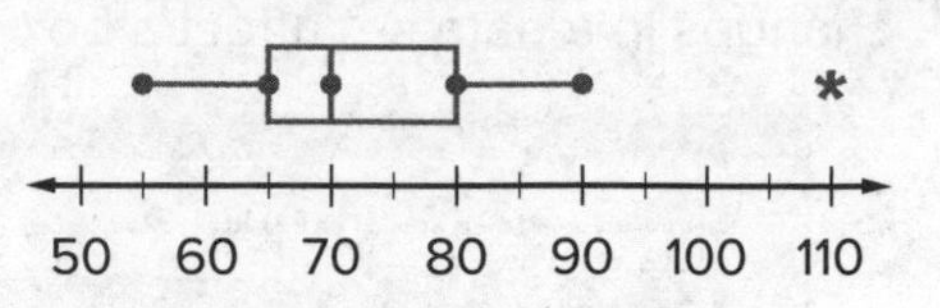

The distribution is not symmetric.

The data contain an outlier, 110 people, indicated by the asterisk.

The median is 70 people. This means that for half of the days, the daily attendance at the fitness club was below 70, and for half of the days, the daily attendance was above 70.

The interquartile range is 80–65, or 15. This means that the middle 50% of the data vary by 15.

The left box is the shortest. This means that 25% of the data is between 65 and 70 people, and these data values are closer together than the data values in the other box or whiskers.

Check

The average gas mileage for various sedans is shown in the box plot. Describe the distribution of the data, including any symmetry, outliers, measures of center, and measures of variation.

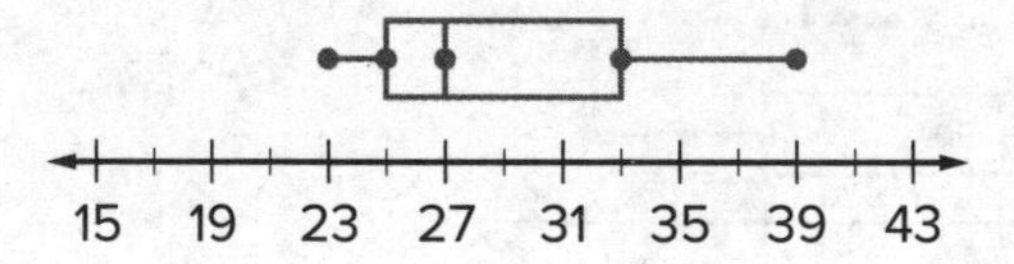

Go Online You can complete an Extra Example online.

Apply Travel

The histogram shows the distances a volleyball team travels to their games. One player claimed that because the peak of the distribution is from 20-24 miles, that the team traveled 20-24 miles more than 50% of the time. Is the player correct?

Distance Traveled

Frequency: 0, 1, 2, 3, 4, 5, 6, 7, 8, 9, 10

10–14 15–19 20–24 25–29 30–34

Distance (mi)

1 What is the task?

Make sure you understand exactly what question to answer or problem to solve. You may want to read the problem three times. Discuss these questions with a partner.

First Time Describe the context of the problem, in your own words.
Second Time What mathematics do you see in the problem?
Third Time What are you wondering about?

2 How can you approach the task? What strategies can you use?

3 What is your solution?

Use your strategy to solve the problem.

4 How can you show your solution is reasonable?

Write About It! Write an argument that can be used to defend your solution.

Talk About It!

Give an example of how the data may have been gathered. Could this have affected the results? Explain your reasoning.

Check

The histogram shows the heights of the students in Mrs. Sanchez's class. What percent of the students are taller than 55 inches? Round to the nearest tenth if necessary.

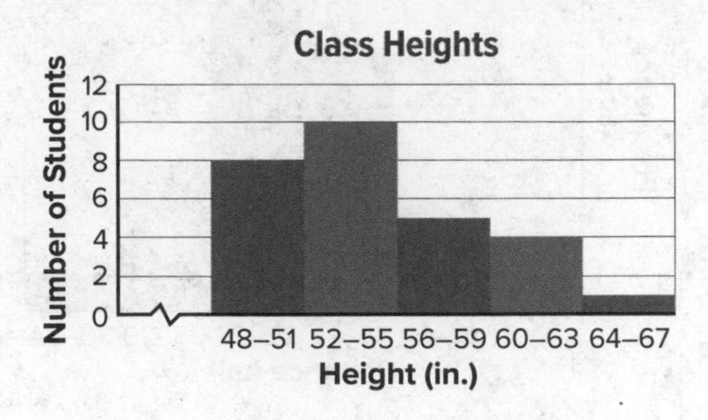

Go Online You can complete an Extra Example online.

Foldables It's time to update your Foldable, located in the Module Review, based on what you learned in this lesson. If you haven't already assembled your Foldable, you can find the instructions on page FL1.

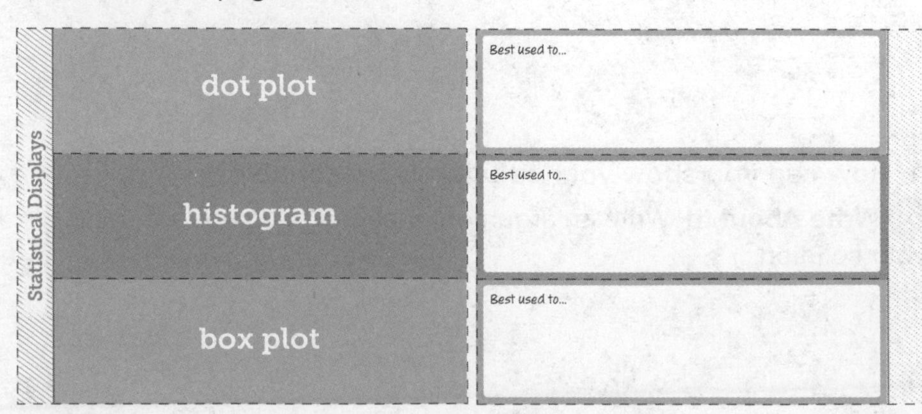

Name ______________________ Period ____________ Date ____________

Practice

Go Online You can complete your homework online.

1. The dot plot shows the number of televisions owned by the families in a neighborhood. Choose the appropriate measure of center and variation. Then use the measures to describe the data set. **(Example 1)**

Number of Televisions

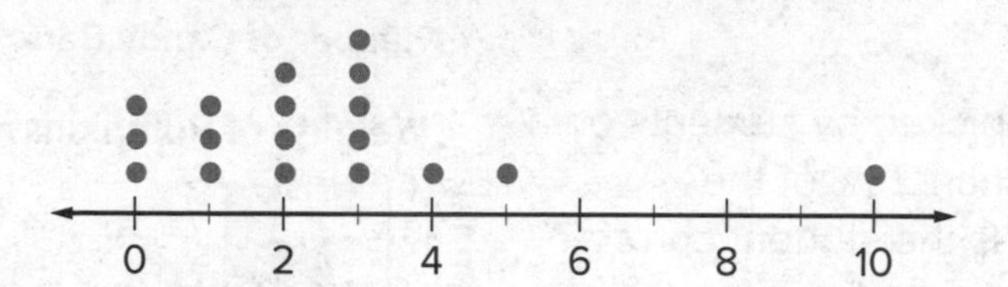

2. The dot plot shows the number of miles run by various sixth grade students. Choose the appropriate measure of center and variation. Then use the measures to describe the data set. **(Example 1)**

Miles Run

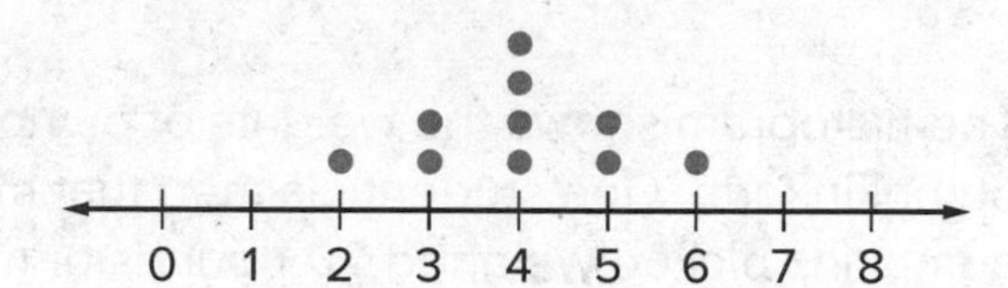

3. The histogram shows the dollars pledged by supporters of an animal shelter. Use clusters, gaps, peaks, outliers, and symmetry to describe the shape of the distribution. **(Example 2)**

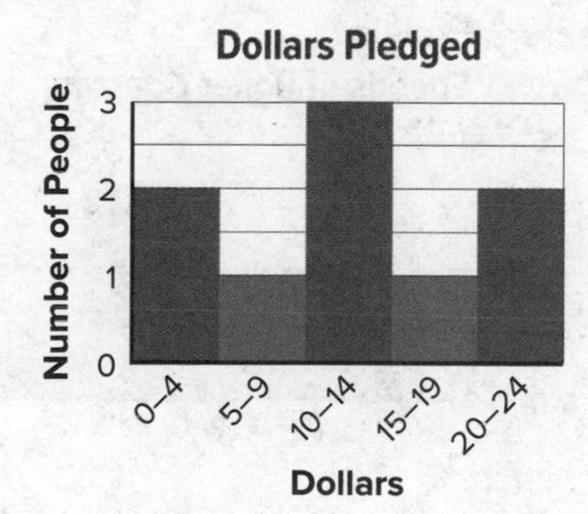

4. The box plot shows the amount of money, in dollars, Olivia saved during various months. Find the median and the measures of variation. Then describe the data. **(Example 3)**

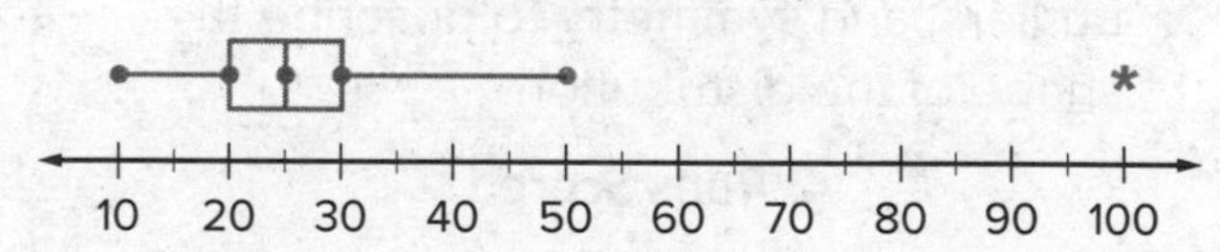

Test Practice

5. Multiple Choice The box plot shows the ticket prices, in dollars, of various concerts. What is the median, interquartile range, and range of the data, in that order?

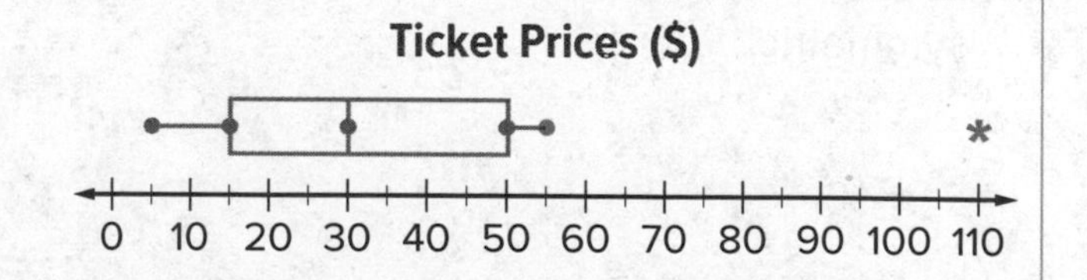

Ⓐ 30; 35; 50

Ⓑ 30; 40; 105

Ⓒ 30; 15; 50

Ⓓ 30; 35; 105

Apply

6. The histogram shows the number of candy bars each player on a football team sold. One player claimed that more than 50% of the players sold 90 or more candy bars. Is the player correct? Write an argument that can be used to defend your solution.

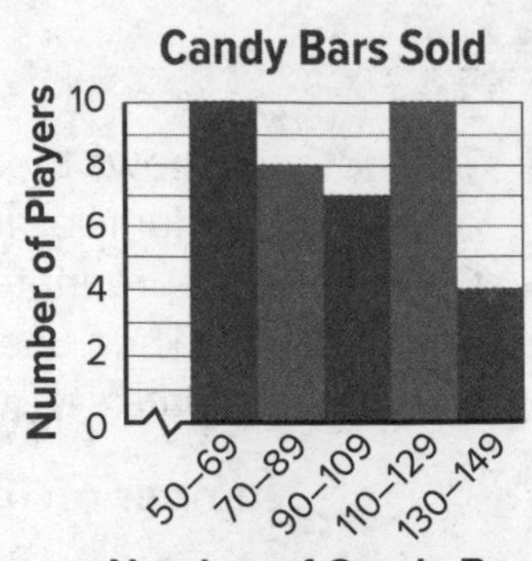

7. The histogram shows the weights of pumpkins picked by students on a pumpkin farm. One student claimed that more than 25% of the pumpkins picked weighed 20 pounds or more. Is the student correct? Write an argument that can be used to defend your solution.

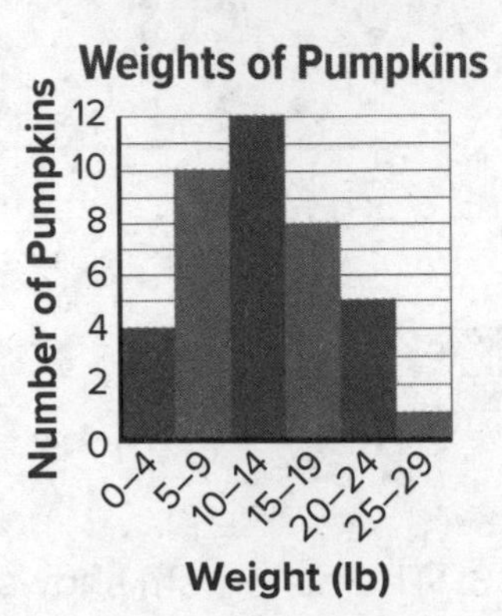

8. MP **Be Precise** The dot plot shows the number of runs scored by a baseball team for last season. Use clusters, gaps, peaks, outliers, and symmetry to describe the shape of the distribution.

Runs Scored

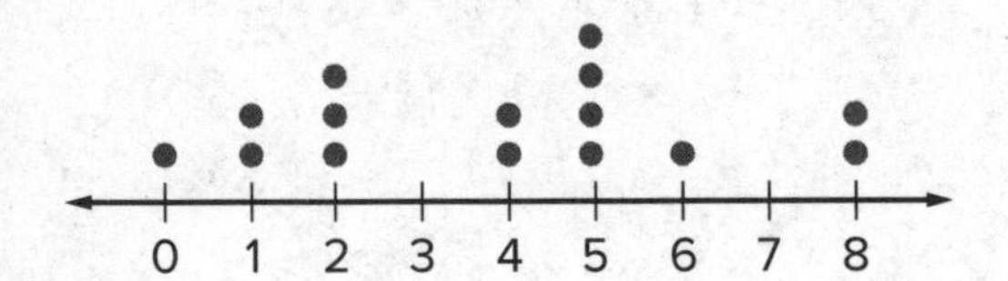

9. MP **Justify Conclusions** According to the histogram, do more than 50% of the roller coasters have a speed of 70 mph or greater? Explain.

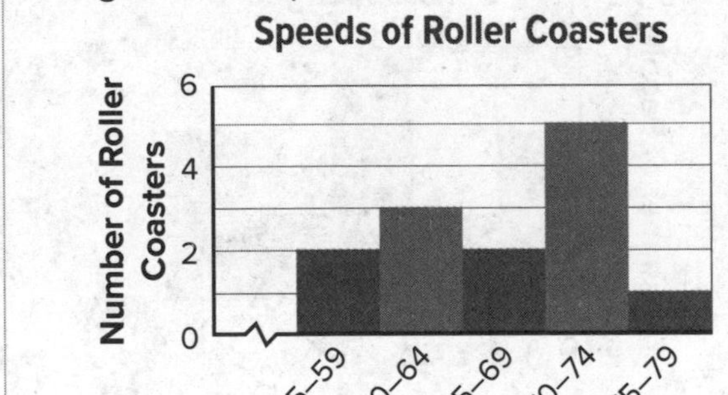

10. Create Draw a dot plot that is not symmetric.

11. MP **Persevere with Problems** If a box plot's distribution is symmetric, which measure of center and measures of spread are most appropriate to use?

Review

Foldables Use your Foldable to help review the module.

Statistical Displays

What measures of center or measures of variation can be found using a dot plot?

What measures of center or measures of variation can be found using a histogram?

What measures of center or measures of variation can be found using a box plot?

Rate Yourself!   

Complete the chart at the beginning of the module by placing a checkmark in each row that corresponds with how much you know about each topic after completing this module.

Write about one thing you learned.

Write about a question you still have.

Reflect on the Module

Use what you learned about statistical measures to complete the graphic organizer.

Essential Question

Why is data collected and analyzed and how can it be displayed?

How are the mean and median helpful in describing data?		
	Mean	**Median**
Definition		
When is it appropriate to use?		
How does an outlier affect it?		

How can data be displayed?			
	Dot Plot	**Histogram**	**Box Plot**
Definition			
Explain how to describe the data.			

Name ______________________ Period ____________ Date ____________

Test Practice

1. Multiselect Which of the following are statistical questions? Select all that apply. **(Lesson 1)**

- ☐ How many countries make up the continent of Africa?
- ☐ How many televisions does the typical family own?
- ☐ How many Major League Baseball teams are there?
- ☐ How many U.S. National Parks are there?
- ☐ How many states has the average student visited?
- ☐ How many students are in the average sixth grade class?

2. Open Response Scott kept track of how long he watched television for five days, and recorded the data in the table. What is the difference between the mean and median length of the time Scott spent watching television? Explain. **(Lesson 3)**

Day	1	2	3	4	5
Time (min)	60	30	45	90	60

3. Open Response The average annual amounts of rainfall for several U.S. cities are given in the table. **(Lesson 4)**

City	Rainfall (in.)
Atlanta	49.7
Baltimore	41.9
Chicago	36.9
Denver	15.6
Houston	49.8
Phoenix	8.2

What are the range and inter quartile range of the data?

4. Multiple Choice Jessica surveyed her teammates using the statistical question, *How many siblings do you have*? The results are shown in the table. **(Lesson 2)**

Number of Siblings		
1	2	4
2	3	0
1	1	3
1	1	2
5	3	3

A. Which dot plot best represents the situation?

Ⓐ
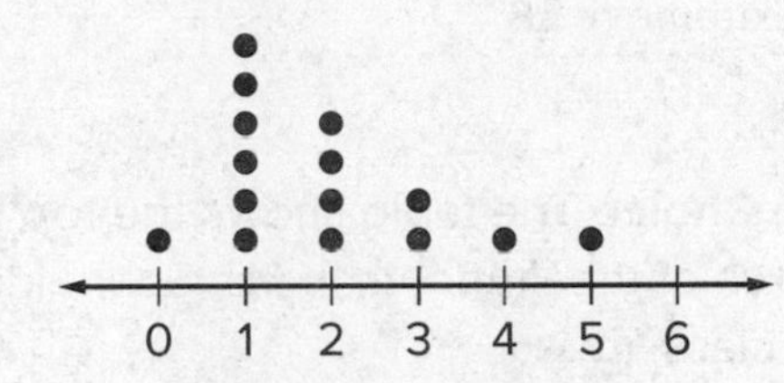

Ⓑ
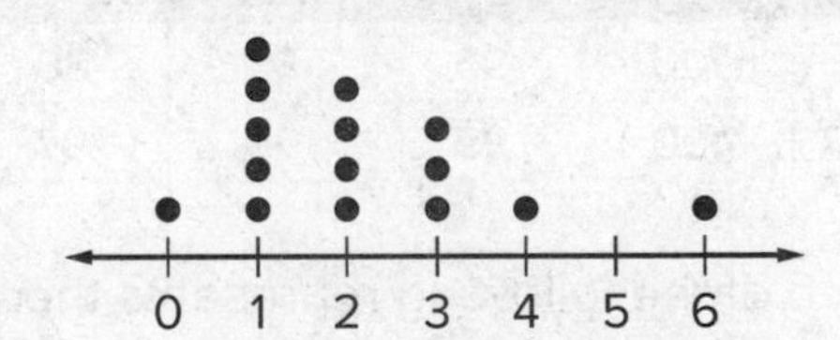

Ⓒ
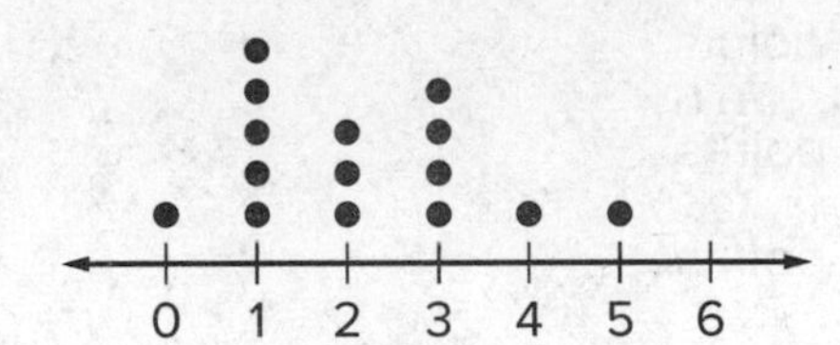

Ⓓ
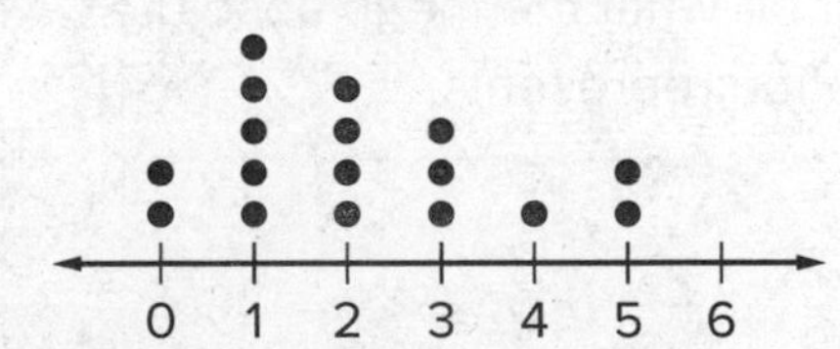

B. What is the greatest number of siblings out of all of her teammates? What is the least number of siblings out of all of her teammates?

5. Table Item The ages of the current students attending an art class at a local community center are shown in the box plot. Consider the parts of the box plot and indicate which of the parts are correctly named. **(Lesson 4)**

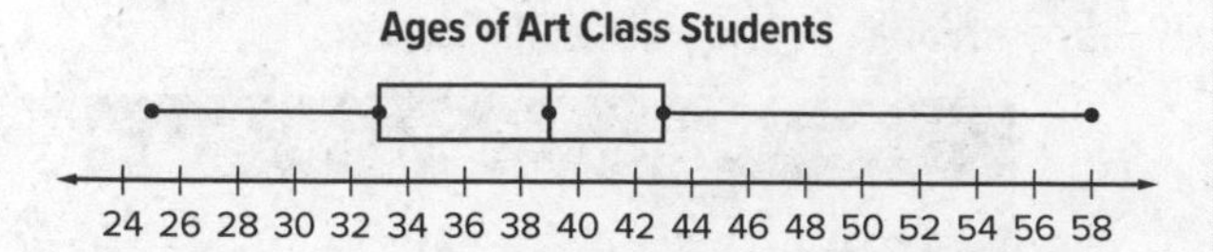

	Correct	Incorrect
Lower Extreme = 24		
Median = 39		
$Q_1 = 33$		
$Q_3 = 44$		
Upper Extreme = 58		

6. Multiple Choice The table shows the top ten test scores of the students in Ms. Schneider's science class. **(Lesson 5)**

Test Scores				
102	100	95	93	88
96	100	99	90	97

A. Which of the following represents the mean absolute deviation of the data?

Ⓐ 3.2 points

Ⓑ 3.4 points

Ⓒ 3.6 points

Ⓓ 4.1 points

B. Describe what the mean absolute deviation represents.

7. Multiselect The heights, in feet, of various trees in the park are 32, 10, 70, 40, 34, 44, and 36. Identify any outliers in the data set. Select all that apply. **(Lesson 6)**

☐ 10 feet

☐ 34 feet

☐ 36 feet

☐ 40 feet

☐ 70 feet

8. Open Response The histogram shows the distances Jerome's co-workers have to commute to work each morning. What percent of his co-workers travel more than 10 miles to work? Round to the nearest percent. **(Lesson 7)**

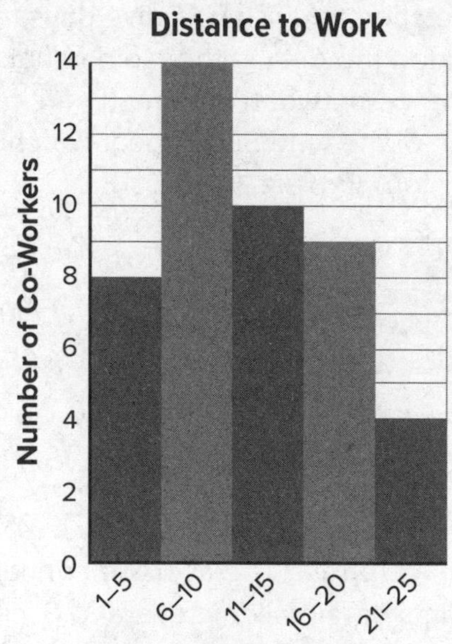

Foldables Study Organizers

What Are Foldables and How Do I Create Them?

Foldables are three-dimensional graphic organizers that help you create study guides for each module in your book.

Step 1 Go to the back of your book to find the Foldable for the module you are currently studying. Follow the cutting and assembly instructions at the top of the page.

Step 2 Go to the Module Review at the end of the module you are currently studying. Match up the tabs and attach your Foldable to this page. Dotted tabs show where to place your Foldable. Striped tabs indicate where to tape the Foldable.

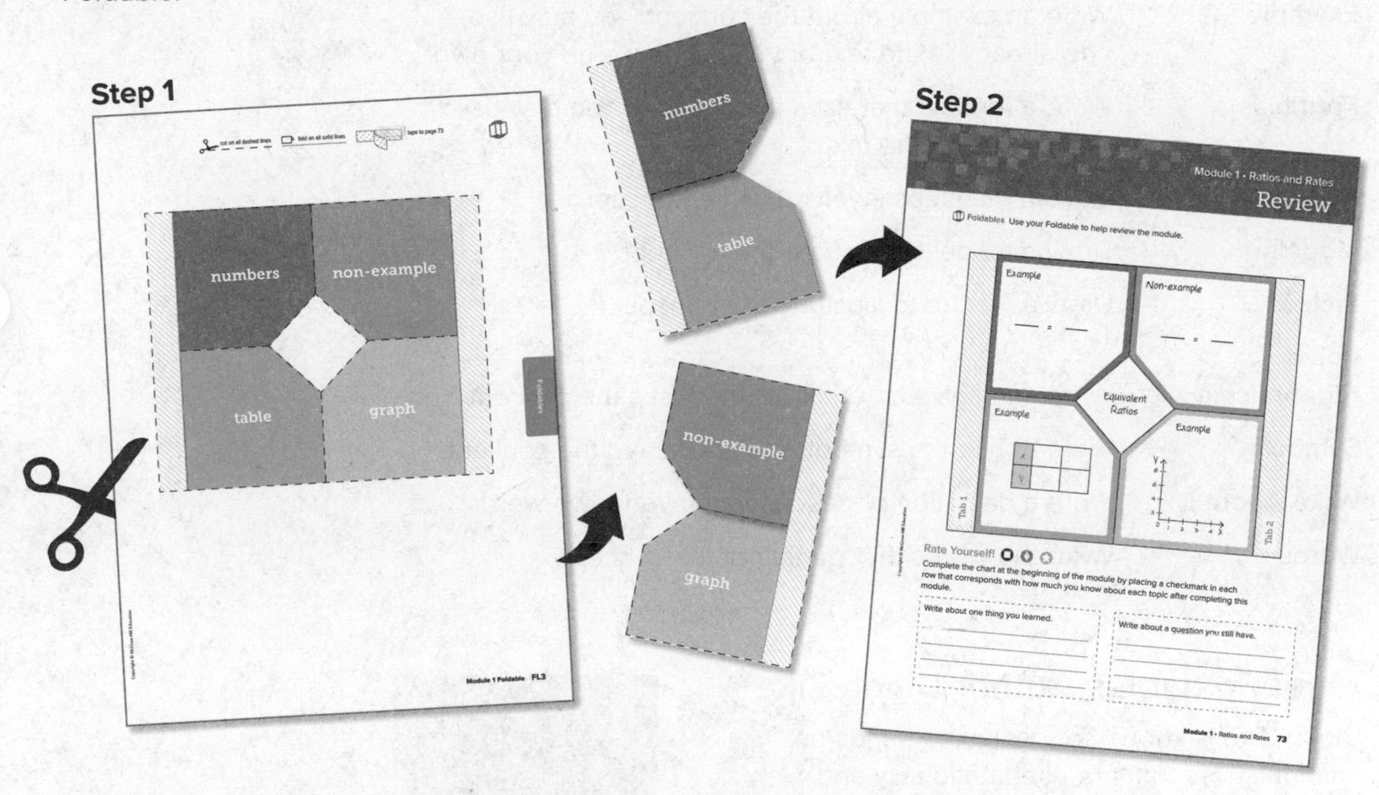

How Will I Know When to Use My Foldable?

You will be directed to work on your Foldable at the end of selected lessons. This lets you know that it is time to update it with concepts from that lesson. Once you've completed your Foldable, use it to study for the module test.

How Do I Complete My Foldable?

No two Foldables in your book will look alike. However, some will ask you to fill in similar information. Below are some of the instructions you'll see as you complete your Foldable. **HAVE FUN** learning math using Foldables!

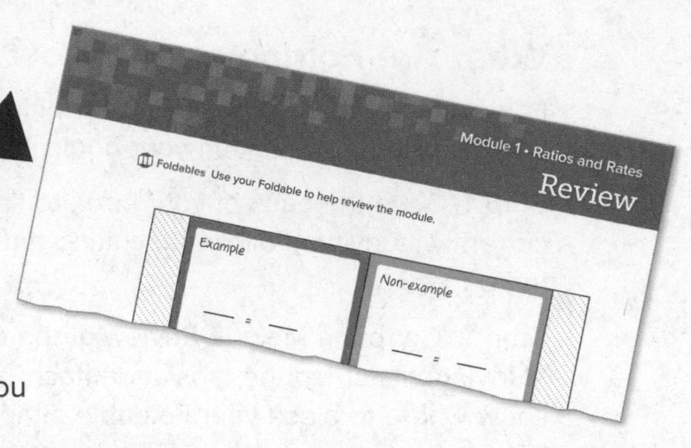

Instructions and What They Mean

Best Used to...	Complete the sentence explaining when the concept should be used.
Definition	Write a definition in your own words.
Description	Describe the concept using words.
Equation	Write an equation that uses the concept. You may use one already in the text or you can make up your own.
Example	Write an example about the concept. You may use one already in the text or you can make up your own.
Formulas	Write a formula that uses the concept. You may use one already in the text.
How do I ...?	Explain the steps involved in the concept.
Models	Draw a model to illustrate the concept.
Picture	Draw a picture to illustrate the concept.
Solve Algebraically	Write and solve an equation that uses the concept.
Symbols	Write or use the symbols that pertain to the concept.
Write About It	Write a definition or description in your own words.
Words	Write the words that pertain to the concept.

Meet Foldables Author Dinah Zike

Dinah Zike is known for designing hands-on manipulatives that are used nationally and internationally by teachers and parents. Dinah is an explosion of energy and ideas. Her excitement and joy for learning inspires everyone she touches.

cut on all dashed lines | fold on all solid lines | tape to page 329

Properties of Addition

Commutative	Associative	Identity
+	+	+
×	×	×
Commutative	Associative	Identity

Properties of Multiplication

Foldables

cut on all dashed lines

fold on all solid lines

tape to page 329

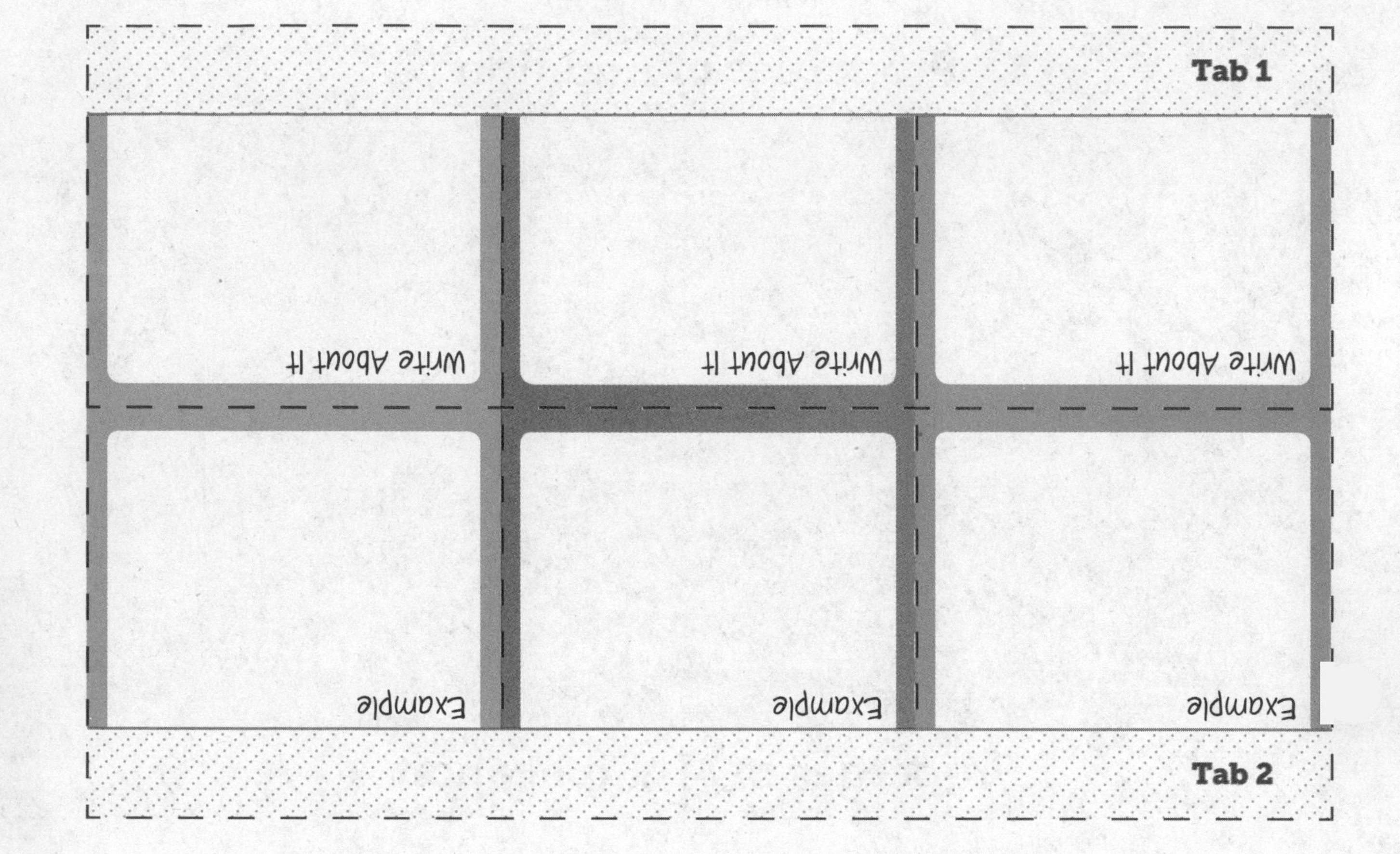

cut on all dashed lines

fold on all solid lines

tape to page 391

equations

Models

Symbols

addition (+)

Models

Symbols

subtraction (−)

Models

Symbols

multiplication (×)

Foldables

cut on all dashed lines

fold on all solid lines

tape to page 391

Tab 4

Write About It

Tab 3

Write About It

Tab 2

Write About It

Tab 1

Write About It

cut on all dashed lines fold on all solid lines tape to page 429

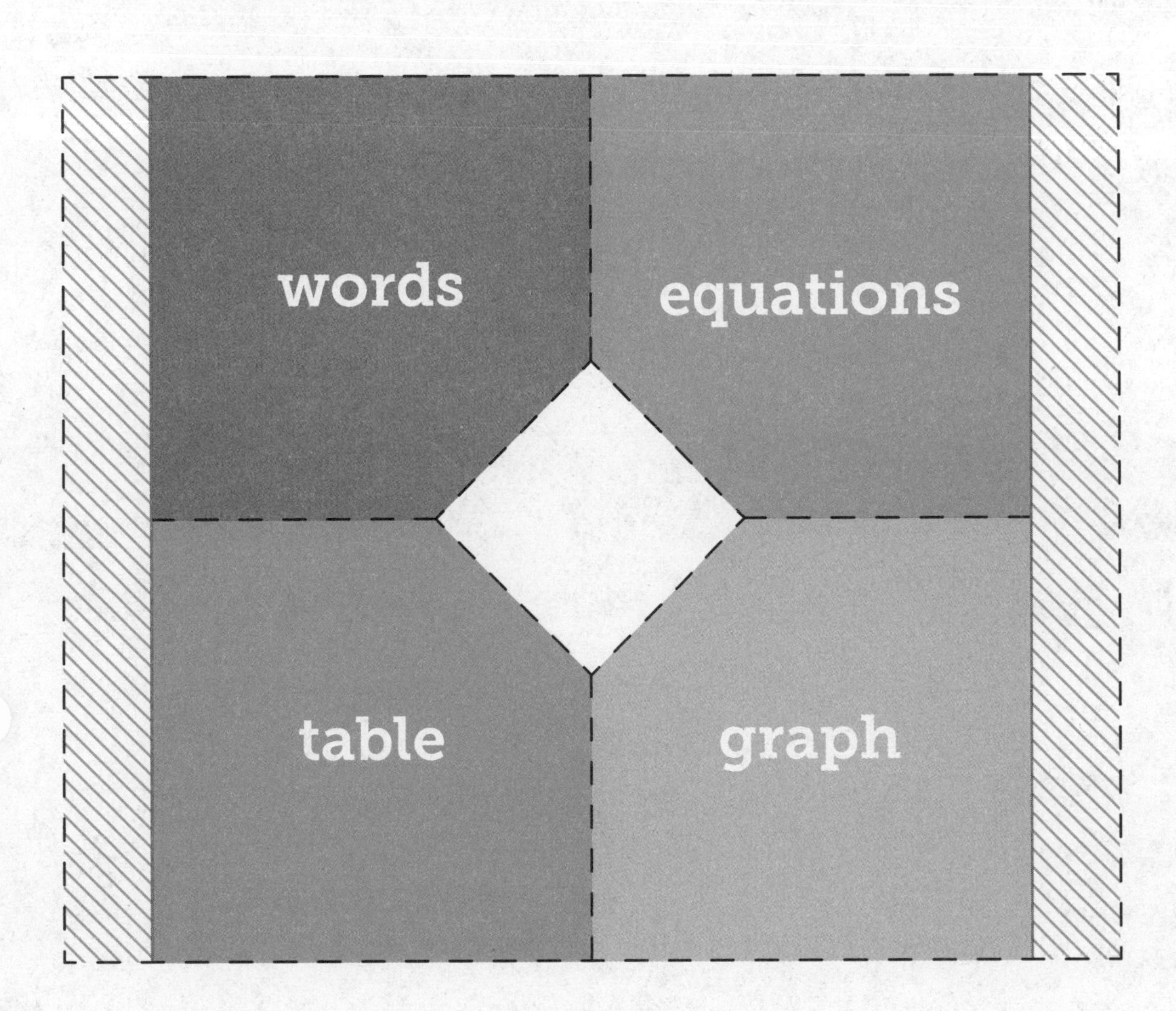

Foldables

cut on all dashed lines

fold on all solid lines

tape to page 429

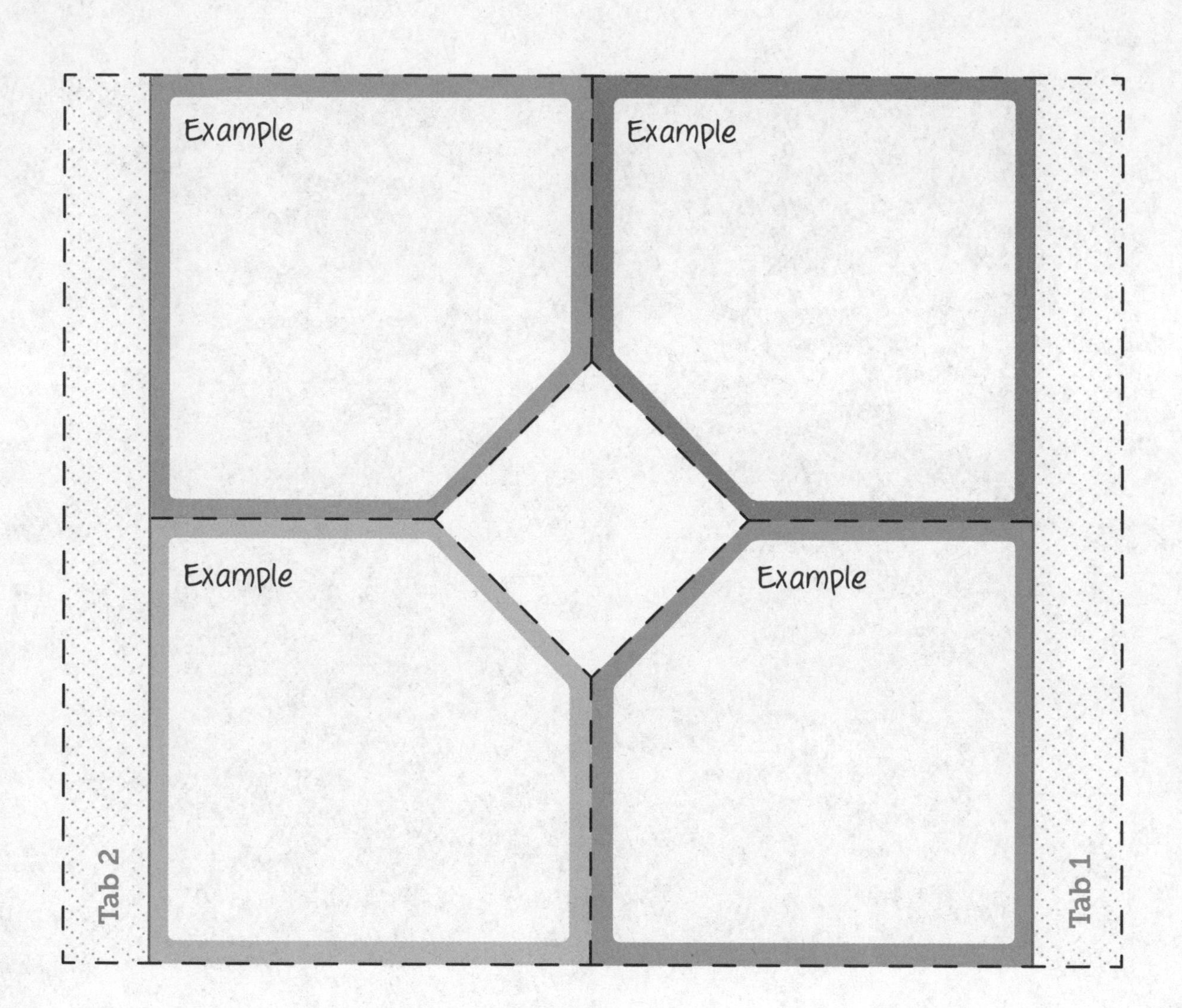

cut on all dashed lines

fold on all solid lines

tape to page 479

Area

parallelograms	triangles	trapezoids

Foldables

cut on all dashed lines   fold on all solid lines

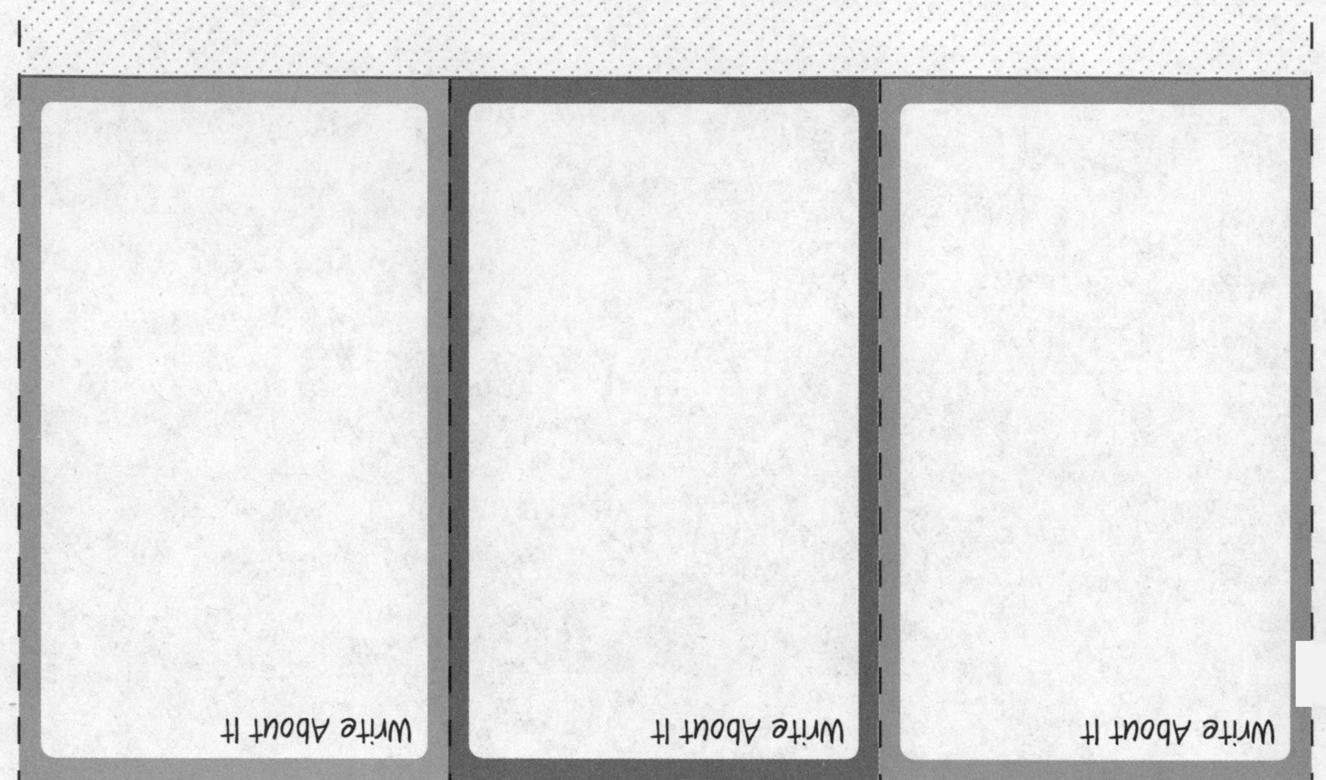

cut on all dashed lines   fold on all solid lines   tape to page 531

volume

surface area

Foldables

cut on all dashed lines

fold on all solid lines

tape to page 531

Tab 1

Formulas

Model

Real-World Examples

Tab 2

cut on all dashed lines fold on all solid lines tape to page 593

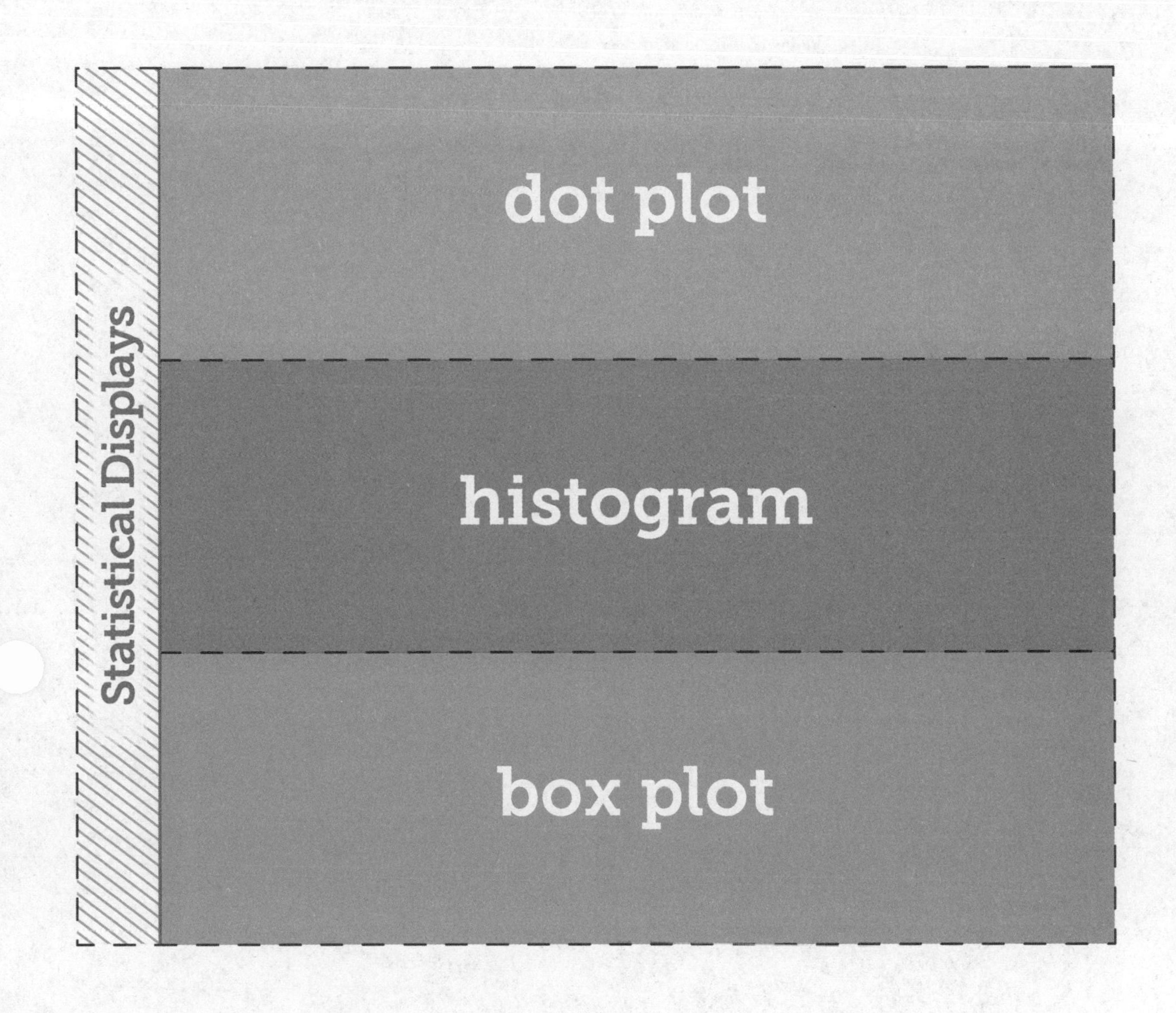

Foldables

cut on all dashed lines

fold on all solid lines

tape to page 593

Best used to...

Best used to...

Best used to...

Glossary

The eGlossary contains words and definitions in the following 14 languages:

Arabic	**English**	**Hmong**	**Russian**	**Urdu**
Bengali	**French**	**Korean**	**Spanish**	**Vietnamese**
Brazilian Portuguese	**Haitian Creole**	**Mandarin**	**Tagalog**	

English / Español

A

absolute value **(Lesson 4-2)** The distance between a number and zero on a number line.

valor absoluto Distancia entre un número y cero en la recta numérica.

Addition Property of Equality **(Lesson 6-8)** If you add the same number to each side of an equation, the two sides remain equal.

propiedad de adición de la igualdad Si sumas el mismo número a ambos lados de una ecuación, los dos lados permanecen iguales.

algebra **(Lesson 5-3)** A mathematical language of symbols, including variables.

álgebra Lenguaje matemático que usa símbolos, incluyendo variables.

algebraic expression **(Lesson 5-3)** A combination of variables, numbers, and at least one operation.

expresión algebraica Combinación de variables, números y, por lo menos, una operación.

analyze **(Lesson 10-1)** To use observations to describe and compare data.

analizar Usar observaciones para describir y comparar datos.

area **(Lesson 8-1)** The measure of the interior surface of a two-dimensional figure.

área La medida de la superficie interior d una figura bidimensional.

Associative Property **(Lesson 5-7)** The way in which numbers are grouped does not change the sum or product.

propiedad asociativa La forma en que se agrupan tres números al sumarlos o multiplicarlos no altera su suma o producto.

average **(Lesson 10-3)** The sum of two or more quantities divided by the number of quantities; the mean.

promedio La suma de dos o más cantidades dividida entre el número de cantidades; la media.

B

base **(Lesson 8-1)** Any side of a parallelogram or any side of a triangle.

base Cualquier lado de un paralelogramo o cualquier lado de un triángulo.

base **(Lesson 9-1)** One of the two parallel congruent faces of a prism.

base Una de las dos caras paralelas congruentes de un prisma.

base (Lesson 5-1) In a power, the number used as a factor. In 10^3, the base is 10. That is, $10^3 = 10 \times 10 \times 10$.

base En una potencia, el número usado como factor. En 10^3, la base es 10. Es decir, $10^3 = 10 \times 10 \times 10$.

bases (Lesson 8-3) The bases of a trapezoid are the two parallel sides.

bases Las bases de un trapecio son los dos lados paralelos.

benchmark percent (Lesson 2-5) A common percent used when estimating part of a whole.

porcentaje de referencia Porcentaje común utilizado para estimar parte de un todo.

box plot (Lesson 10-4) A diagram that is constructed using five values.

diagrama de caja Diagrama que se construye usando cinco valores.

C

cluster (Lesson 10-7) Data that are grouped closely together.

agrupamiento Conjunto de datos que se agrupan.

coefficient (Lesson 5-3) The numerical factor of a term that contains a variable.

coeficiente El factor numérico de un término que contiene una variable.

common factor (Lesson 5-5) A number that is a factor of two or more numbers.

factor común Un número que es un factor de dos o más números.

Commutative Property (Lesson 5-7) The order in which numbers are added or multiplied does not change the sum or product.

propiedad commutativa La forma en que se suman o multiplican dos números no altera su suma o producto.

congruent (Lesson 8-2) Having the same measure.

congruente Ques tienen la misma medida.

congruent figures (Lesson 8-2) Figures that have the same size and same shape; corresponding sides and angles have equal measures.

figuras congruentes Figuras que tienen el mismo tamaño y la misma forma; los lados y los ángulos correspondientes con igual medida.

constant (Lesson 5-3) A term without a variable.

constante Un término sin una variable.

coordinate plane (Lesson 1-3) A plane in which a horizontal number line and a vertical number line intersect at their zero points.

plano de coordenadas Plano en que una recta numérica horizontal y una recta numérica vertical se intersecan en sus puntos cero.

cubic units (Lesson 9-1) Used to measure volume. Tells the number of cubes of a given size it will take to fill a three-dimensional figure.

unidades cúbicas Se usan para medir el volumen. Indican el número de cubos de cierto tamaño que se necesitan para llenar una figura tridimensional.

D

data (Lesson 10-1) Information, often numerical, which is gathered for statistical purposes.

datos Información, con frecuencia numérica, que se recoge con fines estadísticos.

defining the variable (Lesson 5-3) Choosing a variable and deciding what the variable represents.

definir la variable Elegir una variable y decidir lo que representa.

dependent variable (Lesson 7-1) The variable in a relation with a value that depends on the value of the independent variable.

variable dependiente La variable en una relación cuyo valor depende del valor de la variable independiente.

distribution (Lesson 10-7) The arrangement of data values.

distribución El arreglo de valores de datos.

Distributive Property (Lesson 5-6) To multiply a sum by a number, multiply each addend by the number outside the parentheses.

propiedad distributiva Para multiplicar una suma por un número, multiplica cada sumando por el número fuera de los paréntesis.

dividend (Lesson 3-1) The number that is divided in a division problem.

dividendo El número que se divide en un problema de división.

Division Property of Equality (Lesson 6-4) If you divide each side of an equation by the same nonzero number, the two sides remain equal.

propiedad de igualdad de la división Si divides ambos lados de una ecuación entre el mismo número no nulo, los lados permanecen iguales.

divisor (Lesson 3-1) The number used to divide another number in a division problem.

divisor El número utilizado para dividir otro número en un problema de división.

double number line (Lesson 1-2) A double number line consists of two number lines, in which the coordinated quantities are equivalent ratios.

línea doble Una línea numérica doble consta de dos líneas numéricas, en las cuales las cantidades coordinadas son proporciones equivalentes.

dot plot (Lesson 10-2) A diagram that shows the frequency of data on a number line. Also known as a line plot.

diagrama de puntos Diagrama que muestra la frecuencia de los datos sobre una recta numérica.

E

equals sign (Lesson 6-1) A symbol of equality, $=$.

signo de igualdad Símbolo que indica igualdad, $=$.

equation (Lesson 6-1) A mathematical sentence showing two expressions are equal. An equation contains an equals sign, $=$.

ecuación Enunciado matemático que muestra que dos expresiones son iguales. Una ecuación contiene el signo de igualdad, $=$.

equivalent expressions (Lesson 5-7) Expressions that have the same value, regardless of the values of the variable(s).

expresiones equivalentes Expresiones que poseen el mismo valor, sin importer los valores de la(s) variable(s).

equivalent ratios (Lesson 1-2) Ratios that express the same relationship between two quantities.

razones equivalentes Razones que expresan la misma relación entre dos cantidades.

evaluate (Lesson 5-2) To find the value of an algebraic expression by replacing variables with numbers.

evaluar Calcular el valor de una expresión algebraica sustituyendo las variables por número.

exponent (Lesson 5-1) In a power, the number that tells how many times the base is used as a factor. In 5^3, the exponent is 3. That is, $5^3 = 5 \times 5 \times 5$.

exponente En una potencia, el número que indica las veces que la base se usa como factor. En 5^3, el exponente es 3. Es decir, $5^3 = 5 \times 5 \times 5$.

F

face (Lesson 9-1) A flat surface of a prism or pyramid.

cara Una superficie plana de un prisma o pirámide.

factoring the expression (Lesson 5-6) The process of writing numeric or algebraic expressions as a product of their factors.

factorizar la expresión El proceso de escribir expresiones numéricas o algebraicas como el producto de sus factores.

first quartile (Lesson 10-4) The first quartile is the median of the data values less than the median.

primer cuartil El primer cuartil es la mediana de los valores menores que la mediana.

G

gap (Lesson 10-7) An empty space or interval in a set of data.

laguna Espacio o intervalo vacío en un conjunto de datos.

graph (Lesson 1-3) To place a dot on a number line, or on the coordinate plane at a point named by an ordered pair.

graficar Colocar una marca puntual en una línea numérica, o en el plano de coordenadas en el punto que corresponde a un par ordenado.

greatest common factor (GCF) (Lesson 5-5) The greatest of the common factors of two or more numbers.

máximo común divisor (MCD) El mayor de los factores comunes de dos o más números.

guess, check, and revise strategy (Lesson 6-1) A strategy used to solve a problem which involves narrowing in on the correct answer using educated guesses.

adivinar, comprobar y revisar la estrategia Una estrategia utilizada para resolver un problema que implica el estrechamiento en la respuesta correcta usando conjeturas educadas.

H

height (Lesson 8-1) The height of a parallelogram is the perpendicular distance between the base and its opposite side.

altura La altura de un paralelogramo es la distancia perpendicular entre la base y su lado opuesto.

height (Lesson 8-2) The height of a triangle is the perpendicular distance from the base to the opposite vertex.

altura La altura de un triángulo es la distancia perpendicular de la base al vértice opuesto.

height (Lesson 8-3) The height of a trapezoid is the perpendicular distance between the two bases.

altura La altura de un trapecio es la distancia perpendicular entre las dos bases.

histogram (Lesson 10-2) A type of bar graph used to display numerical data that have been organized into equal intervals.

histograma Tipo de gráfica de barras que se usa para exhibir datos que se han organizado en intervalos iguales.

I

Identity Properties (Lesson 5-7) Properties that state that the sum of any number and 0 equals the number and that the product of any number and 1 equals the number.

propiedades de identidad Propiedades que establecen que la suma de cualquier número y 0 es igual al número y que el producto de cualquier número y 1 es iqual al número.

independent variable (Lesson 7-1) The variable in a relationship with a value that is subject to choice.

variable independiente Variable en una relación cuyo valor está sujeto a elección.

inequality (Lesson 6-6) A mathematical sentence indicating that two quantities are not equal.

desigualdad Enunciado matemático que indica que dos cantidades no son iguales.

integer (Lesson 4-1) Any number from the set {..., −4, −3, −2, −1, 0, 1, 2, 3, 4, ...} where ... means *continues without end.*

entero Cualquier número del conjunto {..., −4, −3, −2, −1, 0, 1, 2, 3, 4, ...} donde ... significa que *continúa sin fin.*

interquartile range (IQR) (Lesson 10-4) A measure of variation in a set of numerical data, the interquartile range is the distance between the first and third quartiles of the data set.

rango intercuartil (RIQ) El rango intercuartil, una medida de la variación en un conjunto de datos numéricos, es la distancia entre el primer y el tercer cuartil del conjunto de datos.

interval (Lesson 10-2) The difference between successive values on a scale.

intervalo La diferencia entre valores sucesivos de una escala.

inverse operations (Lesson 6-2) Operations which undo each other. For example, addition and subtraction are inverse operations.

operaciones inversas Operaciones que se anulan mutuamente. La adición y la sustracción son operaciones inversas.

Inverse Property of Multiplication (Lesson 3-3) A property that states that the product of a number and its multiplicative inverse is 1.

propiedad inversa de la multiplicación Una propiedad que indica que el producto de un número y su inverso multiplicativo es 1.

lateral face (Lesson 9-4) Any face that is not a base.

cara lateral Cualquier superficie plana que no sea la base.

least common multiple (LCM) (Lesson 5-5) The smallest whole number greater than 0 that is a common multiple of each of two or more numbers.

mínimo común múltiplo (mcm) El menor número entero, mayor que 0, múltiplo común de dos o más números.

like terms (Lesson 5-3) Terms that contain the same variable(s) to the same power.

términos semejantes Términos que contienen la misma variable o variables elevadas a la misma potencia.

M

mean (Lesson 10-3) The sum of the numbers in a set of data divided by the number of pieces of data.

media La suma de los números en un conjunto de datos dividida entre el número total de datos.

mean absolute deviation (MAD) (Lesson 10-5) A measure of variation in a set of numerical data, computed by adding the distances between each data value and the mean, then dividing by the number of data values.

desviación media absoluta (DMA) Una medida de variación en un conjunto de datos numéricos que se calcula sumando las distancias entre el valor de cada dato y la media, y luego dividiendo entre el número de valores.

measures of center (Lesson 10-3) Numbers that are used to describe the center of a data set. These measures include the mean and median.

medidas del centro Numéros que se usan para describir el centro de un conjunto de datos. Estas medidas incluyen la media, la mediana y la moda.

measures of variation (Lesson 10-4) A measure that is used to describe the variability, or spread, of a data set.

medidas de variación Medida que se utiliza para describir la variabilidad o la dispersión de un conjunto de datos.

median (Lesson 10-3) A measure of center in a set of numerical data. The median of a list of values is the value appearing at the center of a sorted version of the list, or the mean of the two central values, if the list contains an even number of values.

mediana Una medida del centro en un conjunto de datos numéricos. La mediana de una lista de valores es el valor que aparece en el centro de una versión ordenada de la lista, o la media de los dos valores centrales si la lista contiene un número par de valores.

Multiplication Property of Equality (Lesson 6-5) If you multiply each side of an equation by the same nonzero number, the two sides remain equal.

propiedad de multiplicación de la igualdad Si multiplicas ambos lados de una ecuación por el mismo número no nulo, lo lados permanecen iguales.

multiplicative inverses (Lesson 3-3) Any two numbers that have a product of 1.

inversos multiplicativos Cualquier dos números que tengan un producto de 1.

N

negative integer (Lesson 4-1) A number that is less than zero. It is written with a − sign.

entero negativo Número que es menor que cero y se escribe con el signo −.

net (Lesson 9-2) A two-dimensional figure that can be used to build a three-dimensional figure.

red Figura bidimensional que sirve para hacer una figura tridimensional.

numerical expression (Lesson 5-2) A combination of numbers and operations.

expresión numérica Una combinación de números y operaciones.

O

opposites (Lesson 4-2) Two integers are opposites if they are represented on the number line by points that are the same distance from zero, but on opposite sides of zero. The sum of two opposites is zero.

opuestos Dos enteros son opuestos si, en la recta numérica, están representados por puntos que equidistan de cero, pero en direcciones opuestas. La suma de dos opuestos es cero.

order of operations **(Lesson 5-2)** The rules that tell which operation to perform first when more than one operation is used.

1. Simplify the expressions inside grouping symbols.
2. Find the value of all powers.
3. Multiply and divide in order from left to right.
4. Add and subtract in order from left to right.

orden de las operaciones Reglas que establecen cuál operación debes realizar primero, cuando hay más de una operación involucrada.

1. Ejecuta todas las operaciones dentro de los símbolos de agrupamiento.
2. Evalúa todas las potencias.
3. Multiplica y divide en orden de izquierda a derecha.
4. Suma y resta en orden de izquierda a derecha.

ordered pair **(Lesson 1-3)** A pair of numbers used to locate a point on the coordinate plane. The ordered pair is written in the form (*x*-coordinate, *y*-coordinate).

par ordenado Par de números que se utiliza para ubicar un punto en un plano de coordenadas. Se escribe de la forma (coordenada *x*, coordenada *y*).

origin **(Lesson 1-3)** The point of intersection of the *x*-axis and *y*-axis on a coordinate plane.

origen Punto de intersección de los ejes axiales en un plano de coordenadas.

outlier **(Lesson 10-6)** A value that is much greater than or much less than the other values in a set of data.

valor atípico Dato que se encuentra muy separado de los otros valores en un conjunto de datos.

P

parallelogram **(Lesson 8-1)** A quadrilateral with opposite sides parallel and opposite sides congruent.

paralelogramo Cuadrilátero cuyos lados opuestos son paralelos y congruentes.

part-to-part ratio **(Lesson 1-1)** A ratio that compares one part of a group to another part of the same group.

proporción de parte a parte Una proporción que compara una parte de un grupo con otra parte del mismo grupo.

part-to-whole ratio **(Lesson 1-1)** A ratio that compares one part of a group to the whole group.

proporción de parte a total Una proporción que compara una parte de un grupo con todo el grupo.

peak **(Lesson 10-7)** The most frequently occurring value in a line plot.

pico El valor que ocurre con más frecuencia en un diagrama de puntos.

percent **(Lesson 2-1)** A ratio, or rate, that compares a number to 100.

por ciento Una relación, o tasa, que compara un número a 100.

positive integer **(Lesson 4-1)** A number that is greater than zero. It can be written with or without a + sign.

entero positivo Número que es mayor que cero y se puede escribir con o sin el signo +.

powers **(Lesson 5-1)** A number expressed using an exponent. The power 3^2 is read *three to the second power,* or *three squared.*

potncias Números que se expresan usando exponentes. La potencia 3^2 se lee *tres a la segunda potencia* o *tres al cuadrado.*

prism **(Lesson 9-1)** A three-dimensional figure with at least three rectangular lateral faces and top and bottom faces parallel.

prisma Figura tridimensional que tiene por lo menos tres caras laterales rectangulares y caras paralelas superior e inferior.

pyramid **(Lesson 9-4)** A three-dimensional figure with at least three triangular sides that meet at a common vertex and only one base that is a polygon.

pirámide Una figura de tres dimensiones con que es en un un polígono y tres o mas caras triangulares que se encuentran en un vértice común.

Q

quadrants **(Lesson 4-5)** The four regions in a coordinate plane separated by the *x*-axis and *y*-axis.

cuadrantes Las cuatro regiones de un plano de coordenadas separadas por el eje *x* y el eje *y*.

quartiles **(Lesson 10-4)** Values that divide a data set into four equal parts.

cuartiles Valores que dividen un conjunto de datos en cuatro partes iguales.

quotient **(Lesson 3-1)** The result when one number is divided by another.

cociente El resultado cuando un número es dividido por otro.

R

range **(Lesson 10-4)** The difference between the greatest number and the least number in a set of data.

rango La diferencia entre el número mayor y el número menor en un conjunto de datos.

rate **(Lesson 1-7)** A special kind of ratio in which the units are different.

tasa Un tipo especial de relación en el que las unidades son diferentes.

ratio **(Lesson 1-1)** A comparison between two quantities, in which for every *a* units of one quantity, there are *b* units of another quantity.

razón Una comparación entre dos cantidades, en la que por cada *a* unidades de una cantidad, hay unidades *b* de otra cantidad.

ratio table **(Lesson 1-2)** A collection of equivalent ratios that are organized in a table.

table de razones Una colección de proporciones equivalentes que se organizan en una tabla.

rational number **(Lesson 4-4)** A number that can be written as a fraction.

número racional Número que se puede expresar como fracción.

reciprocals **(Lesson 3-3)** Any two numbers that have a product of 1. Since $\frac{5}{6} \times \frac{6}{5} = 1$, $\frac{5}{6}$ and $\frac{6}{5}$ are reciprocals.

recíproco Cualquier par de números cuyo producto es 1. Como $\frac{5}{6} \times \frac{6}{5} = 1$, $\frac{5}{6}$ y $\frac{6}{5}$ son recíprocos.

rectangular prism **(Lesson 9-1)** A prism that has rectangular bases.

prisma rectangular Una prisma que tiene bases rectangulares.

reflection **(Lesson 4-6)** The mirror image produced by flipping a figure over a line.

reflexión Transformación en la cual una figura se voltea sobre una recta. También se conoce como simetría de espejo.

regular polygon **(Lesson 8-4)** A polygon with all congruent sides and all congruent angles.

polígono regular Un polígono con todos los lados congruentes y todos los ángulos congruentes.

S

scaling (Lesson 1-2) The process of multiplying each quantity in a ratio by the same number to obtain equivalent ratios.

homotecia El proceso de multiplicar cada cantidad en una proporción por el mismo número para obtener relaciones equivalentes.

second quartile (Lesson 10-4) Another name for the median, or the center of a set of numerical data.

segundo cuartil Otro nombre para la mediana, o el centro de un conjunto de datos numéricos.

simplest form (Lesson 5-4) The status of an expression when it has no like terms and no parentheses.

forma más simple El estado de una expresión cuando no tiene términos iguales y no hay paréntesis.

slant height (Lesson 9-4) The height of each lateral face of a pyramid.

altura oblicua Altura de cada cara lateral de un pirámide.

solution (Lesson 6-1) The value of a variable that makes an equation true.

solución Valor de la variable de una ecuación que hace verdadera la ecuación.

solve (Lesson 6-1) To replace a variable with a value that results in a true sentence.

resolver Reemplazar una variable con un valor que resulte en un enunciado verdadero.

statistical question (Lesson 10-1) A question that anticipates and accounts for a variety of answers.

cuestión estadística Una pregunta que se anticipa y da cuenta de una variedad de respuestas.

statistics (Lesson 10-1) Collecting, organizing, and interpreting data.

estadística Recopilar, ordenar e interpretar datos.

Subtraction Property of Equality (Lesson 6-2) If you subtract the same number from each side of an equation, the two sides remain equal.

propiedad de sustracción de la igualdad Si sustraes el mismo número de ambos lados de una ecuación, los dos lados permanecen iguales.

surface area (Lesson 9-2) The sum of the areas of all the surfaces (faces) of a three-dimensional figure.

área de superficie La suma de las áreas de todas las superficies (caras) de una figura tridimensional.

survey (Lesson 10-1) A question or set of questions designed to collect data about a specific group of people, or population.

encuesta Pregunta o conjunto de preguntas diseñadas para recoger datos sobre un grupo específico de personas o población.

symmetric distribution (Lesson 10-7) Data that are evenly distributed.

distribución simétrica Datos que están distribuidos.

T

term (Lesson 5-3) Each part of an algebraic expression separated by a plus or minus sign.

término Cada parte de un expresión algebraica separada por un signo más o un signo menos.

third quartile (Lesson 10-4) The third quartile is the median of the data values greater than the median.

tercer cuartil El tercer cuartil es la mediana de los valores mayores que la mediana.

three-dimensional figure (Lesson 9-1) A figure with length, width, and height.

figura tridimensional Una figura que tiene largo, ancho y alto.

trapezoid (Lesson 8-3) A quadrilateral with one pair of parallel sides.

trapecio Cuadrilátero con un único par de lados paralelos.

triangular prism (Lesson 9-3) A prism that has triangular bases.

prisma triangular Prisma con bases triangulares.

U

unit price (Lesson 1-7) The cost per unit of an item.

precio unitario El costo por unidad de un artículo.

unit rate (Lesson 1-7) A rate in which the first quantity is compared to 1 unit of the second quantity.

tasa unitaria Una tasa en la que la primera cantidad se compara con 1 unidad de la segunda cantidad.

unit ratio (Lesson 1-6) A ratio in which the first quantity is compared to 1 unit of the second quantity.

razón unitaria Una relación en la que la primera cantidad se compara con 1 unidad de la segunda cantidad.

V

variable (Lesson 5-3) A symbol, usually a letter, used to represent a number.

variable Un símbolo, por lo general, una letra, que se usa para representar un número.

volume (Lesson 9-1) The amount of space inside a three-dimensional figure. Volume is measured in cubic units.

volumen Cantidad de espacio dentro de una figura tridimensional. El volumen se mide en unidades cúbicas.

X

x-axis (Lesson 1-3) The horizontal line of the two perpendicular number lines in a coordinate plane.

eje x La recta horizontal de las dos rectas numéricas perpendiculares en un plano de coordenadas.

x-coordinate (Lesson 1-3) The first number of an ordered pair. The x-coordinate corresponds to a number on the x-axis.

coordenada x El primer número de un par ordenado, el cual corresponde a un número en el eje x.

Y

y-axis (Lesson 1-3) The vertical line of the two perpendicular number lines in a coordinate plane.

eje y La recta vertical de las dos rectas numéricas perpendiculares en un plano de coordenadas.

y-coordinate (Lesson 1-3) The second number of an ordered pair. The y-coordinate corresponds to a number on the y-axis.

coordenada y El segundo número de un par ordenado, el cual corresponde a un número en el eje y.

Index

A

B

C

N

O

P

Q

R

S

T

Lesson 5-1 Powers and Exponents, Practice Pages 267–268

1. 4^3 **3.** 15^4 **5.** $\left(\frac{1}{3}\right)^7$ **7.** 3,125 **9.** 10,000 **11.** $\frac{8}{125}$ **13.** 3.375 **15.** 0.064 **17.** 7^4; 2,401 **19.** 1,024 cells **21.** Sample answer: The student used the exponent as the base. The base should be 2 and the exponent is 3. The power evaluated should be $2 \times 2 \times 2 = 8$. **23.** Sample answer: Exponential form is repeated multiplication of a common factor.

Lesson 5-2 Numerical Expressions, Practice Pages 275–276

1. 7 **3.** 46 **5.** 23 **7.** 436 **9.** $\frac{3}{5}$ or 0.6 **11.** Sample expression: $(6 \times 1.49) + (2^2) + (3 \times 3.50)$; $23.44 **13.** $8(1.25 + 0.85)$; $8(1.25) + 8(0.85)$ **15.** 294 muffins **17.** Sample answer: The student did not follow the order of operations. The student added first before dividing. The division should have been performed first. $42 + 6 \div 2 = 42 + 3$ or 45 **19.** Sample answer: Frankie and his two sisters each order a hamburger, a fruit cup, and a bottled water for lunch. A hamburger costs $3, a fruit cup costs $0.75, and a bottled water costs $1.25.; $3^2 + (3 \times 0.75) + (3 \times 1.25)$; $15

Lesson 5-3 Write Algebraic Expressions, Practice Pages 285–286

1. terms: 4*e*, 7*e*, 5, 2*e*; like terms: 4*e*, 7*e*, 2*e*; coefficients: 4, 7, 2; constant: 5 **3.** terms: 4, 4*y*, *y*, 3; like terms: 4*y*, *y*; 4, 3; coefficients: 4, 1; constants: 4, 3 **5.** Sample answer: Let *q* represent the number of questions on the first test; $q - 12$ **7.** Sample answer: Let *y* represent the number of yards; $\frac{1}{3}y$ **9.** Sample answer: Let *c* represent the cost of a pizza; $\frac{1}{4}c + 2.5$ **11.** Sample answer: Let *c* represent the number of classes; $35 + 20c$ **13.** Sample answer: Let ℓ represent the length of one of the equal sides; $\ell + \ell + 1.5\ell$ **15.** Sample answer: $2x + 8 + x + 6$; like terms: 2*x*, *x*; 8, 6; coefficients: 2, 1; constants: 8, 6 **17.** $8 + 0.25c$

Lesson 5-4 Evaluate Algebraic Expressions, Practice Pages 293–294

1. 6 **3.** 4 **5.** $\frac{26}{5}$ or $5\frac{1}{5}$ **7.** 2 **9.** 2 **11.** 24.2 ft^2 **13.** 13 **15.** $80 **17.** Sample answer: The student replaced the variables with the incorrect values. The correct value should be $4(2) + 3$, or 11. **19.** Sample answer: If $a = 2$, then $a + 10 = 12$; $(15 + 5) - 8 = 12$

Lesson 5-5 Factors and Multiples, Practice Pages 303–304

1. 6 **3.** 9 **5.** 14 **7.** 20 visits **9.** 12 **11.** 5 flowers **13.** $65 **15.** Sample answer: The bottom row of the factor trees may not show the factors listed in order from least to greatest. I can use the Commutative Property to write the factors in order from least to greatest. **17.** false; Sample answer: 25 and 50; 50 is a multiple of 25, 50 is the LCM, and 50 is the greater number. The LCM is the greater of the two numbers.

Lesson 5-6 Use the Distributive Property, Practice Pages 313–314

1. $3x + 24$ **3.** $27 + 9x$ **5.** 40 **7.** $16(1 + 3)$ **9.** $13(2 + 3)$ **11.** $6(4 + x)$ **13.** $5x + 120$ **15.** $5.40 **17.** Sample answer: $8\left(4\frac{3}{4}\right) = 8\left(4 + \frac{3}{4}\right)$ **19.** no; Sample answer: The Distributive Property combines addition and multiplication. The expression $2(6x)$ is one term with three factors and does not contain addition. $2(6x)$ is equal to $12x$.

Lesson 5-7 Equivalent Algebraic Expressions, Practice Pages 327–328

1. equivalent **3.** not equivalent **5.** $8x + 3$
7. $4x^2 + 7x + 10$ **9.** $10x + 8$ **11.** $76.4x + 8$
13. Sample answer: $2y^2 + y^2 + y + y + \frac{1}{2}$
15. Sample answer: $3x + 0$ and $3x$

Module Review Pages 331–332

1. $5 \times 5 \times 5$; 125 **3a.** $(2 \times 0.75) + (5 \times 1.79) + (3 \times 3)$ **3b.** 19.45 **5.** terms: $8p$, $6q$, 5, $9q$, $12p$; like terms: $8p$ and $12p$, $6q$ and $9q$; coefficients: 8, 6, 9, 12; constant: 5 **7.** C **9.** C
11.

	Equivalent	Not Equivalent
$5x + 2$ and $4x + 1 + x + 2$		X
$(8y + 4x + 4y + 5)$ and $4(3y + x) + 5$	X	
$y^2 + 4y + 5 + 3y$ and $y^2 + y + 5$	X	

Lesson 6-1 Use Substitution to Solve One-Step Equations, Practice Pages 339–340

1. 6 **3.** 23 **5.** 4 **7.** 22 **9.** 8 headbands
11. 7 batches **13.** Sample answer: Jack had $7.50. His mother gave him his allowance at the end of the week. Now Jack has $16. Solve the equation $7.5 + x = 16$ to find how much money his mother gave him. **15.** Sample answer: $x + 1$ is an algebraic expression and is not equal to a specific value. So, there are no restrictions placed on the value of x. $x + 1 = 2$ is an algebraic equation. Each side of an algebraic equation must be equal, so x can only be equal to one value. In this case, $x = 1$.

Lesson 6-2 One-Step Addition Equations, Practice Pages 349–350

1. Sample answer: $320 + c = 647.5$
3. Sample answer: $m + 19.5 = 38.25$
5. 6 **7.** $3\frac{1}{2}$ **9.** 13.25 **11.** Sample equation: $8.99 + 2(5.75) + 2(1.15) + 3.45 + x = 35$; $8.76
13. The value of b must be decreased by 1.
15. 4, 5, 6

Lesson 6-3 One-Step Subtraction Equations, Practice Pages 357–358

1. Sample answer: $c - 17 = 64$ **3.** Sample answer: $f - 1\frac{1}{4} = 1\frac{1}{2}$ **5.** 29 **7.** $10\frac{8}{9}$ **9.** 72.35
11. $667.92 **13.** yes; Sample answer: Solve the equation $x - 7 = 18$ to find the height of Devon's rocket. Devon's rocket reached a height of $18 + 7$ or 25 yards. Since $25 > 23$, Devon's rocket reached a height greater than 23 yards
15a. Sample answer: Today's high temperature is 64°F. This is 9°F less than yesterday's high temperature. What was yesterday's high temperature? **15b.** $x - 9 = 64$ **15c.** 73°F

Lesson 6-4 One-Step Multiplication Equations, Practice Pages 367–368

1. Sample answer: $46.75p = 374$
3. Sample answer: $2.6t = 18.2$ **5.** 2 **7.** $\frac{8}{9}$
9. 6.58 **11.** caramel popcorn; 38 Calories
13. no; Sample answer: Solve the equation $52.5x = 367.50$ to find the number of weeks she needs to save. She needs to save for 7 weeks. Since $7 > 6$, she will not have enough money in 6 weeks. **15.** yes; Sample answer: If you solve each equation you get a value of $x = \frac{1}{9}$. If you replace x with $\frac{1}{9}$ for each equation it makes the equation true. So, $\frac{1}{3} = 3 \times \frac{1}{9}$ or $\frac{1}{3}$ and $\frac{1}{3} \div \frac{1}{9} = 3$.

Lesson 6-5 One-Step Division Equations, Practice Pages 375–376

1. Sample answer: $c \div 284.5 = 6$ **3.** Sample answer: $d \div 5.25 = 3$ **5.** 48 **7.** $\frac{8}{3}$ or $2\frac{2}{3}$
9. 48.852 **11.** cheese crackers: 112.5 oz; pretzels: 227.5 oz; 115 oz

13. 200 miles; Sample answer: Write and solve the division equation $\frac{m}{5} = 40$; 5×40 is 200. So, $m = 200$ miles. **15.** 186 in.; Sample answer: The length of the actual car c divided by 24, the scale, equals the length of the model car: $\frac{c}{24} = 7.75$; So, $c = 186$ in.

Lesson 6-6 Inequalities, Practice Pages 489–490

1. $a \geq 75$

3.

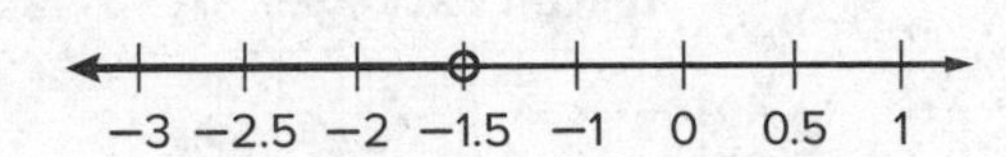

5.

$-1 \quad -\frac{4}{5} \quad -\frac{3}{5} \quad -\frac{2}{5} \quad -\frac{1}{5} \quad 0 \quad \frac{1}{5} \quad \frac{2}{5} \quad \frac{3}{5} \quad \frac{4}{5} \quad 1$

7. 4, 5 **9.** $\frac{1}{4}, \frac{1}{3}$ **11.** Jessica can buy no more than 6 tickets. **13.** China, Maria; $8.25h \geq 74.50$ **15.** Sample answer: More than 2,500 people attended the game; $x > 2{,}500$ **17a.** 4 **17b.** 12 **17c.** 6

Module 6 Review Pages 493–494

1. D **3.** $m + 225 = 478.50$; $225 + m = 478.50$ **5.** $c - 6.50 = 12.99$ **7.** B **9.** $\frac{y}{11}(11) = 28(11)$; $y = 308$

11.

$-1\frac{1}{3} \quad -1 \quad -\frac{2}{3} \quad -\frac{1}{3} \quad 0 \quad \frac{1}{3} \quad \frac{2}{3} \quad 1 \quad 1\frac{1}{3}$

13. 13 teammates; 12 teammates

Lesson 7-1 Relationships Between Two Variables, Practice Pages 403–404

1.

Input, Cost of Pizza ($), p	Rule $p + 3.50$	Output, Total Cost ($), c
9.75	9.75 + 3.50	13.25
12.00	12.00 + 3.50	15.50
14.50	14.50 + 3.50	18.00

3.

Input, Cost of Sundae ($), s	Rule $s - 0.75$	Output, Total Cost ($), c
2.79	2.79 − 0.75	2.04
3.55	3.55 − 0.75	2.80
4.25	4.25 − 0.75	3.50

5.

Input, Number of Pies, p	Rule $9.50p$	Output, Total Cost ($), c
2	9.50(2)	19.00
3	9.50(3)	28.50
5	9.50(5)	47.50

7.

Original Price ($), p	Rule, $p - 15$	Total Cost ($), c
65	65 − 15	50
73	73 − 15	58
79	79 − 15	64

She could buy the pair that originally cost $65 or the pair that originally cost $73.

9.

Input, x	Rule, $2x - 2.5$	Output, y
5	2(5) − 2.5	7.5
6.5	2(6.5) − 2.5	10.5
8	2(8) − 2.5	13.5

11. $14.25; Sample answer: In the equation $c = 2.75p + 1.5d$, replace p with 3 and d with 4 and then simplify. $c = 2.75 \times 3 + 1.5 \times 4$ or 14.25.

Lesson 7-2 Write Equations to Represent Relationships Represented in Tables, Practice Pages 413–414

1. $c = 7t$ **3.** $c = 4g + 2$ **5.** $c = 15m + 10$ **7.** $y = \frac{1}{3}x + 3$ **9.** Sample answer: The student switched the coefficient and the constant. The coefficient is 12 and the constant is 20. The equation should be $c = 12h + 20$.

Selected Answers

Lesson 7-3 Graphs of Relationships, Practice Pages 421–422

1.

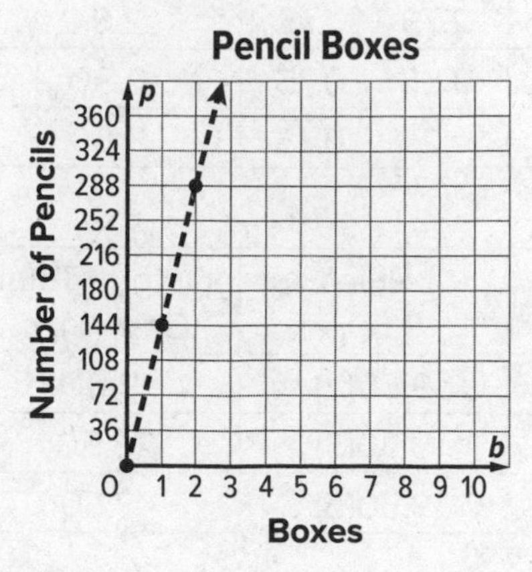

3. $c = 4d + 8$ **5.** 2 more hours **7.** Sample answer: The student switched the coefficient and constant. The correct equation is $s = 5w + 10$. **9.** Sample answer: a straight line through the origin; (0, 0), (2, 1), (4, 2)

Lesson 7-4 Multiple Representations, Practice Pages 427–428

1a. $c = 8t + 2.5$

1b.

Number of Tickets, t	Total Cost (\$), c
1	10.50
2	18.50
3	26.50
4	34.50

1c.

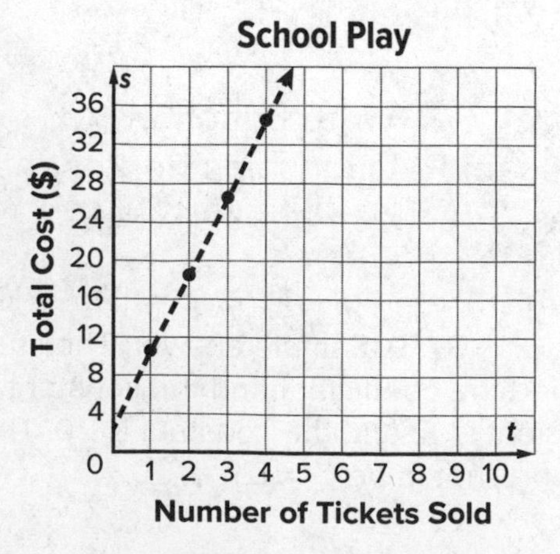

3. $e = 6p$ **5.** $p = 5b + 5$; 55 points
7. no; Sample answer: The graphs of the lines will never meet other than at zero hours.

Module 7 Review Pages 431–432

1. 5.1 **3.** $c = 6b$ **5.** \$61

7a.

Number of Packages, p	Number of Biscuits, b
0	0
1	8
2	16
3	24

7b.

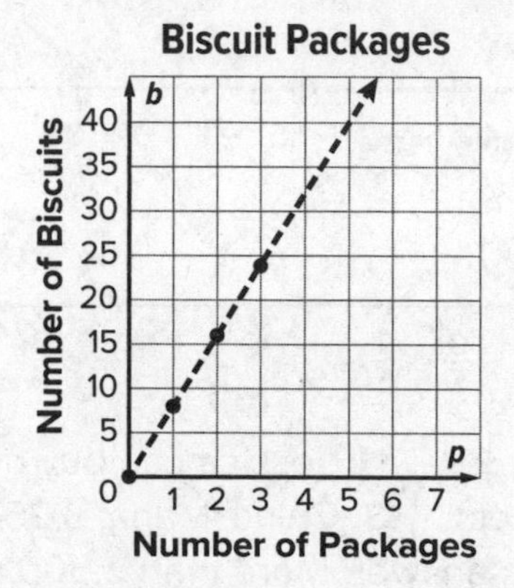

9a. $t = 10s$;

Number of Subscriptions, s	Total Earnings, t
1	10
2	20
3	30

9b.

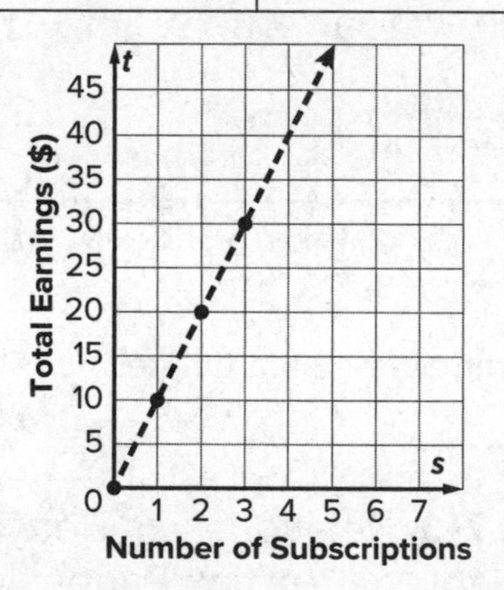

Lesson 8-1 Area of Parallelograms, Practice Pages 441–442

1. 50.02 in^2 **3.** $b = 7$ m **5.** 13.6 in^2 **7.** 41 tiles **9.** 480 mm^2 **11.** Sample answer: Because the area of a parallelogram is found by multiplying the base and height, the area of each of the three parallelograms would be 20 square units, because $5 \times 4 = 20$.

Lesson 8-2 Area of Triangles, Practice Pages 449–450

1. 15 yd^2 **3.** $11\frac{3}{8}$ ft^2 **5.** $b = 9$ km **7.** 144.5 cm^2 **9.** $13.58 **11.** Sample answer: The formula for the area of a triangle is $A = \frac{1}{2}bh$, not bh. $\frac{1}{2}(17)h = 68$; $h = 8$ m **13.** no; Sample answer: The area of her lawn is 125 ft^2 because the area of a triangle is $A = \frac{1}{2}bh$. So, $\frac{1}{2}(25 \times 10) = 125$.

Lesson 8-3 Area of Trapezoids, Practice Pages 461–462

1. 105 cm^2 **3.** 36 in^2 **5.** 155,000 km^2 **7.** 4 in. **9.** The cost of the patio is $1,443.75. Since this is less than $1,500, Greta has budgeted enough money. **11.** Start with the area formula: $A = \frac{1}{2}h(b_1 + b_2)$. Multiply each side by 2: $2A = h(b_1 + b_2)$. Multiply each side by $\frac{1}{h}$: $\frac{2A}{h} = b_1 + b_2$. Subtract b_1 from each side: $\frac{2A}{h} - b_1 = b_2$, or $b_2 = \frac{2A}{h} - b_1$. **13.** 4 cm and 12 cm

Lesson 8-4 Area of Regular Polygons, Practice Pages 467–468

1. 31.82 in^2 **3.** 281.92 cm^2 **5.** $120.03 **7.** 473.2 cm^2 **9.** 492 in^2; the base length of each triangle is 80 ÷ 10 or 8 in. So, $10\left(\frac{1}{2} \times 8 \times 12.3\right) = 492$.

Lesson 8-5 Polygons on the Coordinate Plane, Practice Pages 477–478

1. 38 units **3.** 22 units **5.** 9 square units **7.** Space A; The monthly rental price of Space A is $4,720. The monthly rental price of Space B is $4,756. $4,720 is less than $4,756. **9.** Sample answer: (3, 4) and (3, 7) **11.** Sample answer: The student subtracted 10 − 7 and 7 − 2 to find lengths 3 and 5. The student should have subtracted 7 − 1 and 10 − 2 to find lengths 6 and 8. The perimeter is 28 units.

Module 8 Review Pages 481–482

1. 12 cm^2 **3.** 45 in^2 **5.** 7 **7a.** Sample answer: I would decompose the hexagon into two trapezoids. **7b.** 261 cm^2 **9.** 26

Lesson 9-1 Volume of Rectangular Prisms, Practice Pages 493–494

1. 15 ft^3 **3.** 4 m **5.** 6 ft **7.** medium dumpster **9.** Sample answer: The classmate switched the volume measurement and h in the formula. The correct value for the height is 0.5 centimeter. **11.** 90 in^3; Sample answer: The volume of the pan is 9 × 5 × 3 or 135 cubic inches. Multiply that by two-thirds to find the volume that is filled with batter. $134 \times \frac{2}{3} = 90$.

Lesson 9-2 Surface Area of Rectangular Prisms, Practice Pages 503–504

1.

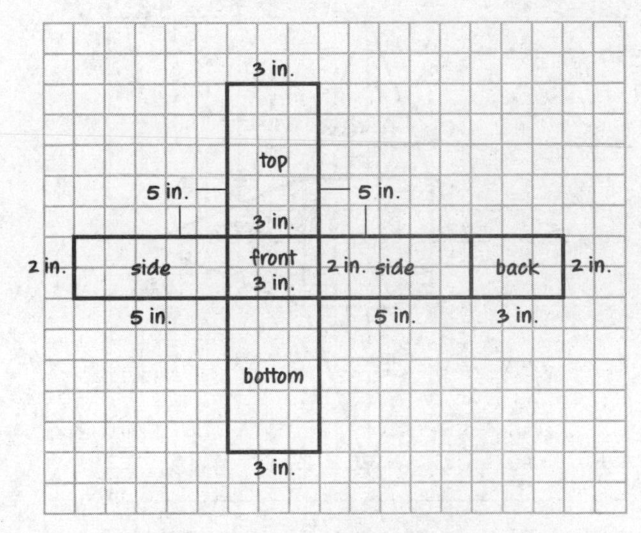

3. 1,120 cm^2 **5.** Sample answer: Let ℓ = length, w = width, and h = height; $S.A. = 2\ell w + 2\ell h + 2wh$ **7.** Block A: 94 in^2; 60 in^3; Block B: 104 in^2; 60 in^3; Block B has a greater surface area. No, the volumes of Blocks A and B are the same.

Selected Answers

Lesson 9-3 Surface Area of Triangular Prisms, Practice Pages 515–516

1.

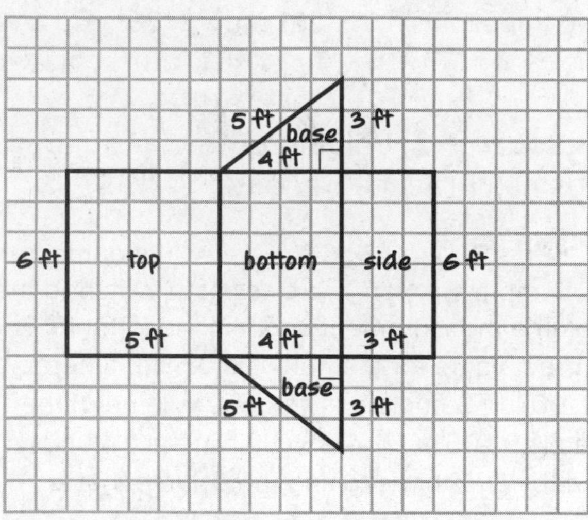

3. 811.2 m^2 **5.** \$106.79 **7.** 75 ft^2

Lesson 9-4 Surface Area of Pyramids, Practice Pages 529–530

1.

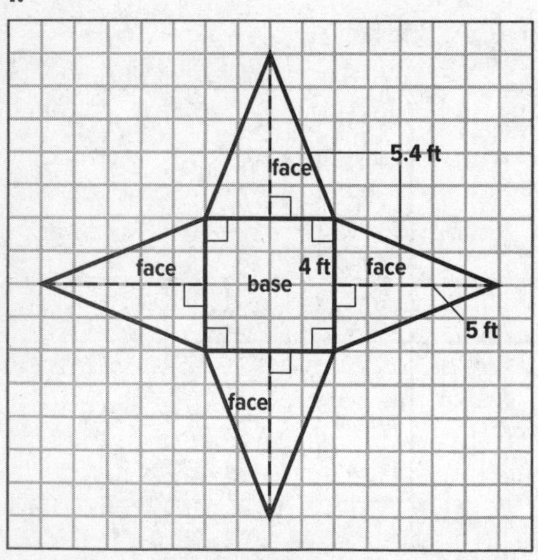

3. 56 ft^2 **5.** \$1.76 **7.** 11.5 yd **9.** 8.7 ft

Module 9 Review Pages 533–534

1. D **3.** 8 cm **5.** The net will be made up of 6 parts, representing the top, bottom, front, back, and both sides of the rectangular prism; Two parts of the net will have dimensions 4 in. by 11 in.; Two parts of the net will have dimensions 4 in. by 9 in. **7.** B

Lesson 10-1 Statistical Questions, Practice Pages 541–542

1. not a statistical question **3.** statistical question

5.

Number of Siblings	Number of Responses
0-1	10
2-3	7
4-5	2
6 or more	1

Sample answer: Half of the students have 0 or 1 siblings.

7.

Number of Sports	Number of Responses
1	4
2	7
3	2
4	1

Sample answer: Half of the students that responded play 2 sports. **9.** Sample answers: How many smartphones does a typical family own?; In what year was the cell phone invented?; The first question is a statistical question because it anticipates a variety of responses. The second question is not a statistical question because it does not anticipate a variety of responses. **11.** yes; Sample answer: Of the 10 families, 3 own one tablet and 4 own two tablets. Since 3 + 4 is 7 and 7 is close to 10, this is a reasonable conclusion.

Lesson 10-2 Dot Plots and Histograms, Practice Pages 567–568

1.

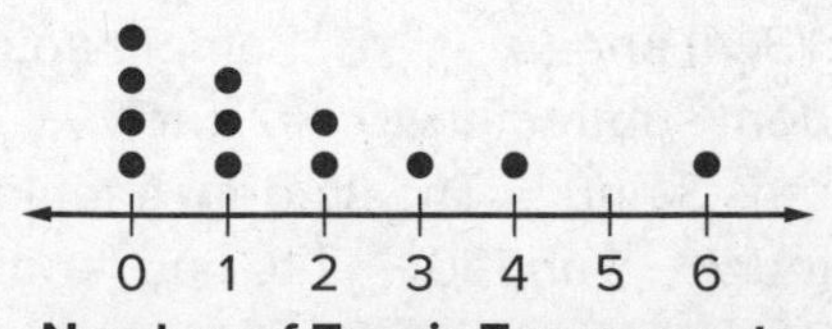

Sample answer: Of the 12 players on Chris's team, some played in as few as 0 and as many as 6 tournaments. Most players played in 1 or fewer tournaments.

3.

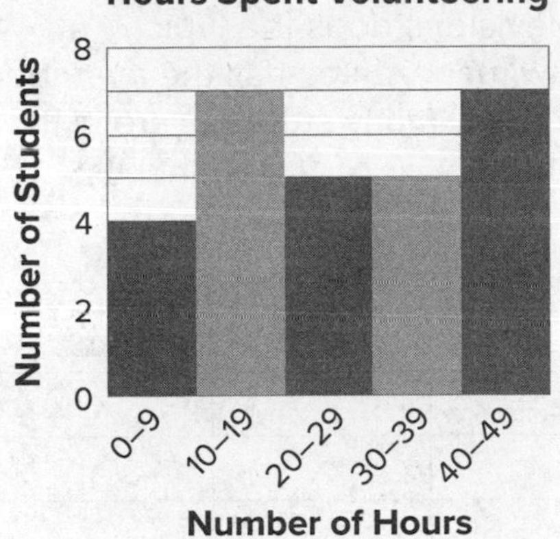

5. 3 video games **7.** Sample answer: Most students have 3 or fewer siblings. The most common number of siblings is 2 or 3. **9.** false; Sample answer: Dot plots display individual data values. Histograms display data by equal intervals, not individual data values.

Lesson 10-3 Measures of Center, Practice Pages 559–560

1. 58 cans **3.** $195 **5.** 23 E-mails
7. 53 points **9.** the mean; Sample answer: The mean of the data is 31 minutes and the median is 25.5 minutes. Since Kenny wants to use a measure that represents a greater number of minutes spent practicing, he should choose the greater of the two measures, the mean.
11. Sample answer: Shoe sizes of the Holden family: 8, 10, 7, 9, and 6. **13.** Sample answer: The student found the median of the data set. The mean of data set is 18 texts.

Lesson 10-4 Interquartile Range and Box Plots, Practice Pages 567–568

1. The data vary by a range of 32 apps. The middle half of the data values vary by 4 apps.
3. Sample answer: The ages range from about 36 years to about 71 years. The middle half of the data range from about 50 years to about 60 years. Because the boxes are shorter than the whiskers, there is less variation among the middle half of the data. Having less variation means there is a greater consistency among the middle 50% of the data than in either whisker.

5. The data vary by a range of $14. The middle half of the data values vary by $6. **7.** false; Sample answer: A box plot does not show individual data values so you cannot find the mean of the data from a box plot alone. **9.** no; Sample answer: Each section of the box plot represents 25% of the total values. This means that each whisker and each box represents the same amount of data values. The length of each section depends on the spread of the data.

Lesson 10-5 Mean Absolute Deviation, Practice Pages 573–574

1. 5; Sample answer: The average distance for each value from the mean is 5 days.
3. Bears: 1.84; Saints: 1.44; Sample answer: The mean absolute deviation of the number of wins is greater for the Bears than for the Saints. The data values for the Saints are closer to the mean. **5.** 11.67 Calories **7.** 4.5 miles per gallon; 7 data values **9.** Sample answer: The term absolute refers to the absolute value of a number, which is the distance a number is from 0 on a number line and distance is always positive. To deviate means to vary or change. So, the mean absolute deviation of a data set is the average (mean) distance from each data value to the mean, which is a description of how the data values deviate or vary from the mean.

Lesson 10-6 Outliers, Practice Pages 581–582

1. 60 minutes **3.** 4 boxes and 56 boxes are both outliers **5.** mean with outlier: $\approx$ 47.6, mean without outlier: $\approx$ 44.2; median with outlier: 45, median without outlier: 44.5; The median best describes the center. **7.** mean with outlier: $\approx$ 21.8, mean without outlier: $\approx$ 17.6; median with outlier: 19, median without outlier: 18; The median best describes the center because the mean was affected the most by the outlier.
9. Sample answer: age, in years, of people attending a picnic: 4, 32, 34, 40, 45, and 72
11. Sample answer: An outlier may make the mean significantly greater or less than the mean would be without the outlier.

Lesson 10-7 Interpret Graphical Displays, Practice Pages 591–592

1. median and interquartile range; The median is 2. This means the data are centered on 2 televisions. The spread of the data around the center is 2 televisions. **3.** Sample answer: The shape of the distribution is symmetric. There is a peak from 10–14 dollars. There are no gaps, clusters, or outliers. **5.** D **7.** no; Sample answer: There were 6 pumpkins that weighed 20 pounds or more, out of 40 total pumpkins picked. $\frac{6}{40}$ = 15%, which is less than 25%, so the student was not correct. **9.** no; Sample answer: There are a total of 13 roller coasters. There are 6 roller coasters that have speeds 70 mph or greater. $\frac{6}{13}$ is about 46.2%. 46.2% is less than 50%. **11.** mean; mean absolute deviation.

Module 10 Review Pages 595–596

1. How many televisions does the typical family own?; How many states has the average student visited?; How many students are in the average sixth grade class? **3.** 41.6 inches; 34.1 inches.

5.

	Correct	Incorrect
Lower Extreme = 24		X
Median = 39	X	
Q_1 = 33	X	
Q_3 = 44		X
Upper Extreme = 58	X	

7. 10 feet; 70 feet